GEOCHEMISTRY - EARTH'S SYSTEM PROCESSES

GEOCHEMISTRY - EARTH'S SYSTEM PROCESSES

Editor

Lester Hanks

scitus academics

Geochemistry - Earth's System Processes
Edited by **Lester Hanks**

ISBN: 978-1-68117-592-8
Library of Congress Control Number: 2016946447

SCITUS Academics LLC,
616, Corporate Way, Suite 2, 4766,
Valley Cottage, NY 10989

www.scitusacademics.com

Notice

Printed in United States of America on Acid Free Paper

PREFACE

Geochemistry is the science that uses the tools and principles of chemistry to explain the mechanisms behind major geological systems such as the Earth's crust and its oceans. The realm of geochemistry extends beyond the Earth, encompassing the entire Solar System and has made important contributions to the understanding of a number of processes including mantle convection, the formation of planets and the origins of granite and basalt. The Earth is part of a closely related family of heavenly bodies, our Solar System, that formed simultaneously. Hence, the realm of geochemistry extends beyond the Earth to encompass the entire Solar System. The goals of geochemistry are thus no different from those of other fields of earth science; just the approach differs. The contributions of geochemistry to this advance have been simply enormous. Geochemists study the composition, structure, processes, and other physical aspects of the Earth. They examine the distribution of chemical elements in rocks and minerals, as well as the movement of these elements into soil and water systems. There is a wealth of information buried in the liquids, gases, and mineral deposits of rock. Understanding the chemical composition of rocks tells oil companies where to drill for oil; enables scientists to put together broad-based theories about the way the Earth is changing; helps environmental management companies decide how to dispose of a toxic or hazardous substance; and steers mining companies toward use of natural resources with a minimum.

Geochemistry - Earth's System Processes brings together the knowledge from a variety of topics within the field of geochemistry. The audience for this book consists of a multitude of scientists such as physicists, geologists, technologists, petroleum engineers, volcanologists, geochemists and government agencies.

TABLE OF CONTENTS

1

CR(VI) OCCURRENCE AND GEOCHEMISTRY IN WATER FROM PUBLIC-SUPPLY WELLS IN CALIFORNIA

John A. Izbicki [a,*], Michael T. Wright [a], Whitney A. Seymour [a], R. Blaine McCleskey [a], Miranda S. Fram [a], Kenneth Belitz [a], Bradley K. Esser [b]

[a] US Geological Survey, United States

[b] Lawrence Livermore National Laboratory, United States

ABSTRACT

Hexavalent chromium, Cr(VI), in 918 wells sampled throughout California between 2004 and 2012 by the Groundwater Ambient Monitoring and Assessment-Priority Basin Project (GAMA-PBP) ranged from less than the study reporting limit of 1 microgram per liter (μg/L) to 32 μg/L. Statewide, Cr(VI) was reported in 31 percent of wells and equaled or exceeded the recently established (2014) California Maximum Contaminant Level (MCL) for Cr(VI) of 10 μg/L in 4 percent of wells. Cr(VI) data collected for regulatory purposes overestimated Cr(VI) occurrence compared to spatially-distributed GAMA-PBP data. Ninety percent of chromium was present as Cr(VI), which was detected more frequently and at higher concentrations in alkaline (pH ≥ 8), oxic water; and more frequently in agricultural and urban land uses compared to native land uses. Chemical, isotopic (tritium and carbon-14), and noble-gas data show high Cr(VI) in water from wells in alluvial aquifers in the southern California deserts result from long groundwater-residence times and geochemical reactions such as silicate weathering that increase pH, while oxic conditions persist. High Cr(VI) in water from wells in alluvial aquifers along the west-side of the Central Valley results from high-chromium in source rock eroded to form those aquifers, and areal recharge processes (including irrigation return) that can mobilize chromium from the unsaturated zone. Cr(VI) co-occurred with oxyanions having similar chemistry, including vanadium, selenium, and uranium. Cr(VI) was positively correlated with nitrate, consistent with increased concentrations in areas of agricultural land use and mobilization of chromium from the unsaturated zone by irrigation return.

KEYWORDS

Groundwater; Chromium; Trace elements; Drinking water; California

1. INTRODUCTION

Chromium is the 21st most abundant element in the earth's crust (Emsley, 2001), with an average concentration between 100 and 180 milligrams per kilogram (mg/kg) (Nriagu and Niebor, 1988, Reimann and de Caritat, 1998 and Hitchon et al., 1999). Chromium concentrations are higher in basaltic and ultramafic rocks which have average concentrations of about 200 and 2400 mg/kg, respectively (Nriagu and Niebor, 1988 and Reimann and de Caritat, 1998), than in granitic rocks which have an average concentration of 10 mg/kg (Reimann and de Caritat, 1998). However, even granitic rocks have sufficient chromium to yield high Cr(VI) concentrations in groundwater under the proper geochemical conditions (Izbicki et al., 2008a).

Natural occurrence of chromium in alkaline, oxic groundwater as hexavalent chromium, Cr(VI), has been reported since the mid-1970s (Robertson, 1975 and Robertson, 1991), and is commonly associated with the presence of ultramafic rocks and serpentinites in ophiolite complexes (Oze et al., 2007). These rocks are relatively common along the western coast of North America (Oze et al., 2004a and Oze et al., 2004b) with serpentine exposed over approximately 2860 square kilometers (km^2) in California (Kruckeberg, 1984). These rocks may potentially impact groundwater geochemistry over a greater area through erosion, transport, and weathering (Morrison et al., 2009 and

Smith et al., 2014). In California, naturally-occurring Cr(VI) in water from wells has been reported in the Mojave Desert (Ball and Izbicki, 2004, Izbicki et al., 2008a and Izbicki et al., 2012), the west-side of the Central Valley (Chung et al., 2001, Dawson et al., 2008, Morrison et al., 2009, Morrison et al., 2015 and Mills et al., 2011; Manning et al., 2015), and the Coast Ranges (Gonzalez et al., 2005 and Steinpress, 2005) (Fig. 1). Reported concentrations in these areas can exceed the U.S. Environmental Protection Agency Maximum Contaminant Level (MCL) for total chromium of 100 micrograms per liter (μg/L) (US Environmental Protection Agency, 2009).

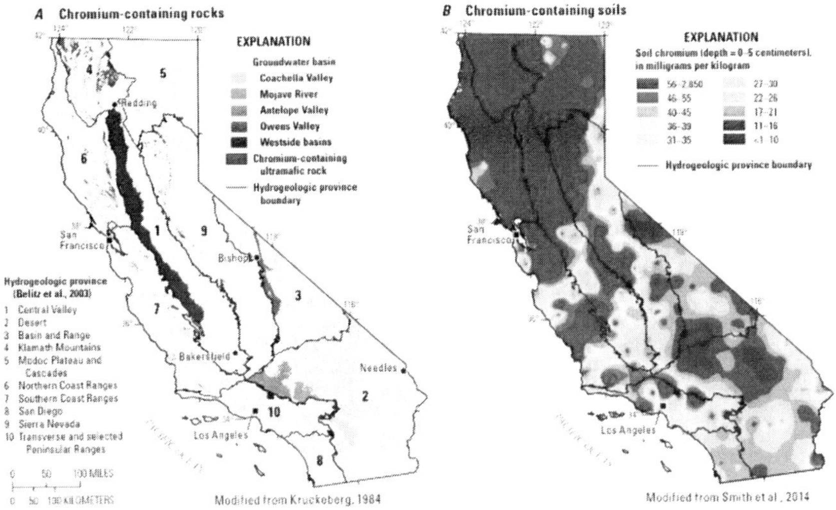

Figure 1. Areal extent of chromium-containing rocks and soils in California.

In addition to naturally-occurring sources, chromium has a wide range of industrial uses (Nriagu and Niebor, 1988), and Cr(VI) is a common industrial contaminant. Prior to widespread reports of naturally-occurring Cr(VI), chromium in groundwater was presumed to be associated primarily with anthropogenic contamination (Hem, 1959). Chromium (VI) is highly soluble and mobile in alkaline, oxic groundwater; and contamination can migrate for kilometers with only limited attenuation (Perlmutter et al., 1963, Blowes, 2002 and Izbicki et al., 2012). The wide-spread industrial use of chromium and the multiplicity of sources in urban areas can combine to create regional-scale contamination, such as contamination observed in alluvial aquifers underlying parts of the Los Angeles metropolitan area (U.S. Environmental Protection Agency, 2008 and California State Water Resources Control Board, 2014a).

Chromium (VI) is a carcinogen if inhaled (Daugherty, 1992 and ATSDR, 2012), and may be a carcinogen if ingested (Sedman et al., 2006 and Beaumont et al., 2008). Concern over chronic exposure to low-levels of Cr(VI) in drinking water prompted establishment of a California MCL for Cr(VI) of 10 μg/L (California State Water Resources Control Board, 2014b). The purpose of this paper is to identify areas in California where groundwater used for public drinking water supply have concentrations greater than the California MCL and to identify geologic, hydrologic, geochemical, and other factors that contribute to the occurrence of Cr(VI). This study was done as part of the California State Water Resources Control Board Groundwater Ambient Monitoring and Assessment-Priority Basin Project (GAMA-PBP).

1.1. Geochemistry of chromium

The most common chromium-containing mineral, chromite ($FeCr_2O_4$), is highly resistant to weathering. Chromium also is commonly substituted within minerals such as amphiboles, garnets, micas, pyroxenes and spinels (Reimann and de Caritat, 1998 and Motzer, 2005). These minerals are more easily weathered than chromite and, where present, may contribute considerable chromium to the environment (Morrison et al., 2009 and Rajapasha et al., 2013).

In most minerals, chromium is present in the +3 oxidation state (Nriagu and Niebor, 1988). Cr(III), an essential micronutrient for humans, is strongly sorbed to mineral surfaces and sparingly soluble under acidic conditions (Kotas and Stasicka, 2000) (Fig. 2). Oxidation of Cr(III) to Cr(VI) occurs in the presence of Mn oxides (Schroeder and Lee, 1975; Eary and Rai, 1987;Guha et al., 2000). Although Cr(VI) is strongly sorbed by Fe oxides at acidic and near neutral pH, it is not strongly sorbed in alkaline or high-ionic strength water (Rai and Zachara, 1984). Under these conditions, Cr(VI) may desorb and enter solution most commonly as the oxyanion CrO_4^{2-} (Fig. 2), or as neutral complexes with calcium and magnesium ($CaCrO_4^0$ and $MgCrO_4^0$, respectively) (Lelli, 2014). Microbially-mediated reduction of Cr(VI) to Cr(III) occurs in the presence of organic material and other reducing agents (Rai and Zachara, 1984). In some cases, reduction of Cr(VI) to Cr(III) also can occur abiotically (Beller et al., 2014). Depending on pH, the reduced chromium may be rapidly sorbed and removed from solution (Mayer and Schick, 1981, Rai and Zachara, 1984 and Charlet and Manceau, 1992).

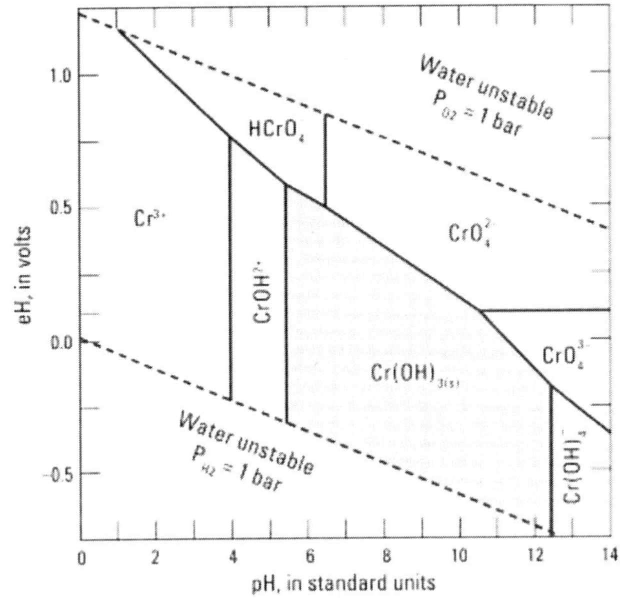

EXPLANATION

Insoluble

Total chromium 1 micromole per liter at 25 degrees Celsius, with no complexing ions present. P_{O_2}, P_{H_2}, partial pressure of oxygen and hydrogen, respectively.

Figure 2. Aqueous forms of chromium.

In simplified conceptual models of the natural-occurrence of chromium in groundwater (Kotas and Stasicka, 2000, Izbicki et al., 2008a and Scancar and Milacic, 2014), the abundance of Cr(III) and Cr(VI) on the surfaces of mineral grains and in solution is determined by: 1) the concentration of chromium in primary minerals, 2) the weathering rates of those minerals, 3) the abundance and reactivity of manganese oxides that convert Cr(III) to Cr(VI), and 4) the abundance and reactivity of reductants that allow microbially-mediated or abiotic reduction of

Cr(VI) to Cr(III). In general, the rate of reduction of Cr(VI) to Cr(III) is at least an order of magnitude greater than the rate of oxidation of Cr(III) to Cr(VI) (Scancar and Milacic, 2014), so that little Cr(VI) is present on aquifer materials in most settings. However, in settings having an abundance of easily weathered chromium-containing minerals and reactive Mn oxides, Cr(VI) may occur naturally in large concentrations (Morrison et al., 2009 and Mills et al., 2011). In alkaline soils and unsaturated deposits, Cr(VI) on the surfaces of mineral grains can desorb and enter solution where it can be mobilized by natural or artificial recharge (Izbicki et al., 2008b, Izbicki et al., 2008c and Mills et al., 2011). In saturated deposits, if the pH of groundwater increases along groundwater flowpaths, as a result of silicate weathering or other processes, Cr(VI) can accumulate in alkaline groundwater over long time periods as long as oxic conditions persist (Izbicki et al., 2008a).

1.2. Hydrogeologic setting

For the GAMA-PBP, Belitz et al. (2003) divided California into 10 hydrogeologic provinces (Fig. 1). (For the purposes of this paper, these provinces were combined into 8 provinces as discussed in the "Methods" section.) Each hydrogeologic province represents a large area having similar hydrologic, geologic, and climatic conditions. The provinces having the largest areal extent are the Desert (including the Basin and Range), Coast Ranges (combined Northern and Southern Coast Ranges), Sierra Nevada, and Central Valley provinces; while the Desert and Central Valley provinces have the largest areal extent of groundwater basins. The largest numbers of public-supply wells, and the largest population dependent on groundwater as a source of supply, are in the Central Valley, Transverse and Selected Peninsular Ranges, and Coast Ranges provinces (Belitz et al., 2015; Table 1).

Ultramafic and serpentine rock containing chromium occurs in the Klamath Mountains, Coast Ranges, and in the foothills along the west-slope of the Sierra Nevada (Fig. 1A) (Kruckeberg, 1984, Oze et al., 2004b, Oze et al., 2004a and Morrison et al., 2009). These rocks have eroded to form alluvial deposits that compose aquifers in the Coast Ranges, and the west-side of the Central Valley provinces. Although ultramafic and serpentine rock also are along the east-side of the Central Valley, alluvium in the east-side is dominated by alluvium eroded from granitic rock in the higher altitudes of the Sierra Nevada. Chromium-containing mafic rocks also occur to a lesser extent along the western margin of the Desert hydrogeologic province (Izbicki et al.,

2008a). The combination of erosion, transport, and weathering of chromium-containing rock has resulted in high-chromium concentrations in soils in northern California and parts of southern California (Smith et al., 2014) (Fig. 1B). In contrast, chromium concentrations in soils within most of the Desert (including the Basin and Range), and San Diego hydrogeologic provinces are low.

The lowest groundwater recharge rates are in the Desert, including parts of the Basin and Range hydrogeologic provinces. Under present-day climatic conditions precipitation in these areas is less than 150 millimeters per year, and groundwater throughout much of the area was recharged many thousands of years ago (Kulongoski et al., 2003,Kulongoski et al., 2005 and Izbicki, 2007). Alluvial materials that compose groundwater basins within these areas are often comparatively non-reactive, and in some areas dissolved oxygen may persist in groundwater for many thousands of years (Izbicki. et al., 2004 and Izbicki and Michel, 2004). Older groundwater also is present in deeper aquifers elsewhere in California, but materials that compose these aquifers are more reactive and dissolved oxygen may not be present in older groundwater (Izbicki, 1996, Izbicki and Martin, 1997 and Izbicki et al., 2003).

2. METHODS

2.1. Data sources and field procedures

Between 2004 and 2012, the GAMA-PBP assessed nearly the entire groundwater resource used for public drinking-water supply in California (Belitz et al., 2015). Areas of the state that contain public-wells were divided into 87 areas, which were subsequently grouped into 35 study units to facilitate sample collection and data interpretation. Each study unit consisted of an alluvial groundwater basin, subbasin, or group of groundwater basins (California Department of Water Resources, 2003) or a hard-rock area outside of defined groundwater basins (Belitz et al., 2003). Detailed descriptions of the study units, and well selection criteria are given in U.S. Geological Survey series reports published for each of the 35 study units (U.S. Geological Survey, 2015).

Table 1. Hydrogeologic provinces, population, and public-supply wells sampled for hexavalent chromium, Cr(VI), in California as part of the Groundwater Ambient Monitoring and Assessment-Priority Basin Project (GAMA-PBP), 2004 to 2012. [Area, number of groundwater basins, and number of public-supply wells from Belitz et al. (2003). Population from U.S. Census data, 2000. Hydrologic province in italics if data combined from two provinces. km^2 is square kilometers; –, data from province combined into larger province for purposes of this study.]

Hydrogeologic province	Area, in km^2	Number of groundwater basins	Area of groundwater basins, in km^2	Population	Number of public-supply wells		Number of public-supply wells sampled for Cr(VI)	
					Inside basins	Outside basins	Inside basins	Outside basins
Central Valley	53,000	36	53,000	53,07,000	5360	0	254	0
Desert	*1,17,000*	*141*	*72,000*	*12,80,000*	*1500*	*80*	205	0
Desert	81,000	96	56,000	12,16,000	1240	60	–	–
Basin and range	36,000	45	16,000	64,000	260	20	–	–
Klamath Mountains	23,000	7	300	75,000	20	110	0	0
Modoc Plateau and Cascades	39,000	53	21,000	1,12,000	240	40	0	0
Coast Ranges	*80,000*	*153*	*19,000*	*80,22,000*	*2510*	*760*	128	8
Northern Coast Ranges	38,000	79	6000	14,23,000	770	280	–	–
Southern Coast Ranges	42,000	74	13,000	65,99,000	1740	480	–	–
San Diego Drainages	10,000	25	1000	34,30,000	180	120	37	13
Sierra Nevada	66,000	22	2000	7,15,000	330	1170	36	96
Transverse and Selected Peninsular Ranges	22,000	33	8000	1,48,99,000	2720	800	108	33

Numbers of samples does not include 343 samples collected and analyzed for Cr(VI) that were later determined to have been contaminated.

To ensure a statistically representative spatial distribution of wells sampled as part of the GAMA-PBP, the study units within the GAMA-PBP were divided into equal-area grids (Scott, 1990). Statewide, the median grid-cell size was 25 km^2, with a median of 20 cells per study unit (Belitz et al., 2015). The objective was to sample one public-supply well (or other well representative of the groundwater resource used for public supply) per grid cell (grid wells). Additional non-grid wells were sampled to provide additional data for the purpose of understanding specific groundwater quality issues. These non-grid wells included wells screened at depths shallower or deeper than the screen depth intervals of public-supply wells.

More than 2400 wells were sampled as part of the GAMA-PBP (Belitz et al., 2015). Samples from 918 of these wells are discussed in this paper for Cr(VI). Because not all GAMA-PBP wells were sampled for Cr(VI), Cr(VI) data have a less rigorous statewide spatial distribution than most other GAMA-PBP data. For example, water from wells within in the Klamath Mountains and Modoc Plateau and Cascades provinces was not analyzed for Cr(VI). In addition, wells in some study units within other hydrogeologic provinces were not analyzed for Cr(VI). As a consequence, data from several hydrologic provinces (Northern and Southern Coast Ranges; and Desert and Basin and Ranges hydrologic provinces) were combined to ensure sufficient Cr(VI) data for analyses.

Samples were collected by U.S. Geological Survey field personnel according to U.S. Geological Survey protocols (U.S. Geological Survey, variously dated). Field parameters (temperature, dissolved oxygen, pH, alkalinity, and specific conductance) were measured at the time of collection, and samples were preserved in the field as required for specific analyses (Appendix A).

Samples for Cr(VI) analyses were processed in the field using exchange resins to remove Cr(III) (Ball and McCleskey, 2003a and Ball and McCleskey, 2003b). After processing, a filtered sample containing Cr(VI) and Cr(III) and the field processed sample containing only Cr(VI) were acidified to pH less than 2.0 using hydrochloric acid and shipped to the U.S. Geological Survey National Research Laboratory in Boulder, Colo. for analyses of chromium (http://wwwbrr.cr. usgs.gov/ projects/ GWC_chemtherm/, accessed November 4, 2014). Cr(III) was determined by difference between the two samples. Field speciation of Cr(VI) allowed for longer holding-times than the 24-h holding time provided by U.S. Environmental Protection Agency (1991) method 218.6 for Cr(VI) preservation and analyses. Most other samples were analyzed at the NWQL in Denver Colo. according to U.S. Geological Survey procedures (http://nwql.usgs.gov/rpt.shtml?pubs, accessed November 4, 2014).

Data collected by the GAMA-PBP are available through the U.S. Geological Survey's National Water Information System (NWIS) (http://waterdata.usgs.gov/nwis), and the California State Water Resources Control Board's GeoTracker GAMA (http://www.waterboards.ca.gov/gama/geotracker_gama.shtml) on-line databases.

Additional data from 1480 public-supply wells within the California State Water Resources Control Board (SWRCB) Division of Drinking Water (DDW) database also were compiled and interpreted as part of this study. Public-supply wells sampled as part of the GAMA-PBP are included in the DDW data. Cr(VI) data for samples collected between 2004 and 2012 (the same time period over which GAMA-PBP samples were collected) were retrieved. For the purposes of this paper, if a well had more than one sample during that time period, the data from the sampling date closest to the date that the GAMA-PBP samples in the same study unit were collected was selected. DDW data were collected by well owners for drinking water compliance purposes and were analyzed by a number of different laboratories certified by the State of California for analyses of Cr(VI). Field parameters were not typically measured, and samples were not generally preserved or filtered in the field. The DDW database does not include data for dissolved oxygen, tritium, or noble gases; thus, only the GAMA-PBP data were used for evaluation of relations between Cr(VI) and explanatory factors. The DDW data for Cr(VI) are not evenly distributed across the state because during the time period of this study, Cr(VI) was classified as an unregulated chemical and data were collected by only a subset of public-supply systems statewide. DDW data are publicly available from the SWRCB's on-line groundwater information system (http://www.waterboards.ca.gov/gama/geotracker_gama.shtml).

2.2. Quality assurance and statistical methods

Data from equipment and field blanks were used for quality assurance and to establish study reporting levels (SRL) for Cr(VI) and trace elements (Olsen et al., 2010 and Davis et al., 2014) (Table S1). Least-squares regression analysis of 91 replicate sample pairs having Cr(VI) concentrations ranging from <1 to 28 μg/L, showed data compared favorably with a R^2 of 0.98 and mean square error of 0.30 μg/L. The average difference between replicate pairs was 4 percent, and in most cases the replicate sample values were identical. As part of this study 1261 samples were collected and analyzed for Cr(VI). Contamination, traced to corroded metal parts in a field pipette used to preserve samples with hydrochloric acid, was identified in 17 of 112 field blank samples collected between September 2004 and December 2005; and resulted in the removal of 343 samples, for a total of 918 samples analyzed for Cr(VI) discussed in this paper. The pipette was replaced with a pipette that did not contain metal parts and contamination issues ceased (Appendix A). Contamination contributed to the less rigorous statewide spatial distribution of Cr(VI) data compared to most other GAMA-PBP data.

Most Cr(VI) data were less than the SRL for this study. Non-parametric descriptive statistics such as medians (and even upper 25-percent quartiles) and associated statistical tests, such as the Median or Wilcoxon rank-sum test (Neter and Wasserman, 1974), are not meaningful for data having a high-occurrence of less than values. As a consequence, differences in Cr(VI) occurrence were evaluated on the basis of the two-sample Kolmogorov–Smirnov test (Taylor and Emerson, 2011) (Appendix A). Co-occurrence of Cr(VI) and other trace elements was evaluated using Kendall's Tau correlation coefficients (Kendall, 1938) and Principle Component Analyses (PCA) (Kshirsagar, 1972 and Gnanadesikan, 1977).

3. RESULTS AND DISCUSSION

3.1. Cr(VI) occurrence

Cr(VI) concentrations in 918 public-supply wells sampled by the GAMA-PBP between 2004 and 2012 ranged from less than the study reporting limit (SRL) of 1 μg/L to 32 μg/L (Fig. 3A). Statewide, Cr(VI) concentrations exceeded the SRL of 1 μg/L in 31 percent of sampled wells, and exceeded the California MCL of 10 μg/L in 4 percent of sampled wells (Fig. 3A). Almost all dissolved chromium, approximately 90 ± 1 percent, was present as Cr(VI) (Fig. 4).

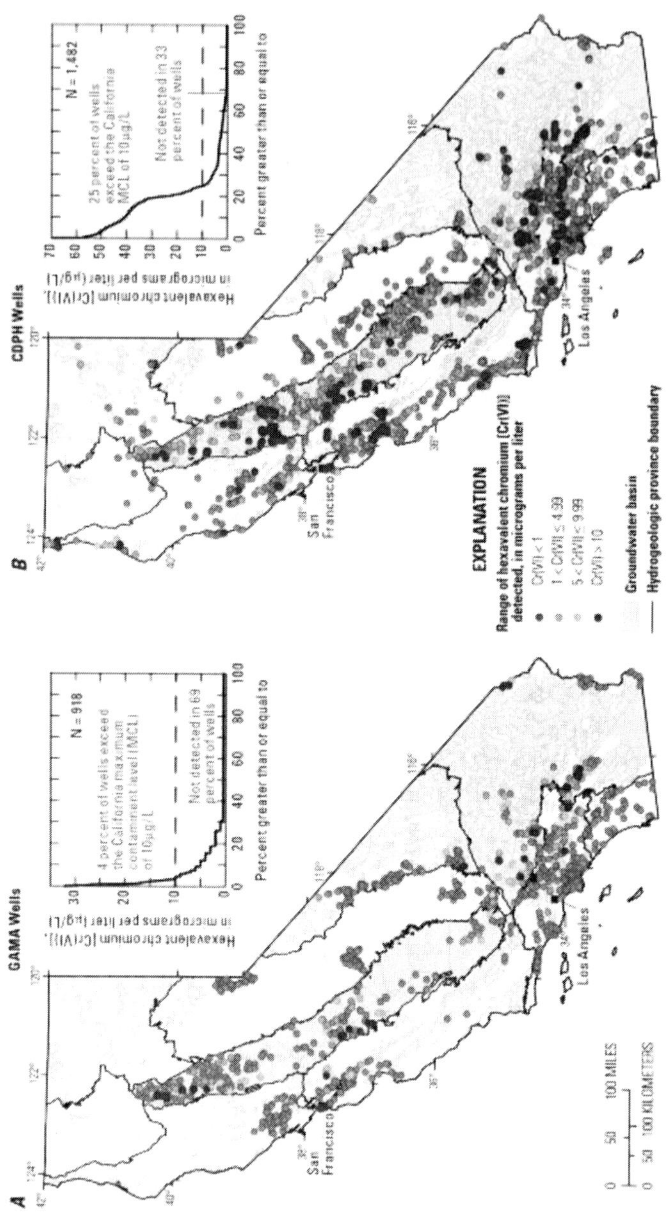

Figure 3. Hexavalent chromium, Cr(VI), concentrations in water from public-supply wells in California: A) sampled as part of the Groundwater Ambient Monitoring and Assessment-Priority Basin Project (GAMA-PBP), 2004–2012, and B) from the California State Water Resources Control Board (SWRCB) Division of Drinking Water (DDW) databases, 2000–2011.

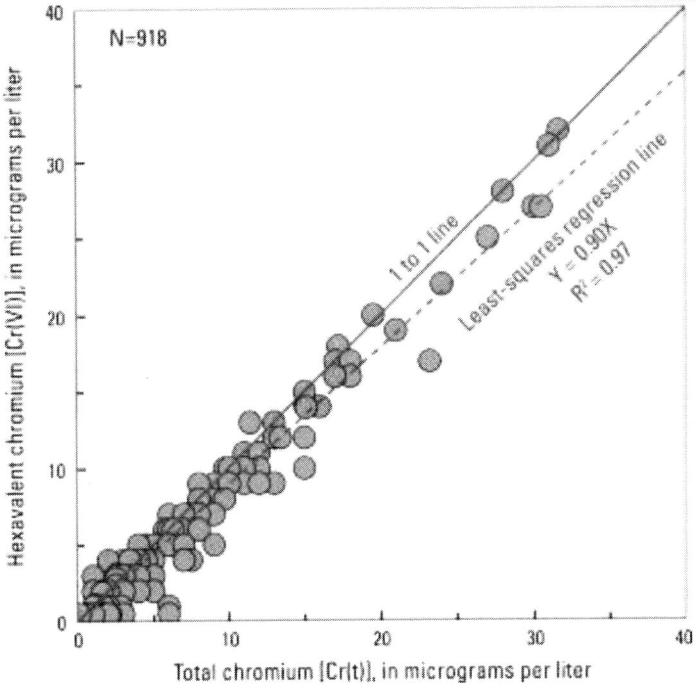

Figure 4. Hexavalent chromium, Cr(VI), as a function of total chromium, Cr(t), in water from public-supply wells in California sampled as part of the Groundwater Ambient Monitoring and Assessment-Priority Basin Project (GAMA-PBP), 2004–2012.

Cr(VI) concentrations in more than 1480 public-supply wells within the DDW database sampled between July 2000 and September 2011 exceeded 1 µg/L in 67 percent of the wells and exceeded the California MCL in 25 percent of wells, with a maximum Cr(VI) concentration of 69 µg/L (Fig. 3B). Similar to the GAMA-PBP data, most Cr(VI) concentrations in excess of the MCL occurred in the Desert and the west-side of the Central Valley. High concentrations also were present in the Transverse and selected Peninsular Ranges near Los Angeles, the eastern part of the San Diego hydrogeologic province, and in areas within the southern part of the Central Valley, and the Northern and Southern Coast Ranges hydrogeologic provinces which were not sampled for Cr(VI) as part of the GAMA-PBP. DDW data show no Cr(VI) concentrations in excess of the California MCL in either the Klamath Mountains or Modoc Plateau and Cascades hydrogeologic provinces (Fig. 3B), despite the presence of chromium-containing rock and soil in those areas (Fig. 1).

Comparison between GAMA-PBP and DDW data using least-squares regression for the same wells sampled less than 3 years apart was good with an R^2 of 0.91 and a Cr(VI) excess of 8 percent in the DDW data (not shown in figures). Some of the Cr(VI) excess in the DDW data may occur because Cr(VI) samples analyzed as part of the GAMA-PBP were filtered and Cr(VI) was separated from Cr(t) in the field; whereas, Cr(VI) samples in the DDW database were not filtered or consistently preserved in the field to prevent changes in Cr(VI) and Cr(III) concentrations. In contrast, a similar comparison between GAMA-PBP and DDW data for samples collected at the same well more than 3 years apart was poor, with an R^2 of 0.58 and a Cr(VI) excess of 30 percent in the DDW data. This difference may result from changing sample collection, preservation, and laboratory methods in the longer period of record represented by the DDW data, or because of trends in Cr(VI) concentrations with time in water from some sampled wells.

DDW data were collected for public-health and regulatory purposes. The higher frequency of Cr(VI) detection and concentrations in excess of the California MCL in the DDW data compared to the GAMA-PBP data, especially in urban areas, may result from clustering of analyses in water from wells in areas known to have natural or anthropogenic Cr(VI) occurrence (Belitz et al., 2003 and Belitz et al., 2015). This clustering creates a spatial bias towards high values. Although Cr(VI) data were less rigorously distributed statewide than most constituents analyzed as part of the GAMA-PBP, they are from representative spatial distributions within individual study units, and are not subject to the same bias associated with sample collection for public-health and regulatory purposes that may over represent areas of known natural or anthropogenic contamination.

In both the GAMA-PBP and DDW data, Cr(VI) concentrations in excess of the California MCL occurred more frequently in the Desert and the west-side of the Central Valley hydrogeologic provinces than in other areas of California, with a few high concentrations in the Coast Ranges and the Transverse and Selected Peninsular Ranges provinces (Fig. 3). The effect of pH, redox, land use, and well construction on Cr(VI) concentrations are discussed to facilitate understanding of differences in Cr(VI) occurrence within different hydrogeologic provinces within California.

3.1.1. Occurrence with pH and redox

Differences in Cr(VI) occurrence with pH data were evaluated by grouping the data into three categories: 1) pH greater than or equal to 8,

2) pH greater than or equal to 7 but less than 8, and 3) pH less than 7. The divisions between the groups were not arbitrary and were established by adjusting the threshold values between the groups at 0.1 unit intervals to maximize the difference in Kendall's Tau correlation coefficients between Cr(VI) and pH within the groups. This approach resulted in statistically significant (although weak) correlation coefficients for pH greater than or equal to 8 of −0.173, for pH greater than or equal to 7 but less than 8 of 0.184, and no significant correlation for pH less than 7.

Despite the negative correlation, Cr(VI) was detected more frequently and at higher concentrations in alkaline water having pH greater than or equal to 8 (Fig. 5); with Cr(VI) present at concentrations greater than the SRL in 45.3 percent of sampled wells and greater than the California MCL in 15 percent of sampled wells. In water from wells having pH greater than or equal to 7 but less than 8, Cr(VI) was present at concentrations greater than the SRL in 39 percent of sampled wells and greater than the California MCL in 4 percent of sampled wells. In water from wells having a pH less than 7, Cr(VI) was present at concentrations greater than the SRL in 8 percent of sampled wells and did not exceed the California MCL in any sampled wells.

The negative correlation between Cr(VI) and pH in water having pH greater than or equal to 8 may result in part from deceases in dissolved oxygen concentrations and increased presence of reduced conditions in older, more alkaline groundwater. If only oxic water from wells is considered there is no significant correlation between Cr(VI) and pH greater than or equal to 8.0. The data are consistent with a threshold pH near 7.5 (Rai and Zachara, 1984, Izbicki et al., 2008c, Izbicki et al., 2008b and Izbicki et al., 2008a), above which desorption of Cr(VI) from the surfaces of mineral grain occurs and most Cr(VI) is present in solution.

Differences in Cr(VI) occurrence by redox status were evaluated by grouping data into two categories, oxic, and reduced. Oxic water had dissolved oxygen concentrations greater than 0.5 mg/L. Reduced water had dissolved oxygen concentrations less than 0.5 mg/L. In the absence of dissolved oxygen data, nitrate (less than 0.5 mg/L as N), manganese (greater than or equal to 0.05 mg/L), or iron concentrations (greater than or equal to 0.1 mg/L) were used to characterize reduced water (McMahon and Chapelle, 2008). A similar approach was used to characterize the redox status of water from wells for analysis of the occurrence of vanadium in California (Wright and Belitz, 2010 and Wright et al., 2014).

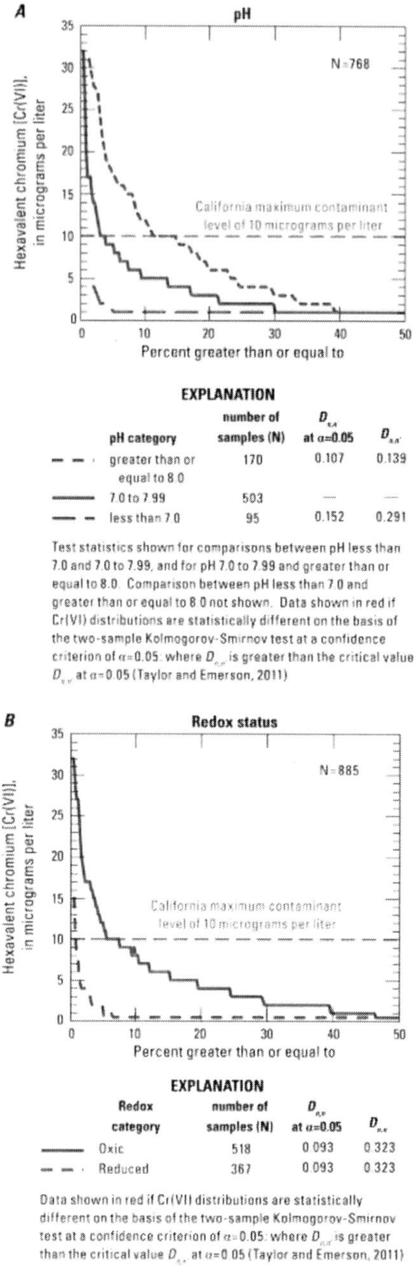

Figure 5. Hexavalent chromium, Cr(VI), as a function of A) pH, and B) redox status, in water from public-supply wells in California sampled as part of the Groundwater Ambient Monitoring and Assessment-Priority Basin Project (GAMA-PBP), 2004–2012.

Statewide, Cr(VI) was detected more frequently and at higher concentrations in oxic water; with Cr(VI) present at concentrations greater than the SRL in 46.4 percent of sampled wells and greater than the MCL in 7.4 percent of sampled wells. In reduced water, Cr(VI) was detected at concentrations greater than the SRL in only 4.9 percent of sampled wells and water from only one well exceeded the MCL of 10 μg/L (Fig. 5B). In 10 of 14 samples in reduced water having reportable Cr(VI) concentrations, nitrate was present at concentrations ranging from 0.7 to 12.7 mg/L as N. In some studies, reduction of Cr(VI) to Cr(III) is reported to increase when nitrate is present (Han et al., 2010); however, other studies (Izbicki, 2008, Izbicki et al., 2008b and Chovanec et al., 2012) show nitrate interferes with microbially mediated reduction of Cr(VI) to Cr(III). Inhibition of Cr(VI) reduction may occur in some settings because many of the same microorganisms that reduce Cr(VI) also reduce nitrate (Chovanec et al., 2012), and these microorganisms may preferentially utilize nitrate, which is commonly present at higher concentrations than Cr(VI).

3.1.2. Occurrence with land use and well construction

Land use within a 500-m radius around wells is a useful predictor of water quality within wells in California (Johnson and Belitz, 2009). Examination of Cr(VI) occurrence in wells showed a statistically significant difference (Kolmolgorov–Smirnov two-sample test at a confidence criterion of $\alpha = 0.05$) between Cr(VI) concentrations in water from wells in natural (undeveloped) land compared to agricultural, and urban land uses (Fig. 6). Cr(VI) occurrence at concentrations greater than the SRL was greater in water from wells in both agricultural and urban land uses (38 and 36 percent, respectively), compared to natural land use (23 percent). However, Cr(VI) occurrence at concentrations in excess of the MCL appears to be greater only in urban land uses (6 percent) compared to natural and agricultural land uses (4 and 3 percent, respectively). There was no statistically significant difference in Cr(VI) occurrence between natural and mixed land uses, possibly because of the smaller number of sampled wells within the mixed land use category. Similar relations were observed between land use and Cr(t) occurrence by Ayotte et al. (2011). The results suggest that although Cr(VI) may be primarily naturally occurring, low-level region-scale contamination may increase Cr(VI) occurrence in public-supply wells in urban areas (U.S. Environmental Protection Agency, 2008 and California State Water Resources Control

Board, 2014a), and processes that alter the hydrology or geochemistry may increase the occurrence of Cr(VI) in agricultural and urban areas.

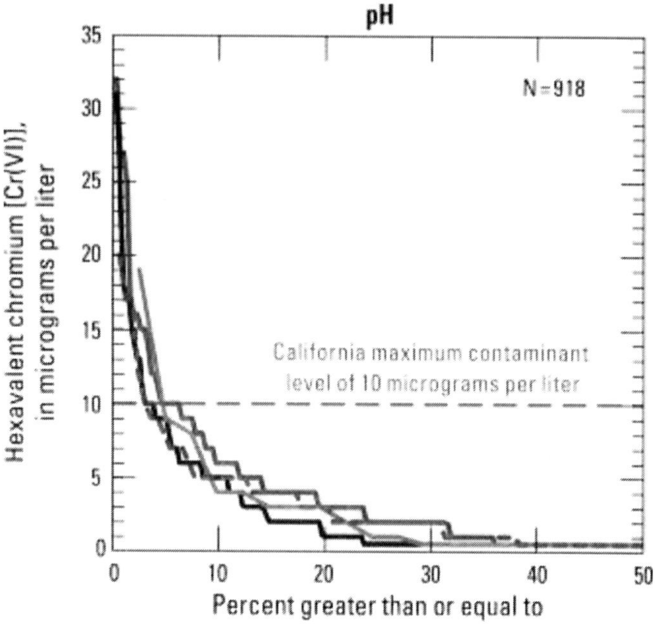

EXPLANATION

Land use	number of samples (N)	$D_{n,n'}$ at α=0.05	$D_{n,n'}$
Natural	401	—	—
Mixed	40	0.226	0.073
Agricultural	168	0.125	0.148
Urban	309	0.103	0.125

Data shown in red if Cr(VI) distributions are statistically different from natural Cr(VI) distribution on the basis of the Kolmogorov-Smirnov D statistic at a confidence criterion of α=0.05: where $D_{n,n'}$ is greater than the critical value $D_{n,n'}$ at α=0.05 (Taylor and Emerson, 2011).

Figure 6. Hexavalent chromium, Cr(VI), as a function of land use in water from public-supply wells in California sampled as part of the Groundwater Ambient Monitoring and Assessment-Priority Basin Project (GAMA-PBP), 2004–2012.

Sampled well depths ranged from 4.5 to more than 650 m below land surface, with a median depth of 122 m below land surface. Additional well-construction data available as part of the GAMA-PBP included top of shallowest opening (screened or perforated interval), and bottom of deepest opening. When only oxic waters are considered, well depth, top of perforated interval, and bottom of perforated interval were significantly positively correlated with Cr(VI) concentrations (Kendall's Tau correlation coefficients of 0.156, 0.070, and 0.078, respectively). For the same samples, pH was similarly positively correlated with well depth, top and bottom of perforated interval (Kendall's Tau correlation coefficients of 0.231, 0.189, and 0.201, respectively). These data suggest that in areas where groundwater is oxic, Cr(VI) concentrations may increase with well depth, and with the depth of top and bottom of perforations, consistent with increases in pH in deeper groundwater and subsequent desorption of Cr(VI) from aquifer materials. In areas where aquifer materials are more reactive, and groundwater is reducing, Cr(VI) may not be present at depth if dissolved oxygen has been consumed.

3.1.3. Occurrence by hydrogeologic province

Cr(VI) occurrence was greater than statewide occurrence in the Desert and Central Valley hydrogeologic provinces with concentrations exceeding the SRL in 40.3 and 45.1 percent of samples and concentrations exceeding the California MCL in 9.7 and 5.9 percent of sampled wells, respectively (Fig. 7A).

Cr(VI) concentrations greater than the California MCL were found in water from wells throughout the Desert (including Basin and Range) hydrogeologic province. However, Cr(VI) concentrations were highest in alluvial aquifers within the Antelope Valley and Coachella Valley study units (Fig. 1) where concentrations exceeded the SRL in 94 and 74 percent of sampled wells, and exceeded the California MCL in 26 and 20 percent of sampled wells, respectively. The chromium-containing Pelona Schist and related schists (Jacobson et al., 2000) that crop out in these areas (not shown on Fig. 1) may be a source of chromium (Izbicki et al., 2008a). Oxic, alkaline groundwater (with dissolved oxygen concentrations greater than 0.5 mg/L in 95 and 83 percent of sampled wells, and median pH's of 8 and 7.8, respectively), coupled with low groundwater recharge rates and long flowpaths through aquifers within the Antelope and Coachella Valley study units, enable chromium weathered from chromium-containing minerals, and oxidized to Cr(VI), to enter and remain in solution. In contrast to other areas in the Desert, Cr(VI) concentrations did not exceed the California MCL for any wells sampled in the Owens Valley study unit and Cr(VI) was present at concentrations greater than the SRL in only 7 percent of sampled wells. The absence of identifiable chromium-containing source rock and more neutral groundwater (median pH 7.4), coupled with higher groundwater recharge rates from the adjacent Sierra Nevada, and shorter groundwater flowpaths may have combined to limit Cr(VI) concentrations in the Owens Valley study unit.

In contrast to the Central Valley as a whole, Cr(VI) was present at concentrations greater than the SRL in 49 percent of sampled wells, and greater than the California MCL in 11 percent of water from wells in study units along the west-side of the Central Valley (Fig. 7B). High Cr(VI) concentrations in the west-side of the Central Valley have been associated with chromium-containing rock that crop out along the valley margin (Fig. 1) (Morrison et al., 2009 and Mills et al., 2011; Manning et al., 2015). Water from sampled wells in the west-side of the Central Valley was oxic with a median dissolved oxygen concentration of 3.1 mg/L, and slightly alkaline with a median pH of 7.6. Concentrations in excess of the California MCL were present in water from only one well along the east-side of the Central valley province, although Cr(VI) occurs more frequently in public-supply wells at concentrations up to 4 μg/L than statewide (Fig. 7B). Cr(VI) concentrations on the east-side of the valley exceeded the SRL in 45 percent of sampled wells. This was greater than the statewide exceedance of 31 percent, and similar to the Cr(VI) exceedance of the SRL on the west-side of the valley (Fig. 7B).

Cr(VI) concentrations were not statistically different from the statewide occurrence in the Coast Ranges, Transverse and selected Peninsular Ranges, and San Diego Drainages hydrogeologic provinces (as determined by the Kolmolgorov-Smirnov two-sample test at a confidence criterion of $\alpha = 0.05$) (Fig. 7). The Coast Ranges province has large areas of chromium-containing rock (Fig. 1), but Cr(VI) concentrations exceed the California MCL in only 3 percent of sampled wells with detections above the SRL in only 18 percent of sampled wells. Cr(VI) concentrations in the Coast Ranges may be limited by the comparatively large percentage of wells in this area having reduced groundwater (about 65 percent) compared to other parts of the state (Table 2). Water from only one well exceeded the MCL in the Transverse Range province and no wells exceeded the California MCL in the San Diego provinces, although Cr(VI) concentrations were greater than the SRL in 18 to 16 percent of sampled wells in these areas (Table 2).

Comparison with DDW data (Fig. 3) shows a number of wells in urban areas near Los Angeles that have Cr(VI) concentrations greater than the California MCL. These wells may have been preferentially sampled and included within the DDW database.

Cr(VI) concentrations in the Sierra Nevada hydrogeologic province were statistically different from statewide concentrations and plot below the statewide distribution (Fig. 7A). The Sierra Nevada has fewer defined groundwater basins than other hydrogeologic provinces within California (Table 1), and about two-thirds of the wells sampled by the GAMA-PBP were in fractured-rock, rather than alluvial aquifers. No areas of known chromium-containing rock (Fig. 1A) were sampled as part of the GAMA-PBP. However DDW data show a number of wells having Cr(VI) greater than 1 μg/L in areas where chromium-containing rock occurs (Fig. 3B), although Cr(VI) concentrations greater than the California MCL were not present. In areas of chromium-containing rock in the Sierra Nevada, such as the Western Metamorphic Belt, water from wells commonly has reduced conditions (Fram and Belitz, 2014), which would limit Cr(VI) occurrence. In most other areas, groundwater flowpaths are commonly short, limiting the potential interactions between groundwater and aquifer material. However, even in areas with oxic conditions, long residence times, and pH values greater than 8 Cr(VI) concentrations are low (Fram and Belitz, 2014).

Cr(VI) data were not collected within the Klamath Mountains, or Modoc Plateau and Cascades hydrogeologic provinces in northern California as part of the GAMA-PBP. The Klamath Mountains have the most extensive occurrence of Cr-containing mafic rocks in California and high-chromium concentrations in surface soils (Fig. 1). Despite the abundance of chromium-containing source material, Cr(t) concentration in water from 39 wells in the Klamath Mountains sampled as part of the GAMA-PBP (Mathany and Belitz, 2014) ranged from less than the SRL of 0.42–6.9 μg/L, with 15 percent of sampled wells greater than or equal to 1 μg/L. Similar to the Sierra Nevada hydrogeologic province, this area also has few defined groundwater basins (Table 1), and all of the wells sampled by GAMA-PBP were in fractured-rock. The wells in this province had (Table 1), and the shallowest median well depth of any hydrogeologic province (Table 2). In addition, many wells are near channels of major rivers and streams and pumping may induce infiltration of surface water from those sources potentially affecting the chemistry of water from the wells. Water from sampled wells was generally oxic and slightly acidic with a median pH of 6.9. Slightly acidic water increases sorption of Cr(VI), potentially limiting the occurrence of Cr(t) and Cr(VI) in water from wells. The Modoc Plateau and Cascades hydrogeologic provinces do not contain extensive areas of Cr-containing mafic rock (Fig. 1A), but the area does have high-chromium concentrations in surface soils (Fig. 1B). Cr(t) concentrations in water from 90 wells in the Modoc Plateau and Cascades sampled as part of the GAMA-

PBP (Shelton et al., 2013) ranged from less than the SRL to 8.9 μg/L, with 40 percent of sampled wells greater than or equal to 1 μg/L. Although water from sampled wells was generally oxic, it was near-neutral pH, with a median pH of 7.2, potentially limiting the occurrence of Cr(VI).

3.2. Other factors affecting Cr(VI) occurrence

In addition to factors such as pH, redox, land use, well construction, and differences in geology and hydrologic conditions within the various hydrogeologic provinces, Cr(VI) occurrence also may be affected by other hydrologic factors including groundwater age and recharge processes. All these factors are linked in environmental systems, and it is often difficult to identify primary and secondary controls on Cr(VI) occurrence. Because Cr(VI) concentrations are higher in the Desert and Central Valley hydrogeologic provinces, differences in groundwater age and recharge processes that affect Cr(VI) occurrence are compared for these two areas.

3.2.1. Groundwater age

As groundwater age increases in oxic alluvial aquifers, pH gradually increases as a result of weathering of primary silicate minerals. As pH becomes increasing alkaline, Cr(VI) and other trace elements may desorb from exchange sites on the surfaces of mineral grains and enter solution (Rai and Zachara, 1984, Izbicki et al., 2008c, Izbicki et al., 2008b and Izbicki et al., 2008a). The effect is most pronounced in large aquifers having long groundwater flowpaths, low recharge rates, and long residence times. In this study, groundwater age in oxic water from wells was evaluated on the basis of tritium and carbon-14 data. Tritium is a radioactive isotope of hydrogen having a half-life of about 12.3 years. Its presence in the environment increased as a result of atmospheric testing of nuclear weapons beginning in about 1952. Water from wells containing measurable tritium is interpreted as containing as least some water recharged after 1952. Carbon-14 is a radioactive isotope of carbon having a half-life of about 5730 years. Unlike tritium, carbon-14 is not part of the water molecule and carbon-14 activities can be affected by reactions other than radioactive decay. For the purposes of this paper, only uncorrected carbon-14 activities that do not account for reactions with aquifer material are presented.

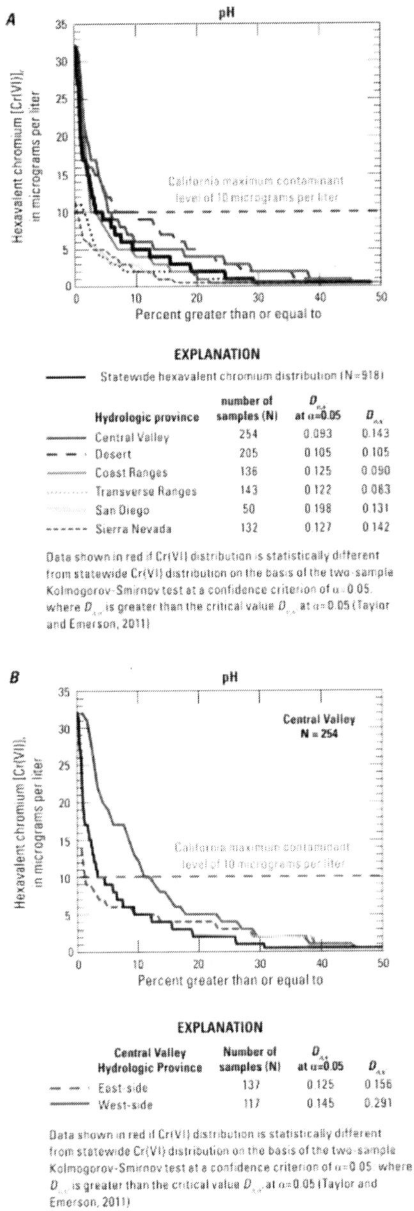

Figure 7. Hexavalent chromium, Cr(VI), as a function of A) hydrogeologic province and B) the east-side and west-side of the Central Valley hydrogeologic province in water from public-supply wells in California sampled as part of the Groundwater Ambient Monitoring and Assessment-Priority Basin Project (GAMA-PBP), 2004–2012.

Table 2. Summary of well-construction, hexavalent chromium (Cr VI), redox, and pH data by hydrogeologic province for public-supply wells sampled for hexavalent chromium, Cr(VI), in California as part of the Groundwater Ambient Monitoring and Assessment-Priority Basin Project (GAMA-PBP), 2004 to 2012. [>, greater than or equal to; <, less than].

Hydrogeologic province	Number of analyses	Top of perforations	Bottom of perforations	Median well depth, in meters	Cr(VI)			Redox status		pH		
					Percent detections, >1 mg/L	Median	Percent >10 mg/L	Percent oxic	Percent reducing	Percent >8	Percent <8 and >7	Percent <7
Statewide	918	52.7	121.9	121.9	29.8	<1	4.5	68	32	20	67	13
Desert, including Basin and Range	205	61.0	121.9	121.9	40.3	<1	9.7	79	21	30	61	9
Central Valley	254	48.8	88.4	86.9	45.1	<1	5.9	65	35	21	72	7
Northern and Southern Coastal Ranges	136	67.1	134.1	130.1	17.6	<1	2.9	37	63	18	72	10
Transverse and selected Peninsular Ranges	141	73.2	168.6	167.6	17.6	<1	0.7	66	34	4	75	21
San Diego Drainages	50	29.6	142.6	134.1	15.6	<1	0	70	30	22	65	13
Sierra Nevada	132	32.6	114.6	111.6	15.8	<1	0	79	14	26	42	32
Klamath Mountains	–	21.0	54.3	*37.2*	*22*		*0*	*95*	*5*	*8*	*38*	*54*
Modoc Plateau and Cascades	–	47.2	82.6	*98.5*	*47*		*0*	*96*	*4*	*12*	*49*	*39*

Water from wells in the Klamath Mountains and Modoc Plateau and Cascades were not analyzed for Cr(VI) as part of the GAMA-PBP. Data in *italics* are for 39 and 90 wells analyzed for Cr(t) as part of the GAMA Program by Mathany and Belitz (2014) and Sheldon et al. (2013), respectively.

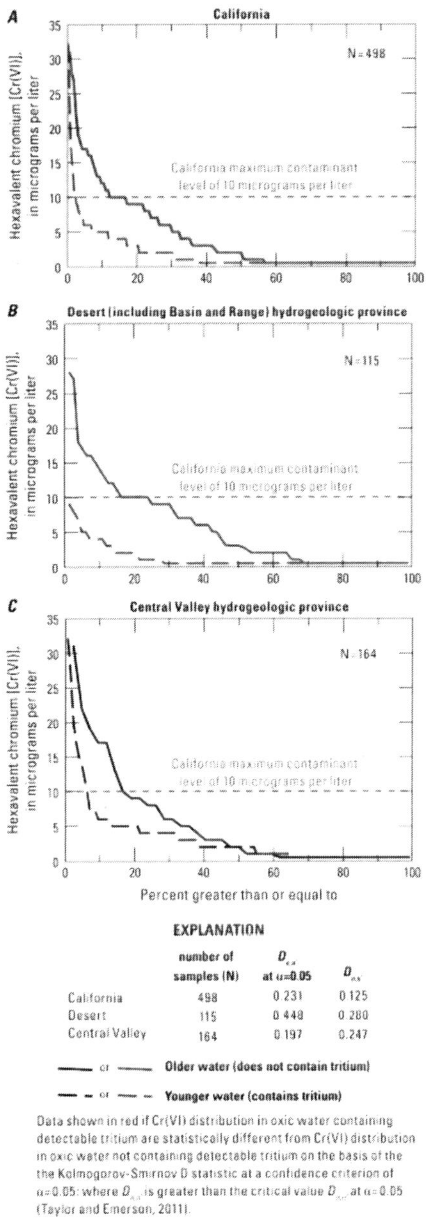

Figure 8. Hexavalent chromium, Cr(VI), as a function of groundwater age in water from public-supply wells in A) California, and the B) the Desert (including Basin and Range), and C) Central Valley Hydrogeologic Provinces sampled as part of the Groundwater Ambient Monitoring and Assessment-Priority Basin Project (GAMA-PBP), 2004–2012.

Statewide, Cr(VI) concentrations were greater in water from wells having only older groundwater that does not contain tritium than from wells having younger groundwater containing tritium (Fig. 8A). Water from 17 percent of wells containing older groundwater that does not contain tritium exceeded the MCL of 10 μg/L for Cr(VI), compared to water from 3 percent of wells having younger water.

In the Desert (including Basin and Range) hydrogeologic province, Cr(VI) concentrations were greater in water from wells having older groundwater than younger groundwater (Fig. 8B). Cr(VI) concentrations in 23 percent of wells yielding only older groundwater not containing tritium exceed the California MCL, compared to wells containing younger groundwater that does not contain tritium. There was a statistically significant negative correlation between Cr(VI) concentrations and carbon-14 activity as percent modern carbon (reflecting increasing Cr(VI) concentrations with groundwater age) in the Antelope Valley study unit within the Desert hydrogeologic province (Kendall's Tau = −0.378; significant at a confidence criterion of α = 0.10). The Antelope Valley study unit has the highest percentage of Cr(VI) concentrations in water from wells in excess of the MCL within California, and the data suggest that Cr(VI) concentrations would continue to increase along groundwater flowpaths as long as groundwater remains oxic. Similar increases in Cr(VI) concentrations with increasing groundwater age have been observed in flowpath studies in alluvial aquifers elsewhere in the Desert hydrogeologic province (Izbicki et al., 2008a).

In the Central Valley hydrogeologic province, Cr(VI) concentrations were not statistically different in water from wells having older and younger groundwater (Fig. 8C). However, Cr(VI) concentrations in water from wells yielding older groundwater that does not containing tritium exceeded the California MCL in about 17 percent of sampled wells, compared to 6 percent of wells containing younger groundwater (Fig. 8C). Because wells sampled as part of the GAMA-PBP are mostly public supply-wells, typically having long screen lengths, many of the samples containing tritium may be mixtures of older and younger water that entered the well from different depths within the aquifer. It also is possible that naturally occurring high-chromium rock eroded to form alluvial aquifers along the west-side of the Central Valley can weather rapidly and influence groundwater Cr(VI) concentrations (Morrison et al., 2009 and Mills et al., 2011). Unlike the Desert hydrogeologic province, areal recharge occurs in the Central Valley province and Cr(VI) in younger groundwater may be associated with groundwater recharge in areas where water moved through thick unsaturated zones and reacted with Cr-containing rock prior to recharge. In addition, the presence of irrigation return water that would have moved through the unsaturated zone was identified on the basis of high-nitrate concentrations in water from some wells having high Cr(VI) concentrations (Mills et al., 2011 and Manning et al., 2015).

3.2.2. Groundwater recharge processes

The effect of differences in groundwater recharge processes on Cr(VI) concentrations in the Desert and Central Valley hydrogeologic provinces was evaluated on the basis of dissolved noble gas concentrations. Noble gasses do not react chemically in water, but their concentration in water is a function of temperature, pressure, and salinity at the time of recharge according to Henry's Law (Stumm and Morgan, 1996). For focused recharge as a result of winter stormflows, dissolved noble-gas concentrations in groundwater may reflect cooler winter recharge temperatures if the unsaturated zone is comparatively thin, typically less than 5–10 m (Bense and Kooi, 2004 and Manning, 2011). For thicker unsaturated zones, water infiltrated from winter stormflows commonly warms to near the average annual temperature of the area prior to recharge. Water recharged as a result of stormflows also commonly has additional gas entrapped during the recharge process, known as excess air (Cey et al., 2008). Dissolved gas in groundwater from excess air is contributed according to the atmospheric proportions of those gases, rather than as a function of temperature, pressure, or salinity according to Henry's Law. For water recharged areally as a result of infiltrating precipitation, or as a result of irrigation return, dissolved-gas concentrations at recharge reflect the temperature at the water table (Cey et al., 2009). For sufficiently thick unsaturated zones, between 5 and 10 m, this temperature approximates the average annual temperature (Stute and Schlosser, 2000). In irrigated areas, dissolved-gas concentrations in recharge from irrigation return may reflect warmer summer (growing season) temperatures. Recharge temperature and excess air were estimated using the computer program NOBLEBOOK (Aeschbach-Hertig et al., 1999 and Aeschbach-Hertig et al., 2000). Input to the program included a suite of noble gas data (including helium, neon, argon, krypton, and xenon) collected as part of the GAMA-PBP (Bryant Jurgens, U.S. Geological Survey, written commun., 2015).

Within the Desert hydrogeologic province there is a wide range in rainy season (October–March) average daily temperatures and annual average temperatures (Fig. 9A). Cooler recharge temperatures in the Desert hydrogeologic province were from wells in the Owens Valley study unit, and to a lesser extent from the floodplain aquifer along the Mojave River within the Mojave River study unit (Fig. 1). As a group, these samples contained tritium, were recently recharged, and had low Cr(VI) concentrations. Otherwise, little relation was apparent for either recharge temperature (Fig. 9A), or excess-air (not shown on Fig. 9A), and Cr(VI) concentrations. This lack of relationship is consistent with recharge from a variety of sources (including sources no longer active under present-day climatic conditions) where Cr(VI) concentrations are controlled by a combination of source abundance and changes in pH and exchange along groundwater flowpaths.

Within the Central Valley hydrogeologic province there is less range and no overlap between rainy season (October–March) average daily temperatures and average annual temperatures (Fig. 9B). Cooler recharge temperatures and lower Cr(VI) concentrations were observed in water from wells along the east side of the Central Valley where recharge from rivers containing winter snowmelt from the higher altitudes of the Sierra Nevada occurs. Similar to the Desert hydrogeologic province, water from these wells contains tritium, was recently recharged, and has generally low Cr(VI) concentrations (Fig. 9B). Younger water containing tritium, warmer recharge temperatures, and higher Cr(VI) concentrations predominates in water from wells along the west-side of the Central Valley. In this area higher Cr(VI) concentrations are consistent with recharge processes through thick unsaturated zones that promote mobilization of Cr(VI) from source chromium-containing materials. Unlike the Desert hydrogeologic province, Cr(VI)-containing groundwater along the west-side of the Central Valley contains tritium, consistent with rapid mobilization of Cr(VI) during recharge from chromium-containing alluvium within the unsaturated zone. Rapid mobilization of Cr(VI) from the unsaturated zone rather than changes in pH and exchange along groundwater flowpaths may be responsible for the high groundwater Cr(VI) concentrations in this area. The presence of Cr(VI) in water from wells in the west-side of the Central Valley having warm recharge temperatures, presumably associated with areal recharge processes, was previously observed by Manning et al. (2015).

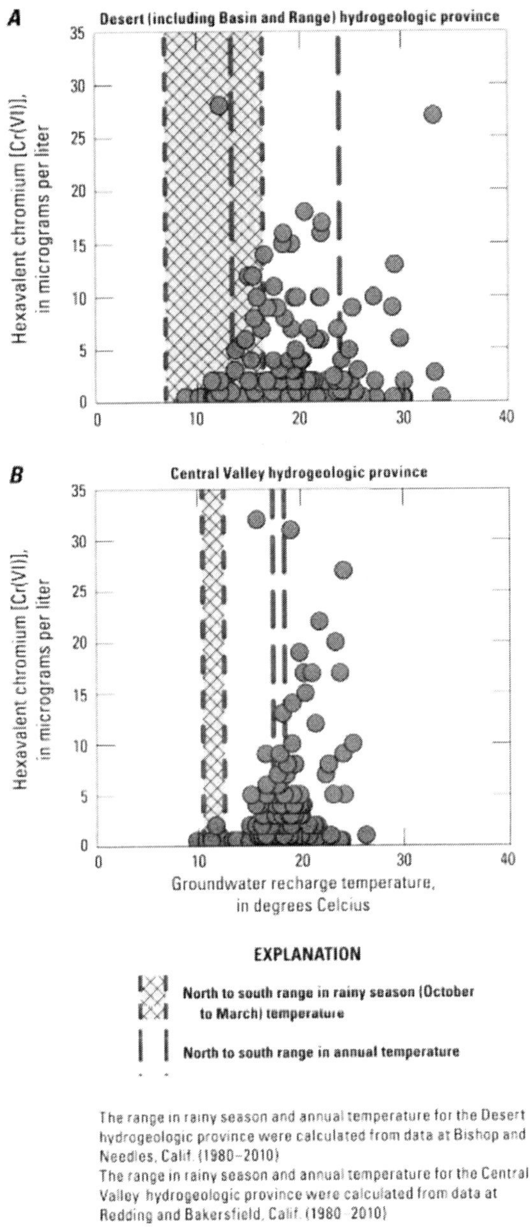

Figure 9. Hexavalent chromium, Cr(VI), as a function of groundwater recharge temperature in A) the Desert (including Basin and Range), and B) Central Valley Hydrogeologic Provinces in water from public-supply wells in California sampled as part of the Groundwater Ambient Monitoring and Assessment-Priority Basin Project (GAMA-PBP), 2004–2012.

3.3. Cr VI co-occurrence with selected trace elements and nitrate

In addition to chromium and Cr(VI), 21 other trace elements (including the reduced and oxidized forms of arsenic, As(III) and As(VI), and of iron, Fe(II) and Fe(III)), were analyzed in water from wells as part of the GAMA-PBP. Kendall's Tau correlation coefficients and Principal Component Analysis (PCA) were used to evaluate similarities and differences in the co-occurrence of selected trace elements with Cr(VI) under different pH and redox conditions.

Kendall's Tau correlation coefficients show significant positive correlations (confidence criterion of $\alpha = 0.05$) between Cr(VI) and the elements vanadium, selenium, uranium and with barium. Most correlation coefficients were comparatively low in magnitude with the largest magnitude correlation, $r = 0.398$, for vanadium (Table S1). With the exception of barium, these elements form oxyanions having similar chemical properties. Vanadium and selenium form soluble oxyanions in the +6 oxidation state, and uranium in the +4 oxidation state. These elements also form comparatively insoluble cations and cationic complexes under reduced conditions. PCA groups these different elements similarly near Cr(VI) (Fig. 10). Barium, while positively correlated with Cr(VI), does not form an oxyanion under alkaline, oxic conditions and does not group closely with Cr(VI) or the other elements (Fig. 10).

Arsenic and its oxyanion, As(V), were not correlated with Cr(VI)in oxic water (Table S1), and were not grouped together in PCA results. In contrast, As(III), which is soluble under reduced conditions, was significantly negatively correlated with Cr(VI). Kendall's Tau correlation coefficients also show significant negative correlations between Cr(VI) and molybdenum, and tungsten. Molybdenum and tungsten in the +6 oxidation state also form oxyanions, but these oxyanions are complex polyatomic ions that differ from Cr(VI) in chemical behavior. Because of the polyatomic nature of molybdenum and tungsten oxyanions, analyses of filtered water samples commonly measure only the dissolved cationic forms. Negative correlation coefficients also were obtained between Cr(VI) and Fe(II), lithium, and boron. Collectively, results of PCA analyses group molybdenum, tungsten, As(III), Fe(II), lithium, and boron separately from Cr(VI), and the (non-polyatomic) oxyanion forming elements vanadium, selenium, and uranium (Fig. 10A). Cadmium, manganese, and nickel also are negatively correlated with Cr(VI). These elements form divalent cations soluble in reduced water but comparatively insoluble in oxic water, and form a separate group on the basis of the PCA results.

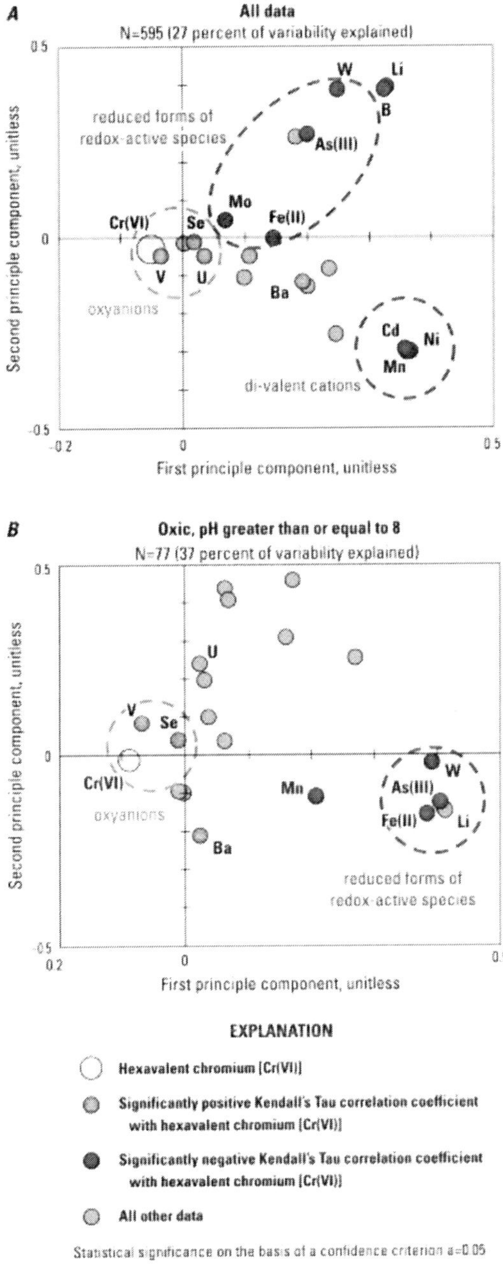

Figure 10. Principal component analysis (PCA) of hexavalent chromium, Cr(VI), and selected trace element in water from wells in California sampled as part of the Groundwater Ambient Monitoring and Assessment-Priority Basin Project (GAMA-PBP), 2004–2012.

Fewer elements were significantly correlated with Cr(VI) under alkaline ($pH \geq 8$) oxic conditions (possibly because fewer data are available) (Fig. 10B). However, the data show the same general pattern; with the oxyanion-forming elements, vanadium and selenium, grouping near Cr(VI), and the reduced forms of redox-active species Fe-II and As-III and tungsten plotting differently. The data are consistent with Cr(VI) co-occurrence with oxyanion-forming elements as a result of similarities in their aqueous chemistry over a range of pH and redox conditions within aquifers, rather than as a result of similarities in their occurrence within source rock. Co-occurrence of these elements may increase as pH increases along groundwater flowpaths, as long as oxic conditions persist.

In addition to trace elements, Cr(VI) also was positively correlated with nitrate (Kendall's correlation coefficient of 0.358). This occurs, in part, because both constituents occur primarily in oxic groundwater, although Cr(VI) may persist in reduced groundwater until nitrate is completely denitrified (Izbicki, 2008, Izbicki et al., 2008b and Chovanec et al., 2012) (Fig. 5B). If only oxic water from wells is considered, the correlation coefficient increases to 0.592. These data are consistent with increased Cr(VI) occurrence in areas of agricultural land use (Fig. 6), a finding that with the noble-gas data (Fig. 9) suggests that areal recharge and recharge associated with irrigation return may increase Cr(VI) concentrations in groundwater as a result of increased contact with chromium containing minerals in the unsaturated zone prior to recharge (Mills et al., 2011 and Manning et al., 2015). Cr(VI) was not correlated with either Ca or Mg, despite a tendency for Cr(VI) to form complexes with these ions (Lelli 204).

4. CONCLUSIONS

Statewide Cr(VI) was detected in 31 percent of public-supply wells sampled as part of the GAMA-PBP, and exceeded the recently established (2014) California Maximum Contaminant Level of 10 μg/L in 4 percent of sampled wells. Almost all dissolved chromium, about 90 percent, was present as Cr(VI). Data collected for public-health and regulatory purposes by the DDW had a higher frequency of detection compared to the spatially distributed data collected as part of the GAMA-PBP, as a result of clustering of analyses in the DDW database from wells in areas known to have natural or anthropogenic Cr(VI) occurrence. Use of public-health or regulatory data to assess the occurrence of Cr(VI), and other trace elements, over large areas may overestimate the occurrence of these constituents compared to data collected using spatially-distributed sample designs (Belitz et al., 2015).

Statewide, about 20 percent of the public-supply wells in California have high concentrations of one or more constituents, and trace-elements, specifically arsenic, manganese, and uranium, were more prevalent in public-supply wells than nitrate or organic compounds (Belitz et al., 2015). Arsenic and uranium occur naturally in rock and alluvium, with whole-rock concentrations almost two orders of magnitude lower than chromium (Nriagu and Niebor, 1988 and Reimann and de Caritat, 1998), but occur commonly in water from public-supply wells, especially in the Central Valley and Desert hydrogeologic provinces because of alkaline, oxic groundwater and hydrologic conditions in these areas (Belitz et al., 2015). (Although arsenic, as As(III), concentration increased in reduced groundwater because arsenic also is soluble under reduced conditions.) Although commonly detected at the SRL of 1 μg/L, the occurrence of Cr(VI) in groundwater was less than what might be expected on the basis of its relative abundance in the continental crust (compared to arsenic and uranium), and the widespread occurrence of chromium-containing rock and soil in California. Cr(VI) was most abundant in aquifers having chromium-containing source rock and alkaline oxic groundwater, especially alluvial aquifers in the Desert and west-side of the Central Valley hydrogeologic provinces.

Within the Desert hydrogeologic province, Cr(VI) was more commonly detected and present at concentrations greater than the California MCL in the Antelope and Coachella Valley study units. In both study units the combination of chromium-containing source rock and alkaline, oxic groundwater contributed to the both the high frequency of Cr(VI) occurrence and the high Cr(VI) concentrations. These data suggest that geologic, hydrologic, and geochemical factors do not individually control Cr(VI) occurrence, but in combination these factors can create a high frequency of occurrence, and high Cr(VI) concentrations. However, Cr(VI) in excess on the California MCL was widely distributed throughout the Desert independent of chromium abundance in geologic materials, and data show increases in pH resulting from silicate weathering in comparatively non-reactive alluvial aquifers contributes to high chromium occurrence in groundwater, as long as dissolved oxygen and oxic conditions persist.

In contrast, within the Central Valley hydrogeologic province, chromium-containing geologic material, coupled with areal groundwater recharge (from both infiltrating precipitation and recharge from irrigation return), facilitate interaction of water with unsaturated sediments prior to groundwater recharge—resulting in increased Cr(VI) occurrence and concentrations in water from public supply wells. However, the presence of chromium-containing material does not solely explain occurrence of Cr(VI) in groundwater. Parts of California having large areas of chromium-containing rock or soil, such as the Klamath Mountains, or Modoc Plateau and Cascades hydrogeologic provinces, do not have high concentrations of Cr(VI) in water from public supply wells (on the basis of California DPHS data) because of hydrologic and geochemical factors. Hydrology can play a role in limiting Cr(VI) concentrations in areas having shallow aquifers with short groundwater flowpaths that limit interaction between groundwater and aquifer materials, especially where wells located near streams and rivers are influenced by nearby surface water. Similarly, Cr(VI) concentrations also are low in both the Desert hydrogeologic province and the east-side of the Central Valley in areas where focused recharge from streams and rivers contribute large amounts of recent groundwater recharge. Rapid recharge of large quantities of water and the relatively young age of water in parts of these systems also limits interaction between groundwater and aquifer materials.

Interpretation of data in this paper is at a large, statewide scale, and is intended to highlight major geologic, hydrologic, and geochemical controls on Cr(VI) occurrence. Additional interpretation may be useful for evaluation of more localized processes that control Cr(VI) occurrence within individual within smaller-scale study units and groundwater basins that compose larger areas.

ACKNOWLEDGMENTS

This study was funded as part of the California State Water Resources Control Board Groundwater Ambient Monitoring and Assessment-Priority Basin Project (GAMA-PBP). The authors thank the large number of participating well owners across the state that provided access to wells for sample collection.

REFERENCES

1. Aeschbach-Hertig, W., Peeters, W., Beyerle, U., Kipfer, R., 1999. Interpretation of dissolved atmospheric noble gases in natural waters. Water Resour. Res. 35 (9), 2779-2792.

2. Aeschbach-Hertig, W., Peeters, W., Beyerle, U., Kipfer, R., 2000. Paleotemperature reconstruction from noble gases in groundwater taking into account equilibration with entrapped air. Nature 405, 1040-1044.

3. ATSDR, September 2012. Toxicological Profile for Chromium. Agency for Toxic Substances and Disease Registry, Public Health Service, US Department of Health and Human Services. http://www.atsdr.cdc.gov/toxprofiles/tp.asp? id¼62&tid¼17 (accessed 03.12.14.).

4. Ayotte, J.D., Gronberg, J.M., Apodaca, L.E., 2011. Trace Elements and Radon in Groundwater across the United States, 1992e2003. U.S. Geological Survey Scientific Investigations Report 2011-5059, 115 p. http://pubs.usgs.gov/sir/2011/ 5059 (accessed 06.11.14.).

5. Ball, J.W., McCleskey, R.B., 2003a. A new cation-exchange method for accurate field speciation of hexavalent chromium. Talanta 61 (3), 305-313.

6. Ball, J.W., McCleskey, R.B., 2003b. A New Cation-exchange Method for Accurate Field Speciation of Hexavalent Chromium. U.S. Geological Survey Water-Resources Investigations Report 03-4018, 17 p. http://wwwbrr.cr.usgs.gov/projects/GWC_ chemtherm/pubs/ WRIR%2003-4018.pdf (accessed 04.11.14.).

7. Ball, J.W., Izbicki, J.A., 2004. Occurrence of hexavalent chromium in ground water in the western Mojave Desert, California. Appl. Geochem. 19 (7), 1123e1135. http://www.sciencedirect.com/ science/article/pii/S088329270400023X (accessed 24.03.15.).

8. Beaumont, J.J., Sedman, R.M., Reynolds, S.D., Sherman, C.D., Li, L., Howd, R.A., Sandy, M.S., Zeise, L., Alexeeff, G.V., 2008. Cancer

mortality in a Chinese population exposed to hexavalent chromium in drinking water. Epidemiology 19 (1), 12-23.

9. Belitz, K., Dubrovsky, N.M., Burrow, K., Jurgens, B., Johnson, T., 2003. Framework for a Ground-water Quality Monitoring and Assessment Program for California. U.S. Geological Survey Water-Resources Investigations Report 03-4166, 78 p. http:// pubs.usgs.gov/ wri/ wri034166/ (accessed 31.10.14.).

10. Belitz, K., Fram, M.S., Johnson, T.D., 2015. Metrics for assessing the quality of groundwater used for public supply, CA, USA: equivalent-population and area. Environ. Sci. Technol. 49 (14), 8330-8338.

11. Beller, H.R., Yang, L., Varadharajan, C., Ruyang, H., Lim, H.C., Karaoz, U., Molins, S., Marcus, M.A., Brodie, A.L., Steefel, C.I., Nico, P.S., 2014. Divergent aquifer biogeochemical systems converge on similar and unexpected Cr(VI) reduction products. Environ. Sci. Technol. 48 (18), 10699-10706.

12. Bense, V.F., Kooi, H., 2004. Temporal and spatial variations of shallow subsurface temperature as a record of lateral variations in groundwater flow. J. Geophys. Res. B Solid Earth 109 (4). B04103 1-13.

13. Blowes, D.W., 2002. Tracking hexavalent Cr in groundwater. Science 295 (5562), 2024-2025.

14. California Department of Water Resources, 2003. California's groundwater. Bulletin 118. http://www.water.ca.gov/ groundwater/bulletin118/index.cfm (accessed 08.04.15.).

15. California State Water Resources Control Board, 2014a. Groundwater Information Sheet: Hexavalent Chromium. http:// www.waterboards.ca.gov/water_issues/ programs/gama/docs/ coc_hexchromcr6.pdf (accessed 03.12.14.).

16. California State Water Resources Control Board, 2014b. Chromium-6 Drinking Water MCL. http://www.waterboards.ca.gov/ drinking_

water/certlic/ drinkingwater/Chromium6.shtml (accessed 03. 12. 14.).

17. Cey, B.D., Hudson, G.B., Moran, J.E., Scanlon, B.R., 2008. Impact of artificial recharge on dissolved noble gases in groundwater in California. Environ. Sci. Technol. 42 (4), 1017-1023.

18. Cey, B.D., Hudson, G.B., Moran, J.E., Scanlon, B.R., 2009. Evaluation of noble gas recharge temperatures in a shallow unconfined aquifer. Ground Water 47 (5), 646-659.

19. Charlet, L., Manceau, A.A., 1992. X-ray absorption spectroscopic study of the sorption of Cr (III) at the oxide-water interface, II. Adsorption, coprecipitation, and surface precipitation on hydrous ferric oxide. J. Colloid Interface Sci. 148 (2), 443-458.

20. Chovanec, P., Sparacino-Watkins, C., Zhang, N., Basu, P., Stolz, J.F., 2012. Microbial reduction of chromate in the presence of nitrate by three nitrate respiring organisms. Front. Microbiol. 3, 146-158.

21. Chung, J.-B., Burau, R.G., Zasoski, R.J., 2001. Chromate generation by chromate depleted subsurface materials. Water Air Soil Pollut. 138, 407-417.

22. Daugherty, M.L., 1992. Toxicity Summary for Chromium. Oak Ridge National Laboratory Chemical Hazard Evaluation and Communication Group, Oak Ridge, TN.

23. Dawson, B.J.M., Bennett, G.L., Belitz, K., 2008. Ground-water Quality Data in the Southern Sacramento Valley, California, 2005 e Results from the California GAMA Program. In: US Geological Survey Data Series 285, 93 p. http://pubs. usgs.gov/ds/285/ (accessed 10.04. 15.).

24. Davis, T.A., Olsen, L.D., Fram, M.S., Belitz, K., 2014. Updated Study R 903 Eporting Levels (SRLs) 904 for Trace-element Data Collected for the California Groundwater Ambient Monitoring and 905 Assessment (GAMA) Priority Basin Project, October 2009eMarch

2013. U.S. Geological Survey Scientific Investigations Report 2014e5105, 52 p. (accessed 04.11.14.).

25. Eary, L.E., Rai, D., 1987. Kinetics of Chromium (III) oxidation to chromium (VI) by reaction with manganese dioxide. Environ. Sci. Technol. 21, 1187-1193.

26. Emsley, John, 2001. Nature's Building Blocks: an AeZ Guide to the Elements. Oxford University Press, Oxford, England, UK, pp. 495-498.

27. Fram, M.S., Belitz, Kenneth, 2014. Status and Understanding of Groundwater Quality in the Sierra Nevada Regional Study Unit, 2008dCalifornia GAMA Priority Basin Project. U.S. Geol. Survey Scientific Investigations Report 2014-5174, 118 p. http://pubs.usgs.gov/sir/2014/5174/ (accessed 21.01.15.).

28. Gnanadesikan, R., 1977. Methods for Statistical Data Analysis of Multivariate Observations. John Wiley and Sons, New York. Gonzalez, A.R., Ndung'n, K., Flegal, A.R., 2005. Natural occurrence of hexavalent chromium in the Aromas Red Sands Aquifer, California. Environ. Sci. Technol. 39 (15), 5505-5511.

29. Guha, H., Saiers, J.E., Brooks, S., Jardine, P., Jayachandran, K., 2000. Chromium transport, oxidation, and adsorption in manganese-coated sand. Contam. Hydrol. 49 (1e2), 311-334.

30. Han, R., Geller, J.T., Yang, L., Brodie, E.L., Chakraborty, R., Larsen, J.T., Beller, H.R., 2010. Physiological and transcriptional studies of Cr(VI) reduction under aerobic and denitrifying conditions by an aquifer-derived Pseudomonad. Environ. Sci. Technol. 44 (19), 7491-7497.

31. Hem, J.D., 1959. Study and Interpretation of the Chemical Characteristics of Natural Water, first ed. U.S. Geological Survey Water-Supply. Paper 1473, 269 p.

32. Hitchon, B., Perkins, E.H., Gunter, W.D., 1999. Introduction to Groundwater Geochemistry. Geoscience Publishing Ltd., Sherwood Park, Alberta, Can., 310 p.

33. Izbicki, J.,A., 1996. Source, Movement, and Age of Groundwater in a California Coastal Aquifer. U.S. Geological Survey Fact Sheet 126-96, 4 p. http://pubs.er. usgs.gov/publication/fs12696 (accessed 12.12. 14.).

34. Izbicki, J.A., Martin, P., 1997. Use of Isotopic Data to Evaluate Recharge and Geologic Controls on the Movement of Groundwater in Las Posas Valley, Ventura County, California. U.S. Geological Survey Water Resources Investigations Report 97- 4035, 12 p. http://pubs.er.usgs.gov/publication/wri974035 (accessed 12.12. 14.).

35. Izbicki, J.A., Borchers, J.W., Leighton, D.A., Kulongoski, J.T., Fields, L., Galloway, D.L., Michel, R.L., 2003. Hydrogeology and Geochemistry of Aquifers Underlying the San Lorenzo and San Leandro Areas of the East Bay Plain, Alameda County, California. U.S. Geological Survey Water-Resources Investigations Report 2002- 4250, 71 p. http://pubs.er.usgs.gov/publication/wri024259 (accessed 12.12. 14.).

36. Izbicki, J.A., Stamos, C.L., Nishikawa, T., Martin, P., 2004. Comparison of groundwater flow model particle-tracking results and isotopic data in the Mojave River ground-water basin, southern California, USA. J. Hydrol. 292, 30-47. http://pubs.er.usgs.gov/ publication/ 70026967 (accessed 12.12.14.).

37. Izbicki, J.A., Michel, R.L., 2004. Movement and Age of Ground Water in the Western Part of the Mojave Desert, Southern California, USA. USGS Water-Resources Investigations Report: 2003-4314, 42 p. http://pubs.er.usgs.gov/publication/ wri034314 (accessed 12.12. 14.).

38. Izbicki, J.A., 2007. Physical and temporal isolation of mountain headwater streams in the western Mojave Desert, southern California. J. Am. Water Resour. Assoc. 43 (1), 26e40. http://pubs.er. usgs.gov/publication/70033123 (accessed 12.12.14.).

39. Izbicki, J.A., 2008. Chromium Concentrations, Chromium Isotopes, and Nitrate in the Unsaturated Zone at the Wateretable Interface, El Mirage, California. Written comm. to the Lahontan Regional Water Quality Control Board, December 24, 2008. http://www.swrcb.ca.gov/lahontan/publications_forms/available_ documents/molycorp_elmirage1208.pdf (accessed 05.11.14.).

40. Izbicki, J.A., Ball, J.W., Bullen, T.D., Sutley, S.J., 2008a. Chromium, chromium isotopes and selected trace elements, western Mojave Desert, USA. Appl. Geochem. 23 (5), 1325e1352. http://ca.water.usgs.gov/news/2008/Chromium-report.pdf (accessed 24.03.15.).

41. Izbicki, J.A., Kulp, T.R., Bullen, T.D., Ball, J.W., O'Leary, D., 2008b. Chromium mobilization from the unsaturated zone (Abs.). In: Proc. 2008 Goldschmidt Conf. Canada, Vancouver. July 13-18, 2008.

42. Izbicki, J.A., Flint, A.L., Stamos, C.L., 2008c. Artificial recharge through a thick, heterogeneous unsaturated zone. GroundWater 46 (3), 475-488.

43. Izbicki, J.A., Bullen, T.D., Martin, P., Schroth, B., 2012. Delta chromium-53/52 isotopic composition of native and contaminated groundwater, Mojave Desert, USA. Appl. Geochem. 27 (4), 841-853. http://ca.water.usgs.gov/pubs/IzbickiEtAl2012-2.pdf (accessed 24.03.15.).

44. Jacobson, C.E., Barth, A.P., Grove, M., 2000. Late Cretaceous protolith age and provenance of the Pelona and Orocopia Schists, southern California: implications for evolution of the Cordilleran margin. Geology 28 (3), 219-222.

45. Johnson, T.D., Belitz, K., 2009. Assigning land use to supply wells for the statistical characterization of regional groundwater quality: correlating urban land use and VOC occurrence. J. Hydrol. 370 (1e4), 100-108. http://ca.water.usgs.gov/ projects/gama/pdfs/ Johnson_2009_1-s2.0-S0022169409001462-main.pdf (accessed 04.11.14.).

46. Kendall, M., 1938. A new measure of rank correlation. Biometrika 30 (1e2), 81e89. Kotas, J., Stasicka, Z., 2000. Chromium occurrence in the environment and methods of its speciation. Environ. Pollut. 107 (3), 263-283.

47. Kruckeberg, A.R., 1984. California serpentines: Flora, Vegetation, Geology, Soils, and Management Problems. University of California Press, Ltd., Berkeley, 180 p.

48. Kshirsagar, A.M., 1972. Multivariate Analysis. Marcel Dekker, New York. Kulongoski, J.T., Hilton, D.R., Izbicki, J.A., 2003. Helium isotope studies in the Mojave Desert, California: implications for groundwater chronology and regional seismicity. Chem. Geol. 202, 95-113. http://pubs.er.usgs.gov/publication/70025023 (accessed 12.12.14.).

49. Kulongoski, J.T., Hilton, D.R., Izbicki, J.A., 2005. Source and movement of helium in the eastern Morongo groundwater Basin: the influence of regional tectonics on crustal and mantle helium fluxes. Geochim. Cosmochim. Acta 69 (15), 3857-3872. http:// pubs.er.usgs.gov/publication/70031458 (accessed 12.12.14.).

50. Lelli, M., Grassi, S., Amadori, M., Franceschini, F., 2014. Natural Cr(VI) contamination of groundwater in the Cecina coastal area and its inner sectors (Tuscany, Italy). Environ. Earth Sci. 71 (9), 3907-3919.

51. Manning, A.H., 2011. Mountain-block recharge, present and past, in the eastern Espanola Basin, New Mexico, USA. Hydrogeol. J. 19, 379-397.

52. Manning, A.H., Mills, C.T., Morrison, J.M., Ball, L.B., 2015. Insights into controls on hexavalent chromium in groundwater provided by environmental tracers, Sacramento Valley, California, USA. Appl. Geochem. 61. http://www. sciencedirect.com/science/article/pii/ S004896970200231O (accessed 16.07.15.).

53. Mathany, T.M., Belitz, K., 2014. Groundwater-quality Data in the Klamath Mountains Study Unit, 2010: Results from the California

GAMA Program. In: U.S. Geological Survey Data Series 803, 82 p. http://pubs.usgs.gov/ds/803/ (accessed 20.01.15.).

54. Mayer, L.M., Schick, L.L., 1981. Removal of hexavalent chromium from estuarine waters by model substrates and natural sediments. Environ. Sci. Technol. 15 (12), 1482-1484.

55. McMahon, P.B., Chapelle, F.H., 2008. Redox processes and water quality of selected principal aquifer systems. GroundWater 46 (2), 259-271.

56. Mills, C.T., Morrison, J.M., Goldhaber, M.B., Ellefsen, K.J., 2011. Chromium(VI) generation in vadose zone soils and alluvial sediments of the southwestern Sacramento Valley, California: a potential source of geogenic Cr(VI) to groundwater. Appl. Geochem. 26 (8), 1488-1501.

57. Morrison, J.M., Goldhaber, M.B., Lee, L., Holloway, J.M., Wamty, R.B., Wolf, R.E., Ranville, J.F., 2009. A regional-scale study of chromium and nickel in soils of northern California, USA. Appl. Geochem. 24 (8), 1500-1511.

58. Morrison, J.M., Goldhaber, M.B., Mills, C.T., Breit, G.N., Hooper, R.L., Holloway, J.M., Diehl, S.F., Ranville, J.F., 2015. Weathering and transport of chromium and nickel from serpentinite in the coast range ophiolite to the Sacramento Valley, California, USA. Appl. Geochem. 62, 72-86. http://www.sciencedirect.com/science/article/pii/S0883292715001407 (accessed 16.07.15.).

59. Motzer, W.E., 2005. In: Guertin, J., Jacobs, J.A., Avakian, C.P. (Eds.), Chemistry, geochemistry, and geology of chromium and chromium compound, Chapt. 2, Chromium (VI) Handbook, 784 p.

60. Neter, J., Wasserman, W., 1974. Applied Linear Statistical Models, vol. 1. Richard D. Irwin Inc., Homewood, IL., 152 p.

61. Nriagu, J.O., Niebor, E. (Eds.), 1988. Chromium in the Natural and Human Environments. John Wiley & Sons, New York, 571 p.

62. Olsen, L.D., Fram, M.S., Belitz, K., 2010. Review of Trace-element Field-blank Data Collected for the California Groundwater Ambient Monitoring and Assessment (GAMA) Program, May 2004eJanuary 2008. U.S. Geological Survey Scientific Investigations Report 2009-5220, 47 p. http://pubs.usgs.gov/sir/2009/5220/pdf/ sir2009 5220.pdf (accessed 04.11.14.).

63. Oze, C., Fendorf, S., Bird, D.K., Coleman, R.G., 2004a. Chromium geochemistry of serpentinite soils. Int. Geol. Rev. 46 (2), 97-126.

64. Oze, C., Fendorf, S., Bird, D.K., Coleman, R.G., 2004b. Chromium geochemistry in serpentinized ultramafic rocks and serpentine soils from the Franciscan Complex of California. Am. J. Sci. 304 (1), 67-101.

65. Oze, C., Bird, D.K., Fendorf, S., 2007. Genesis of hexavalent chromium from natural sources in soil and groundwater. Proc. Natl. Acad. Sci. 104 (16), 6544-6549.

66. Perlmutter, N.M., Liber, M., Frauenthal, H.L., 1963. Movement of Waterborne Cadmium and Hexavalent Chromium Wastes in South Farmingdale, Nassua County, Long Island, pp. C170-C184. US Geological Survey Professional Paper 475C. http://pubs.usgs.gov/pp/0475c/report.pdf (accessed 03.12.14.).

67. Rai, D., Zachara, J.M., 1984. Chemical Attenuation Rates, Coefficients, and Constants in Leachate Migration: Volume 1: a Critical Review. Battelle Pacific Northwest Laboratories, Richland, WA. EA-3336 Volume 1, Research Project 2198e1, variously paged.

68. Rajapasha, A.U., Vithanage, M., Sik, Y., Oze, C., 2013. Cr(VI) formation related to Cr(III)-muscovite and birnessite interactions in ultramafic environments. Environ. Sci. Technol. 47 (17), 9722-9729.

69. Reimann, C., de Caritat, P., 1998. Chemical Elements in the Environment. SpringerVerlag, Berlin, 398 p. Robertson, F.N., 1975. Hexavalent chromium in the ground water, in Paradise Valley, Arizona. GroundWater 13 (6), 516-527.

70. Robertson, F.N., 1991. Geochemistry of Ground Water in Alluvial Basins of Arizona and Adjacent Parts of Nevada, New Mexico, and California. US Geological Survey Professional Paper 1406-C, 90 p. http://pubs.er.usgs.gov/publication/pp1406 (accessed 05.12.14.).

71. Scancar, J., Milacic, R., 2014. A critical overview of Cr speciation analysis basedon high performance liquid chromatography and spectrometric techniques. J. Anal. Atomic Spectrom. 29, 427-443.

72. Schroeder, D.C., Lee, G.F., 1975. Potential transformations of chromium in natural waters. Water Air Soil Pollut. 4, 355e365. Scott, J.C., 1990. Computerized Stratified Random Site-selection Approaches for Design of a Ground-water-quality Sampling Network. U.S. Geological Survey, Water-Resources Investigations Report 90-4101, 109 p. http://pubs.er.usgs.gov/ publication/ wri904101.

73. Sedman, R.M., Beaumont, J., McDonald, T.A., Reynolds, S., Krowech, G., Howd, R., 2006. Review of the evidence regarding carcinogenicity of hexavalent chromium in drinking water. J. Environ. Sci. Health Part C 24 (1), 155-182.

74. Shelton, J.L., Fram, M.S., Belitz, Kenneth, 2013. Groundwater-quality Data in the Cascade Range and Modoc Plateau Study Unit, 2010dResults from the California GAMA Program. In: U.S. Geol. Survey Data Series 688, 124 p. http://pubs. usgs.gov/ds/688/ (accessed 21.01.15.).

75. Smith, D.B., Cannon, W.F., Woodruff, L.G., Solano, F., Ellefsen, K.J., 2014. Geochemical and Mineralogical Maps for Soils in the Conterminous United States. U.S. Geological Survey Open-File Report 2014-1082, 368 p. http://pubs.usgs.gov/of/ 2014/1082/ (accessed 12.12.14.).

76. Steinpress, M.G., 2005. In: Guertin, J., Jacobs, J.A., Avakian, C.P. (Eds.), Naturallyoccurring chromium(VI) in groundwater, Chapt. 3, Chromium (VI) Handbook, 784 p. Stumm, W., Morgan, J.J., 1996. Aquatic Chemistry. Wiley, New York, 1022 p.

77. Stute, M., Schlosser, P., 2000. Atmospheric noble gases. In: Cook, P., Herezeg, A.L. (Eds.), Environmental Tracers in Subsurface Hydrology. Kluwer, Boston, 529 p.

78. U.S. Environmental Protection Agency, 1991. Determination of Dissolved Hexavalent Chromium in Drinking Water, Groundwater and Industrial Wastewater Effluents by Ion Chromatography. http://water.epa.gov/scitech/methods/cwa/ bioindicators/upload/2007_07_10_methods_method_218_6.pdf (accessed 04.11.14.).

79. U.S. Environmental Protection Agency, 2008. Addressing Chromium Contamination in the San Fernando Valley. U.S. Environmental Protection Agency, Region 9, San Francisco, Calif.. http://www.epa. gov/Region9/superfund/chromium/index.html (accessed 03.12.14.).

80. U.S. Environmental Protection Agency, 2009. National Primary Drinking Water Regulations. EPA 816-F-09e004 6 p. http:// water.epa.gov/drink/contaminants/ upload/mcl-2.pdf (accessed 03.12.14.).

81. U.S. Geological Survey, 2015. Groundwater Ambient Monitoring and Assessment Program. http://ca.water.usgs.gov/gama/includes/ GAMA_publications.html (accessed 15.07.15.).

82. U.S. Geological Survey, Variously Dated, National Field Manual for the Collection of Water-quality data: U.S. Geological Survey Techniques of Water-resources Investigations, Book 9, Chaps. A1e A9 http://pubs.water.usgs.gov/twri9A, (accessed 04.11.14).

83. Wright, M.T., Belitz, K., 2010. Factors controlling the regional distribution of vanadium in groundwater. Ground Water 48 (4), 515e525. http://pubs.er.usgs.gov/ publication/70003992 (accessed 05.11.14.).

84. Wright, M.T., Stollenwerk, K.G., Belitz, K., 2014. Assessing solubility controls on vanadium in groundwater, northeastern San Joaquin Valley, CA. Appl. Geochem. 48, 41e52. http://www. sciencedirect. com/science/journal/08832927 (accessed 26.03.15.)

2

SOIL CONTAMINATION NEAR NON-FERROUS METAL SMELTERS: A REVIEW

Vojtěch Ettler

Institute of Geochemistry, Mineralogy, and Mineral Resources, Faculty of Science, Charles University in Prague, Albertov 6, Prague 2, 128 43, Czech Republic

ABSTRACT

This review summarizes over 160 studies focused on soil contamination near non-ferrous metal smelters. The methods of these investigations were examined with an emphasis on the combinations of traditional (geo)chemical approaches with various mineralogical and metal isotope techniques that are particularly helpful for depicting the fate of smelter-derived contamination in the soil. Differences in the distributions and binding of metal(loid)s in smelter-affected soils from temperate and (sub)tropical climatic zones indicate the greater vulnerability of the latter. Prevailing wind direction is a key factor affecting the dispersion of smelter emissions and their subsequent deposition into the soils, with greater importance found especially in arid areas. Whereas the greatest contamination is generally observed in the surface soil layers, downward migration of metal(loid)s in the soil profiles has been documented at numerous sites. Contamination of smelter soils significantly affects both plants and soil organisms, but suitable remediation techniques (such as chemical stabilization of soils by amendments) can be used for reducing the bioavailability of contaminants.

KEYWORDS

Soil; Non-ferrous metal smelting; Pollution; Metal(loid)s

1. INTRODUCTION

The mining and smelting of metal ores are one of the important sources of environmental pollution by metals and metalloids (Nriagu, 1996). According to a global inventory byNriagu and Pacyna (1988), in the late 1980s, about 356–857 × 10^6 kg Pb, and 557–1360 × 10^6 kg Zn were released into the environment annually through mining and smelting activities. More recently, Pirrone et al. (2010) estimated that ore mining and processing is responsible for 13% of global Hg emissions. Despite the fact that Hg, Pb, and Cd emissions from the non-ferrous metal industry decreased in Europe over the last 50 years due to the installation of efficient flue gas cleaning systems (Pacyna et al., 2009), the increasing industrial activities on other continents (Asia, Africa) could significantly affect the global emissions of metals (e.g., Pirrone et al., 2010 and Zhang et al., 2011). For example, Zhang et al. (2011) estimated that up to 2007 cumulative emissions from mining, ore dressing, and smelting activities in China were about 1.62 Mt Pb and 3.32 Mt Zn, with the contribution of the smelting processes accounting for 19% and 27%, respectively. It is known that large-scale non-ferrous metal smelters are important local-to-regional sources of pollution. For instance, Li et al. (2011) reported that the largest Pb/Zn smelter in China, located in Zhuzhou (Hunan), emitted 77.82 tonnes of Cd into the atmosphere in the period 1991–2000, accounting for 95% of total emissions from the city.

Soils represent direct sinks for contaminants emitted to the atmosphere by smelters. Contaminants associated with particulates emitted from mining operations are usually concentrated in the fine fraction (<2 μm), and those from smelting even concentrated in the ultrafine particle fraction (<0.5 μm), which may travel greater distances into the environment (Ettler et al., 2005a, Uzu et al., 2011, Csavina et al., 2011, Csavina et al., 2012, Csavina et al., 2014 and Sorooshian et al., 2012). This was perfectly demonstrated by Hou et al. (2006), who studied Pb concentrations and isotope compositions in soils affected by the Horne copper smelter (Rouyn-Noranda, Canada), and found that the signature of the smelter's emissions was to be found as far as 116 km downwind. However, highly volatile contaminants such as Hg can be deposited in soils to lesser extent. Wu et al. (2014) calculated that a large-scale Pb/Zn smelter in Zhuzhou (China) emitted 105 t of Hg during the period 1960–2011, with only 14% of this amount deposited locally in the soil, with the remainder emitted into the global pool. Historic smelting activities were often responsible for extensive soil contamination having persisted through to modern times. Baron et al. (2006) studied medieval metallurgical sites at the Mont-Lozère Massif (southern France), and estimated from the mass balances and isotope compositions that 95% of Pb in the soils originated from the medieval metallurgical workshops, rendering later pollution sources negligible. Similarly, Kierczak et al. (2013) found that soils in the areas around historic smelters, which were active between the 14th and 16th centuries, are still highly polluted with metal(loids)s (up to 4000 mg/kg Cu, 1500 mg/kg Zn, 300 mg/kg As, and 200 mg/kg Pb) due especially to the centuries-long dissolution of smelter wastes into the soils. They concluded that the legacy of old smelting sites, even those which have not been operational for centuries, must still be considered as a serious environmental problem today; these sites should be properly investigated before any decisions regarding remediation strategies are made (Kierczak et al., 2013).

This review's aim is to summarize the relevant studies (primarily those published in the last 15 years) devoted to the pollution of soils in the vicinity of non-ferrous metal smelters (with the exception of alluvial soils and peats), with a special focus on isotope techniques and advanced mineralogical approaches, which are extremely powerful in tracing pollution, especially in areas where mining and smelting operations are geographically close and standard geochemical approaches are unable to distinguish between these two sources of pollution. Moreover, those smelter sites, which heretofore had been overlooked (e.g., in Africa or China), have begun to be investigated in the last decade, in order to understand not only the extent of the pollution, but also the potential impacts on biota. I am following in the footsteps of an older review paper by Rieuwerts and Farago (1995) dedicated to Pb contamination in smelting and mining environments. Here, I especially emphasize the methodological advances, which have enabled the deciphering of geochemical processes in smelter-contaminated soils. Rieuwerts and Farago (1995) concluded that for polluted dusts and soils, Pb absorption by the human body is apparently lower in mining areas than in smelter areas, probably due to differences in chemical and mineralogical forms of Pb. These facts were later confirmed by comparative studies assessing oral bioaccessibility of contaminants from mining- and smelter-impacted soils (e.g., Davis et al., 1996, Rieuwerts et al., 2000 and Ettler et al., 2012a). Along with the consumption of contaminated food and water, which has greater significance for less developed areas where the population is dependent on local food and water sources, it has been suggested that the ingestion/inhalation of dust re-suspended from contaminated soils as well as small-size aerosols from smelting operations are among the major routes of human exposures to metal(loid)s, resulting in elevated contaminant concentrations in the blood or urine (Banza et al., 2009, Cheyns et al., 2014 and Csavina et al., 2014). Despite the increasing number of investigations focused on potential contaminant bioaccessibilities from smelter soils (e.g., Carrizales et al., 2006, Romero et al., 2008, Roussel et al., 2010, Ettler et al., 2012a, Pelfrêne et al., 2011 and Pelfrêne et al., 2015), this subject remains beyond the scope of the present paper.

2. METHODS OF INVESTIGATION

2.1. Screening geochemical studies

The spatial distribution of contaminants in soils near non-ferrous metal smelters depends on: (i) wind direction (hotspots are commonly elongated according to the prevailing wind direction) (Kříbek et al., 2010, Stafilov et al., 2010, Ettler et al., 2011, Ettler et al., 2014a,Ettler et al., 2014b and Šajn et al., 2013, Fig. 1), (ii) the size of particulates emitted by the smelter (smaller particles can travel longer distances) (Csavina et al., 2011, Csavina et al., 2012 and Csavina et al., 2014 and references therein), and (iii) the chemical and mineralogical composition of the particulates (if soluble forms are present in the dusts, greater downward leaching of contaminants can occur in the soil profiles) (Ettler et al., 2012b).

The highest concentrations of contaminants in soils near smelters occur in the uppermost horizons (e.g., Sterckeman et al., 2000, Ettler et al., 2004 and Cecchi et al., 2008). For this reason, the strategies of sampling for determination of the spatial distribution of contaminants in soils are either based on: (i) topsoils collected in the grid to construct distribution maps of elements (Kříbek et al., 2010 and Stafilov et al., 2010,Fig. 1), or (ii) topsoils collected in transects in different geographical or wind directions (Tembo et al., 2006 and Hou et al., 2006). For topsoils, composite samples are generally prepared: for instance, Kříbek et al. (2010) combined samples from the edges as well as from the central point of a 25 m × 25 m square; whereas Sterckeman et al. (2002)obtained a representative topsoil sample at each point by the mixing of twenty topsoil sub-samples collected within a circle of approximately 20 m in diameter. To assess the spatial distribution of contaminants in soils around a smelter, tens to hundreds of samples should be collected, analysed, and statistically evaluated before being plotted. However, inconsistencies in the definition of "topsoil" still occur in the literature. While only the few uppermost centimetres have been collected in some studies (0–5 cm;Rieuwerts and Farago, 1996b, Martley et al., 2004a, Stafilov et al., 2010 and Kříbek et al., 2010), other investigations were based on a ploughed layer corresponding to a depth of 0–20 or 0–25 cm (Adamo et al., 1996, Sterckeman et al., 2002, Douay et al., 2007, Li et al., 2011 and Qasim et al., 2015). Moreover, the topsoil sampling is often done irrespective of the natural development of the soil horizons; thus, the final sample can be a mixture of different soil materials. For the screening studies, it is convenient to compare topsoil concentrations with regional or local background concentration levels. Based on the large scale investigation in the Zambian Copperbelt, Kříbek et al. (2010) did not only collect topsoils, but also subsurface soil samples from a depth of 80–90 cm using a soil probe; by a comparative approach, these authors were able to discriminate concentration hotspots related to mining and smelting activities, and also natural enrichments derived from the bedrock, and transferred into the overlying mineral soil by weathering.

Another sampling strategy is based on the collection of soil profiles at various distances from the smelter stack and positioned according to the prevailing wind direction (upwind*vs.* downwind). The advantage of this approach is that samples collected in the soil pits can be fully described and characterized according to soil taxonomies (Soil Survey Staff, 2014 and IUSS Working Group WRB, 2014). Two approaches are generally adopted: (i) soil sampling according to nominally selected depths (0–5 cm, 5–10 cm, 10–20 cm, etc.) (Dumat et al., 2001, Martley et al., 2004b and Kříbek et al., 2010), or (ii) soil sampling according to naturally developed horizons down to the bedrock, which is generally preferred because different biogeochemical processes can occur in the individual soil horizons (it is recommended to divide thicker horizons into smaller subsamples; Ettler et al., 2005b, van Oort et al., 2006 and Ettler et al., 2011). Finer-scale sampling (0.5–5 cm segments) was also adopted in some cases to describe fine vertical variations in contaminant concentrations in soil profiles (Fernandez et al., 2008, Klaminder et al., 2008, Ettler et al., 2011 and Ettler et al., 2014a). Moreover, sampling of soil profiles in different landscape settings (agricultural *vs.* forest; permanent pasture *vs.* arable land) provides very useful information on the effects of land use on the processes influencing variability in the contaminant distribution and mobility (Ettler et al., 2005b, Ettler et al., 2014a, van Oort et al., 2006, Fernandez et al., 2007, Fernandez et al., 2010, Chrastný et al., 2012a and Vaněk et al., 2013).

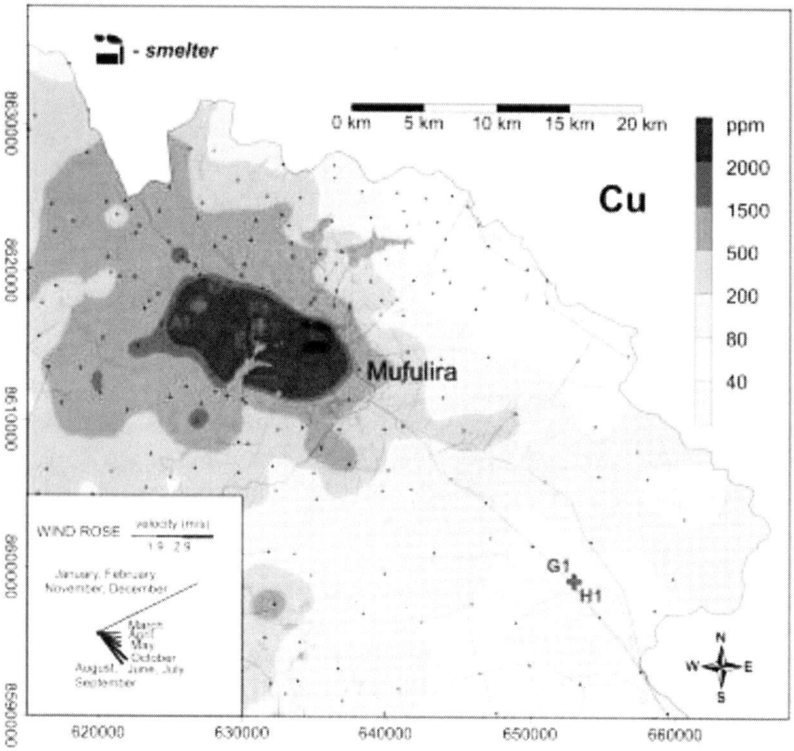

Fig.ure 1. The spatial distribution of Cu in topsoils near a Mufulira copper smelter, Zambian Copperbelt (modified from Kříbek et al. (2010) and Ettler et al. (2014a)). Dots represent the topsoil sampling points, and red crosses the localization of the deep soil pits.

Total contaminant concentrations in soils are generally determined after digestion in mineral acids. Whereas numerous authors use *aqua regia* "pseudo-total" extraction (Martley et al., 2004b, Kříbek et al., 2010 and Šajn et al., 2013); it has been found that especially for tropical soils this extraction might be insufficient, and total digestion in perfluoric (HF) and perchloric ($HClO_4$) is preferred, as was recently demonstrated on Pb from smelter-affected Oxisols (Ettler et al., 2011).

2.2. Chemical extractions and soil mineralogy

A large number of single extractions simulating the "plant-available" or "labile" fraction of metal(loid)s in soils have been proposed and used for smelter-affected soils. The 0.01 M $CaCl_2$ extraction, simulating soil pore water (Dumat et al., 2001, Ettler et al., 2007c and Qasim et al., 2015), and the 0.05 M EDTA extraction simulating the plant-available fraction (Komárek et al., 2006, Chrastný et al., 2012a, Popescu et al., 2013 and Ettler et al., 2014a) have been the most frequently used for metallic contaminants. The popularity of these extraction methods lies in their simplicity, extraction rapidity, comparability with other studies, as well as the availability of soil reference materials with certified (or at least indicative) concentration values for the major metallic contaminants such as Cd, Cr, Cu, Ni, Pb, and Zn (CRM 483 – sewage sludge amended soil, certified by European Community Bureau of Reference [BCR], Standard, Measurements and Testing Programme). In contrast, simple deionized water and 0.1 M Na_2HPO_4 extractions have been suggested for quick estimates of easily exchangeable and specifically sorbed fractions of anionic contaminants (As, Sb) (Ettler et al., 2007b).

Various schemes for sequential extraction procedures (SEP) have also been applied to smelter-affected soils, in order to assess the chemical fractionation of metals (Adamo et al., 1996, Kabala and Singh, 2001, Burt et al., 2003, Ettler et al., 2005b, Knight and Henderson, 2006 and Cecchi et al., 2008) and metalloids (Ettler et al., 2010). In the last decade, BCR SEP based on the work of Rauret et al. (1999) has frequently been used, especially due to its simplicity (a 3-step extraction) and the possibility of accuracy checking using two BCR certified reference materials (CRM 483 – sewage sludge amended soil, and CRM 701 – freshwater sediment) (Yang et al., 2010, Ettler et al., 2011,Chrastný et al., 2012a and Qasim et al., 2015).

Selective extractions, alone, can only hardly predict the binding of contaminants to soil constituents. For this purpose, their combination with mineralogical investigations at various levels of concentration and scale is extremely useful. Basic phase composition obtained by X-ray diffraction (XRD) analysis is routinely used to determine the major soil minerals, including those which can potentially bind contaminants by sorption or co-precipitation (e.g., clay minerals, oxides, and hydroxides) (e.g., Chrastný et al., 2012a and Vaněk et al., 2013). At high contaminant concentrations in soils, even metal(loid)-bearing phases can be directly detected by XRD in bulk samples. Using this kind of analysis, Ettler et al. (2005b) found anglesite ($PbSO_4$) derived from smelter emissions in the most polluted horizon of the forest soil at Příbram (Czech Republic) (total Pb: 35,300 mg/kg); and Burt et al. (2003) detected anglesite, plattnerite (PbO), galena (PbS), quenselite ($PbMnO_2(OH)$), tennantite ($(Cu,Fe)_{12}As_4S_{13}$), arsenosulvanite (Cu_3AsS_4), and chalcopyrite ($CuFeS_2$) in soils near the Anaconda copper smelter (Montana, USA) with 475 mg/kg Pb and 1270 mg/kg Cu.

Scanning electron microscopy (SEM) coupled with energy dispersion spectrometry (EDS) is a useful tool for the visualization and analysis of metal(loid)-bearing particles from bulk soil samples (Henderson et al., 1998, Leguédois et al., 2004 and Cabala and Teper, 2007) or from the dense soil fraction. The dense fraction is often separated by extraction in heavy liquids, such as bromoform or tetrabromethane (densities of 2.89 g/cm^3 and 2.96 g/cm^3, respectively) (Adamo et al., 1996, Morin et al., 1999, Chopin and Alloway, 2007, Ettler et al., 2011 and Konečný et al., 2014), or with a magnet (Lanteigne et al., 2012). Subsequently, specimens can be prepared as individual grains separated under a binocular microscope and attached on double-sided carbon tape (Knight and Henderson, 2006, Chopin and Alloway, 2007 and Cabala and Teper, 2007) or embedded in epoxy resin and prepared as polished (thin) sections (Lanteigne et al., 2014, Ettler et al., 2014a and Konečný et al., 2014).

Smelter-derived particulates from soils are chemically and mineralogically complex materials. They contain: (i) spherically shaped particles representing melt relicts emitted from the smelting process – composed of sulphides, oxides (spinels), or slag-like phases (glass and silicates); and (ii) irregular particles derived from materials not heated up to melting temperatures, which may correspond to unmelted ore, unreacted flux, or fuel materials (Fig. 2) (Adamo et al., 1996, Henderson et al., 1998, Gregurek et al., 1999,Cabala and Teper, 2007, Ettler et al., 2014a and Konečný et al., 2014). Moreover, as smelters are often located close to mining operations, the contamination may be mixed, and angular grains of metal-bearing particles windblown from mine tailing disposal sites are often found in the soil dense fraction (Fig. 2). Additional methods such as Raman spectroscopy, Mössbauer spectroscopy (Lanteigne et al., 2012), or trace element analysis by laser ablation ICP-MS (Lanteigne et al., 2014) can be used for further investigations of the soil particles. At some smelting sites it has been shown that the smelter slag particles are wind-dispersed from the slag dumps and directly deposited into the soils. The SEM/EDS investigations of slag fragments, weathered in soils, have indicated that metallic/sulphide inclusions and glass are the most susceptible to weathering, and could be responsible for the release of metals into soils; nevertheless, this process remains limited to the vicinity of slag disposal sites, and due to the larger sizes of slag particles (>1 mm) does not represent a significant risk for the broader environment (Sobanska et al., 2000, Kierczak et al., 2009 and Chopin and Alloway, 2007). Compared to SEM, higher resolution microscopic methods such as transmission electron microscopy (TEM) coupled to EDS only rarely have been used for visualization and analysis of metal-bearing particles from smelter-affected soils, this probably due to the difficulties with specimen preparation (Buatier et al., 2001).

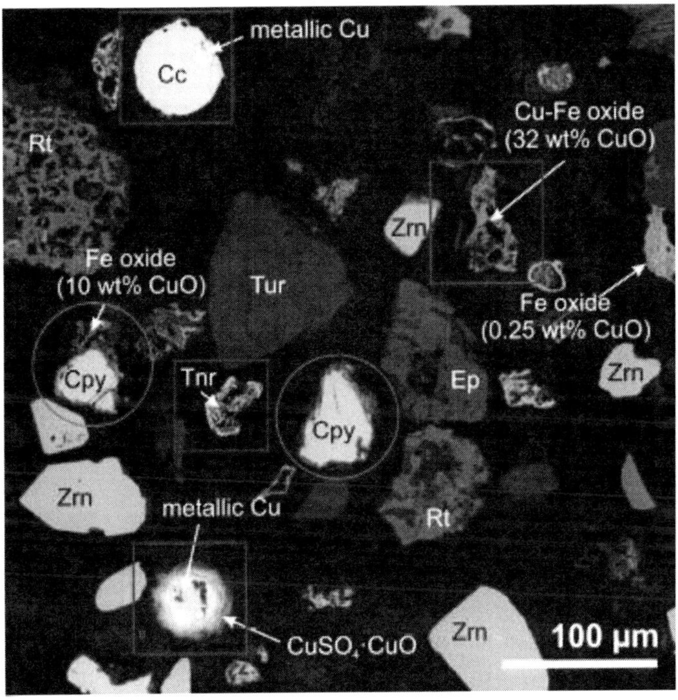

Figure 2. Back-scattered electron (BSE) microphotograph of particles from the dense fraction of the highly polluted soil located 3.6 km downwind from the Mufulira copper smelter, Zambian Copperbelt (total Cu: 8980 mg/kg). Particles derived from smelter dusts are indicated by red squares, green circles indicate particles originating from mine tailings, other particles have a geogenic origin. Abbreviations: Cc – chalcocite (Cu_2S); Cpy – chalcopyrite ($CuFeS_2$); Cu–Fe oxide with 32 wt%, CuO closely corresponding to delafossite ($CuFeO_2$); Ep – epidote ($Ca_2Al_2(Fe,Al)(SiO_4)(Si_2O_7)O(OH)$); Rt – rutile ($TiO_2$); Tnr – tenorite (CuO); Tur – tourmaline (mainly schorl: $NaFe_3Al_6Si_6O_{18}(BO_3)_3(OH)_4$); Zrn – zircon ($ZrSiO_4$) (modified fromKonečný et al., 2014). (For interpretation of the references to colour in this figure legend, the reader is referred to the web version of this article.)

So far, relatively few studies have applied synchrotron-based methods to investigate the solid speciation of contaminants in smelter-contaminated soils. In the pioneering work byMorin et al. (1999), extended X-ray absorption fine structure (EXAFS) spectroscopy on soil samples from the vicinity of Pb/Zn smelters in northern France was used to quantitatively distinguish between Pb binding to humic substances (predominant in wooded soils) and Pb absorption to Fe- and Mn-oxides (predominant in tilled soils). In near-neutral Zn-polluted soils from the same area, EXAFS spectroscopy indicated that the Zn was mainly bound in the clay fraction (<2 μm) either as Zn/Al hydrotalcite ($[Zn,Al](OH)_2(CO_3)_{0.2}\cdot nH_2O$), a layered double hydroxide mineral (Juillot et al., 2003) or as dioctahedral ferruginous smectites (Vespa et al., 2010). In contrast, Roberts et al. (2002)showed by a combination of EXAFS with other mineralogical techniques that in more acidic smelter-impacted soils, the Zn was mainly bound in franklinite ($ZnFe_2O_4$) and sphalerite (ZnS), both emitted during the smelting process, and a minor proportion of Zn was adsorbed by Fe- and Mn-oxides. Nachtegaal et al. (2005) used μ-EXAFS spectroscopy, and confirmed that 10 years after the remediation of a soil polluted by a Belgian Zn-smelter (soil was treated by amendment with cyclonic ash, beringite, compost, and subsequent re-vegetation) about 50–70% of Zn was mainly bound to phyllosilicates. Other techniques such as synchrotron based X-ray microfluorescence (μ-SXRF) imaging were applied to visualize the distributions of contaminants in particles extracted from smelter-affected soils and to find correlations with other elements (Juillot et al., 2003, Nachtegaal et al., 2005 and van Oort et al., 2006). Takaoka et al. (2005) used X-ray absorption near edge structure (XANES) to determine the redox speciation of Sb in smelter soils, and found that Sb occurred exclusively as Sb(V) compounds suggesting that metallic Sb and Sb_2O_3 species, presumably emitted by the smelter, were fully oxidized after their deposition in soils.

2.3. Applications of isotopes

With the increasing use of metal isotopes in environmental geochemistry, as well as advances in isotope measurements using multi-collector ICP-MS, numerous metal isotope systems have become useful for the tracing of contamination sources. Since the review by Rieuwerts and Farago (1995), the development of isotope techniques became one of the major advances in investigation of the origin and fate of contaminants near non-ferrous metal smelters.

It has been demonstrated that Pb isotopes are not fractionated during the smelting process, and variations in Pb isotope compositions of the smelter effluents (slag, fly ash, emissions) can only be attributed to mixing of the Pb sources (ores, recycled Pb) in the smelting furnace charge (Rabinowitz, 2005 and Shiel et al., 2010). Lead isotopes have thus been used to trace the extent of smelter contamination, which can be recognized at large distances, commonly tens of kilometres from the source (e.g., Hou et al., 2006 and Ettler et al., 2011). Lead isotopes can also be used to differentiate between the smelter fingerprint and other potential sources (e.g., petrol combustion, other industrial processes, geogenic sources) (Ketterer et al., 2001, Bacon and Dinev, 2005, Fernandez et al., 2008, Tyszka et al., 2012, Wen et al., 2015, Mihaljevič et al., 2011, Mihaljevič et al., 2015 and Félix et al., 2015); however, changes in the smelting technology, the isotope compositions of the smelted ores, or the isotope composition of other pollution sources can complicate the situation. An interesting application of Pb isotopes was recently demonstrated by Klaminder et al. (2008), who found that O-horizons of boreal forest soils collected near a copper smelter (Rönnskär, Sweden), and subsampled to 0.5-cm thick segments, can be successfully used as a semi-archive for the temporal reconstruction of smelter emissions. If the past atmospheric input recorded in the O-horizons were compensated to leaching losses, the Pb isotope and concentration patterns closely correlated with a similar record in peatbog profiles (Klaminder et al., 2008). A few studies also used Pb isotopes to calculate the downward penetration rate of smelter-derived Pb in the soil profiles. The length of time of the smelter's operation must be known. The fact that soils near smelters are often developed on a mineralized background, and the isotope signature of the processed ore is thus similar to that of the bedrock must also be taken into account for these calculations (Ettler et al., 2004, Prapaipong et al., 2008 and Ettler et al., 2011). Ettler et al. (2004) took advantage of the historical change in smelting technology, where the primary Pb smelting (ore processing) was substituted by the secondary Pb smelting (car battery processing), and calculated, on the basis of changes in Pb isotope compositions of soils and smelter wastes, that the downward penetration rate of Pb, which corresponded to emissions from car battery processing, ranged from 0.3 to 0.36 cm/year. Chrastný et al. (2014) recently verified this hypothesis by incubation and laboratory experiments, and found that such high mobility of Pb can primarily occur in the forest litter. Even higher penetration rates attaining ~1 cm/year were calculated by Prapaipong et al. (2008) for soils near a Pb smelter in Missouri (USA). The Pb isotope compositions of different fractions obtained by selective leaching can show which part of smelter-derived Pb can be mobile. For instance, Ettler et al. (2011)found that in highly polluted Oxisol near a

Cu smelter in the Zambian Copperbelt, the "labile" fraction, corresponding to ~1.5% of total Pb, and exhibited a downward migration at the rate of 1.36 cm/year.

In contrast to Pb, isotope fractionation during the smelting process, leading to isotopically lighter emissions and isotopically heavier slag residues, has been suggested for other metals despite the fact that contradictory results have been reported in the literature. Whereas for Cd, Cloquet et al. (2006), and Shiel et al. (2010) found that smelter emissions were isotopically lighter compared to slag, more recently Chrastný et al. (2015) did not observe any Cd fractionation during the smelting process itself, but suggested that the interaction of Cd with solid soil components after its deposition in soil could be responsible for the variations found in different horizons of forest soils. Similar variable results were observed for Zn isotopes. The first detailed investigations byMattieli et al. (2009) indicated that significant fractionation occurred in the smelting and refining processes in Pb/Zn smelters in the North of France, progressing from the δ^{66}Zn value of sphalerite ore (from +0.88‰ to +0.16‰), to negative values in emissions from the blast furnace (−0.37‰ to −0.66‰) and distillation (−0.35‰). Slag-like particulates collected in proximity to the smelter exhibited positive δ^{66}Zn values (+0.02‰ to +0.19‰), and the authors hypothesized that these were derived from the re-suspension of the slag heaps. In contrast, distant aerosols (PM_{10}) that had negative values (−0.52‰ to −0.02‰) corresponded well to the isotopic signature of the smelter emissions (Mattieli et al., 2009). Similarly, at a Cu smelting site in Slovakia, Bigalke et al. (2010) found that Zn was isotopically lighter in smelter ash (−0.41‰) and in the organic soil horizons near the smelting facility (−0.85‰ to −0.47‰), whereas slag exhibited an isotopically heavier value (+0.18‰). In contrast to these studies, Shiel et al. (2010) reported that Zn isotope fractionation in the Trail Pb–Zn smelting complex (Canada) did not occur to such a large extent; surprisingly, the effluent of the hydrometallurgical process and the fumes were even heavier isotopically (δ^{66}Zn from +0.41‰ to +0.51‰) than the Zn ore concentrate (0.09‰–0.17‰). Relatively little information is known about Cu isotope fractionation in the smelting process. Bigalke et al. (2010) found that δ^{65}Cu values varied little in the wastes and soils near a Cu smelting site, and no source identification was possible. All these observations clearly indicate that isotope fractionation is probably site- and/or technology-specific, and other biogeochemical processes (sorption, biological cycling) may be responsible for the isotope fractionation directly in the soil. In addition, the origin of ore concentrates processed in a smelter may be rather variable over time, and can also be responsible for changes in isotope signatures of the emissions despite the fact that, for example, Zn isotope compositions of sphalerite ores of various origins were found to be remarkably homogeneous with a mean δ^{66}Zn value of +0.16 ± 0.20‰ (Sonke et al., 2008). The applications of diagrams in which isotope signatures of the soils and smelter-derived materials (feed, wastes) are plotted against inverse metal concentration are extremely useful to discern mixing between background and potential pollution sources (Fig. 3a). Moreover, it seems that the multi-isotopic approach, with at least two isotopic systems applied to the feed and effluents from smelting processes and environmental samples (soils, aerosols), will be more extensively used in the future for tracing the contamination near smelters as has recently been successfully demonstrated by Cloquet et al., 2008 and Bigalke et al., 2010, and Wen et al. (2015) (Fig. 3b).

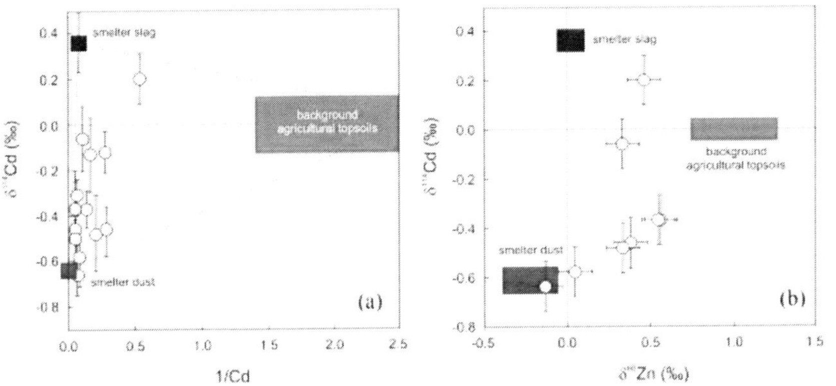

Figure3. (a) Cadmium isotopic compositions (δ^{114}Cd) as a function of inverse Cd concentrations (mg/kg) in soils, and Pb/Zn smelter-derived waste materials from Noyelles-Godault (northern Francc) with estimated compositions for: reference topsoils at 0‰ (green rectangular), polluted topsoils (open circles), dust (red symbol) and slag from a Pb/Zn smelter (black symbol). (b) Cd and Zn isotope compositions (δ^{114}Cd *vs.* δ^{66}Zn) in soils, smelter slag and dust, and theoretical mixing curves calculated according to compositions of three end members (isotopically light smelter dust, isotopically heavy smelter slag, and uncontaminated soils) (redrawn after Cloquet et al., 2006 and Cloquet et al., 2008). (For interpretation of the references to colour in this figure legend, the reader is referred to the web version of this article.)

2.4. Laboratory and field experiments

Laboratory and field (*in situ*) experiments are suitable tools for understanding the processes of contaminant release from smelter particulates after their deposition into soils. For instance, Cornu et al. (2007) used a laboratory pot experiment to study how redox changes influence Cd dynamics, in a re-wetted to 90% water-holding capacity (WHC) agricultural soil, contaminated by smelter fallout. As Cd is generally presumed to be more mobile in polluted soils, compared to other metals, this experiment confirmed a relatively high release of Cd during the initial stages of soil incubation, followed by a progressive decrease related to the establishment of reducing conditions, leading to precipitation of Cd sulphide in the soil (Cornu et al., 2007). Despite the fact that extraction methods have frequently been used to simulate soil solutions and to estimate the "labile" or "easily-extractable" pool of metal(loid)s in contaminated soils, it is always convenient to compare these extraction results with real pore water samples collected in the field.Clemente et al. (2008) sampled pore waters in soils near a former Cu refinery 20 years after cessation of the pollution source's activity, and found contrasting concentrations and mobility for the studied contaminants. Whereas Cu and As concentrations in the soil pore water were much higher closer to the soil surface, elevated concentrations of Zn and Cd were found in the deeper soil layers, indicating their potential vertical mobility in the soil profiles. In contrast, the mobility and concentrations of Pb and Sb in soil pore waters were rather low, even in the uppermost soil horizons, where the highest concentrations of these elements were observed (Clemente et al., 2008). Denaix et al. (2001) found up to 11 mg/l Zn in pore water, collected by zero-tension lysimeters, under the surface soil horizon (11,995 mg/kg Zn) in a Pb/Zn smelter-polluted area. Whereas 95% of Zn and Cd were mainly present in the dissolved form and were highly mobile, 50% of Pb in soil pore water occurred as colloids (Denaix et al., 2001). Citeau et al. (2003) also demonstrated that the Pb colloidal fraction was important, primarily in agricultural and grassland soils; however, the proportion of the colloidal fraction for all three studied contaminants (Pb, Zn, Cd) in a forest soil (podzol) was <10%.

Whereas particulates emitted by the smelter stacks are often composed of highly soluble phases, the slag-like fragments deposited into the soil by wind dispersion are generally less soluble (e.g., Ettler et al., 2005a, Knight and Henderson, 2006, Uzu et al., 2011 and Ettler et al., 2014b). Ettler et al., 2012b, Ettler et al., 2012c and Ettler et al., 2015a recently used laboratory pot experiments with re-wetted acidic soils, where porous polyamide experimental bags (1-μm mesh) filled with Pb smelter fly ash were incubated for 21 days. This approach has been found very useful for the calculation of mass balances and fluxes of individual contaminants released from fly ash into the soil, as well as for possible investigation of mineralogical changes of the fly ash after incubation using XRD, SEM, or TEM. Whereas As, Cd, and Zn were rapidly released from the fly ash into the soils and soil pore water, Pb and Sb were relatively efficiently retained due to formation of less soluble secondary phases, observed by TEM, within the weathered fly ash material (anglesite, $PbSO_4$; and bindheimite, $Pb_2(Sb,Sn)_2O_6(O,OH)$, respectively) (Ettler et al., 2012b and Ettler et al., 2012c). Chrastný et al. (2012b) incubated mixtures of a similar Pb smelter fly ash with litter/humus, and also found that Cd and Zn were mobilized to a higher extent than Pb. The leaching and downward migration of contaminants is even more pronounced in tropical soils, as has recently been demonstrated by laboratory incubation of Cu smelter fly ash in an Oxisol (Ettler et al., 2015a).

A limited number of long-term field investigations of smelter fly ash reactivity in soils exist in the literature. They are generally based on three experimental setups, represented by the following examples:

(i)Copper slag particles attached to a polymethylmethacrylate (PMMA) polymer support were exposed in acidic forest soils over the period of 18 months, and their weathering has been studied by X-ray fluorescence analysis (XRF) and SEM/EDS (Birkefeld et al., 2006). Despite the fact that Cu slag particles were found to be relatively resistant to weathering compared to other materials investigated (PbO and Cu concentrate), comparison with the results of laboratory leaching tests indicated that the faster dissolution rates observed in the field are probably caused by more intensive biological activity (effects of exudates from plant roots and fungi) (Birkefeld et al., 2006).

(ii)Ettler et al. (2012b) carried out *in situ* incubations of Pb smelter fly ash in acidic forests and grassland soils using an experimental bag technique (described above). Their results indicated that up to 66% of fly ash dissolved after 1 year of incubation in soils, and newly formed phases efficiently controlled the release of Pb, which was significantly less leached than other contaminants. In addition, the release of Pb was dependent on soil pH; whereas Cd and Zn were leached to a greater extent irrespective of soil pH, and partitioned more easily into the "labile" soil fractions (Ettler et al., 2012b).

(iii)Yet another type of long-term experiment was conducted by Nowack et al. (2010), who used an artificial contamination approach.

They manually introduced a filter dust from a non-ferrous metal smelter into the topsoil, equipped the experimental site with field lysimeters to collect soil water, and planted the site with trees. They investigated changes in chemical fractionation of contaminants in soils as a function of time (up to 42 months). The authors found that the largest changes in speciation, expressed as the redistribution between the fractions of the sequential extraction, occurred within the first 6 months. Moreover, plant activity kept the metals in a more soluble form, whereas at those experimental plots with no vegetation, a much higher proportion of metals were present in more refractory phases (Nowack et al., 2010).

3. DISTRIBUTION AND MOBILITY OF CONTAMINANTS IN SMELTER SOILS

An overview of those studies dealing with soils polluted by non-ferrous metal smelters in temperate climatic zones; plus those in tropical, subtropical, and arid areas is given inTable 1 and Table 2, respectively. These tables clearly show that several sites have been extensively studied by numerous authors, having unravelled various aspects of the distribution, binding, and (bio)availability of contaminants in soils (Noyelles-Godault/Auby and Mortagne du Nord in northern France, Příbram in the Czech Republic, sites in the African Copperbelt, and a few localities in Canada and Poland). Moreover, a significantly lower numbers of studies have been conducted in (sub)tropical and dry areas, despite the fact that these soils are generally more vulnerable to metal(loid) pollution, because they generally exhibit a low soil organic matter (SOM) content, as well as containing low-activity clays (kaolinite) and less reactive Fe-oxides (hematite), leading to greater mobility of the contaminants (Rieuwerts, 2007). Maximum contaminant concentrations in these soils often exceed regional screening levels (RSLs) of chemical contaminants for residential soils, and in some cases also industrial soils, as proposed by the USEPA (2015): As (0.67 and 3.0 mg/kg), Cd (7.0 and 98 mg/kg), Co (2.3 and 35 mg/kg), Cu (310 and 4700 mg/kg), Hg (elemental – 0.94 and 4 mg/kg), Pb (400 and 800 mg/kg), and Zn (2300 and 35,000 mg/kg), respectively. Canadian soil quality guidelines for the protection of environmental and human health (CCME, 2007) have often been used for polluted tropical soils. For example, Kříbek et al. (2010) in the their screening study, based on 719 topsoil samples across the Zambian Copperbelt, found that 91.23% of soils exceeded the Cu concentration guideline for agricultural/residential soils (63 mg/kg), and that 85.26% of soils exceeded the guideline for commercial/industrial soils (91 mg/kg).

Table 1. Examples of studies devoted to smelter-affected temperate soils.

Country	Locality	Smelting operation	Contaminants (maximum concentrations in mg/kg)	Spatial distribution	Soil profiles	Mineralogy	Extractions	Isotopes	Bioavailability	Reference
France	Noyelles-Godault/ Auby	Pb/Zn smelters	Pb (1900)		* (Topsoils)	*				Morin et al. (1999)
			Pb (4890), Zn (2685), Cd (61.7)		*					Sterckeman et al. (2000)
			Pb (4264), Cu (127), Cd (39.9)		*		*			Dumat et al. (2001)
			Pb (4480), Zn (11,995), Cd (169)		*	* (Colloids)				Denaix et al. (2001)
			Pb (2670), Zn (1865)		*	*				Buatier et al. (2001)
			Zn (1378), Pb (1132), Cd (21) and 15 other elements	*						Sterckeman et al. (2002)
			Zn (4391)		*	*	*			Juillot et al. (2003)
			Pb (1132), Cd (21)	* (Topsoils)				* (Pb, Cd)		Cloquet et al. (2006)
			Cd (21), Zn (n.a.)	* (Topsoils)				* (Cd, Zn)		Cloquet et al. (2008)
			Pb (1815), Cd (11.8)	* (Topsoils)					* (Vegetables)	Douay et al. (2007)
			Zn (6908), Pb (3711), Cd (31.4) and 15 other elements	*	*					Douay et al. (2008a)
			Pb (821), Cd (14.5)		* (Topsoils)				* (Wheat)	Douay et al. (2008b)
			Pb (41,959), Zn (38,763), Cd (2402)	*			*			Douay et al. (2009)
			Zn (1215), Pb (1058), Cd (14.9)		* (Topsoils)				* (Herbs)	Bidar et al. (2009)
			Zn (6908), Pb (3026), Cd (31.4)	* (Topsoils)			*		* (Bioaccessibility)	Roussel et al. (2010)
			Zn (2550)			*				Vespa et al. (2010)
			Zn (1250), Pb (824), Cd (13)	* (Topsoils)			*		* (Bioaccessibility)	Pelfrêne et al. (2011)
			Zn (10,606), Pb (364), Cd (70)		* (Litter)				* (Soil macrofauna)	Codet et al. (2011)
			Zn (7726), Pb (3551), Cd (72)		* (Topsoils)		*		* (bean)	Qasim et al. (2015)
			Zn (7263), Pb (6002), Cd (62.7)	* (Topsoils)					* (Bioaccessibility)	Pelfrêne et al. (2015)
	Mortagne-du-Nord	Pb/Zn smelters	Zn (19,000)	*					* (Soil macrofauna)	Nahmani and Lavelle (2002)
			Zn (17,956), Pb (4720), Cd (109)		*	* (Colloids)				Citeau et al. (2003)
			Zn (2635), Pb (517), Cu (113), Cd (3.39)		*	*				Leguédois et al. (2004)
			Zn (2635), Pb (517), Cu (113), Cd (3.39)		*	*				van Oort et al. (2006)
			Zn (110), Pb (59.7)		*	*	*			Fernandez et al. (2007)
			Pb (69.9)		*		*	* (Pb)		Fernandez et al. (2008)
			Zn (109), Pb (61.5)		*		*			Fernandez et al. (2010)
	Bazoches-les-Gallerandes	Pb smelter (secondary)	Pb (1932), As (28), Sb (67) and other 5 elements		*	*	*			Cecchi et al. (2008)
Czech Republic	Pribram	Pb smelter (primary and secondary)	Pb (37,300), Zn (27,600), Cu (715), Cd (48), As (1640), Sb (980)	* (Transect)						Rieuwerts and Farago (1996a)
			Hg (2.32)	* (Transect)						Rieuwerts and Farago (1996)
			Pb (58,500), Zn (21,900), Cu (2440), Cd (96)	* (Topsoils)						Rieuwerts et al. (1999)
			Pb (4660), Zn (5740), Cu (146), Cd (43.5), As (204), Sb (167)	* (Garden topsoils)		*	*		* (Bioaccessibility)	Rieuwerts et al. (2000)
			Pb (81,041), Zn (1868), Cu (1097), Cd (46.63) and 32 other elements	* (Humus)						Suchara and Sucharová (2004)
			Pb (35,300)		*			* (Pb)		Ettler et al. (2004)
			Pb (35,300)		*	*	*			Ettler et al. (2005b)
			Pb (17,000)		*		*	* (Pb)		Komárek et al. (2006)
			Pb (36,234), Zn (519), Cu (488), Cd (11.9), As (120), Ag (25.1), Se (5.88)		*			* (Pb)	* (Mushrooms)	Komárek et al. (2007a)
			Hg (6.49)		*					Ettler et al. (2007a)
			Sb (705), As (336)		*		*			Ettler et al. (2007b)
			Pb (7331), Cd (98)						* (Maize)	Komárek et al. (2009)
			Sb (2058), As (447)		*		*			Ettler et al. (2010)
			Pb (57,100)		*			* (Pb)		Chrastný et al. (2014)
Canada		Cu/Ni smelters	Ni (2149), Cu (1891)	* (Topsoil)		*	*			Adamo et al. (1996)

	Sudbury (Ontario)		Ni (2150), Cu (1890), Zn (146) and other 5 elements	* (Topsoil)		*	*			Adamo et al. (2002)
			Cu (321), Ni (234), Zn (67.5), Pb (56.0), As (20.0)		* (Transect)				* (Microbes)	Anand et al. (2003)
			Cu (391), Pb (81.7), Ni (38.5) and other 7 elements		*				* (Spruce)	Nkongolo et al. (2008)
			n.a.			*				Lanteigne et al. (2012)
			n.a.			*				Lanteigne et al. (2014)
			Ni (1600), Cu (1330), Pb (176) and other 8 elements		*					Narendrula et al. (2012)
			Ni (1600), Cu (1330), Pb (176) and other 8 elements		*		*		* (Spruce)	Narendrula et al. (2013)
	Rouyn-Noranda (Québec)	Cu smelter	n.a.	* (Transect, humus)		*				Knight and Henderson (2006)
			Pb (869)		* (Transect)			* (Pb)		Hou et al. (2006)
	Flin Flon (Manitoba)	Cu/Zn smelter	Zn (7428), Cu (2670), Pb (1692), As (558), Cd (39), Hg (10) and other 2 el.	* (Transects)		*	*			Henderson et al. (1998)
Poland	Olkusz	Zn/Pb smelter	Pb (147,700), Zn (83,400), Cd (428), Tl (139)		* (Topsoil)	*				Cabala and Teper (2007)
			Zn (9630), Pb (1595), Cd (82.9), Cu (48.1)		* (Transect)					Tozsa et al. (2010)
			Zn (>3000), Pb (>1000), Cd (>20)		*	*	*			Chrastný et al. (2012a)
			Tl (30.1)		*	*	*			Vaněk et al. (2013)
			Cd (153)		*			* (Cd)		Chrastný et al. (2015)
	Legnica and Glogow	Cu smelters	Cu (1710), Pb (460)		*		*			Karczewska et al. (1996)
			Cu (426), Pb (130), Zn (92.3)		*		*			Kabala and Singh (2001)
			Cu (4500), Pb (2146), Zn (1222), As (90)		*			* (Pb)		Tyszka et al. (2012)
	Bukowno	Pb/Zn smelter	Zn (12,400), Pb (3570), Cd (73.2)		*		*			Verner et al. (1996)
	Rudawy Janowickie	Cu smelter (medieval)	Cu (4011), Zn (1503), As (333), Pb (200)		*	*		* (Pb)		Kierczak et al. (2013)
USA	Palmerton (Pennsylvania)	Zn smelter	Zn (32,500), Pb (3600), Cd (780), Cu (700)[a]		*			* (Pb)		Ketterer et al. (2001)
			Zn (6200)		*	*				Roberts et al. (2002)
	Glover (Missouri)	Pb smelter	Pb (1439)		*			* (Pb)	* (Tree biomass)	Prapaipong et al. (2008)
	Anaconda (Montana)	Cu smelter	Cu (1270), Zn (849), Pb (474.5)		*		*			Burt et al. (2003)
	10 sites	Pb smelters (secondary)	Pb (2550)		* (Topsoils)					Eckel et al. (2002)
UK	Avonmouth	Pb/Zn smelter	Zn (3630), Pb (1740), Cu (161), Cd (54.5)		* (Topsoils)				* (Earthworms)	Nahmani et al. (2007)
	Prescot	Cu smelter	Cu (677), Pb (809), As (143), Zn (176)		*		*			Clemente et al. (2008)
South Korea	Janghang	Cu/Pb/Zn smelter	As (907), Pb (590), Cu (441), Zn (61)		* (Topsoils)		*		* (Earthworms)	Bade et al. (2012a)
			As (145)		*		*		* (Bacteria)	Shagol et al. (2014)
			Pb (219), Cu (122), As (18), Zn (67)		* (Topsoils)				* (Fungi)	Krishnamoorthy et al. (2015)
Finland	Harjavalta	Cu/Ni smelter	Cu (2304), Ni (161)		*					Derome and Lindroos (1998)
			Cu (2900), Ni (460)	* (Topsoils)					* (Worms)	Salminen and Haimi (1999)
Japan	Shiga	Sb smelter	Sb (2100)	*	*	*				Takaoka et al. (2005)
Sweden	Rönnskär	Cu smelter	Pb (9700), Hg (1.5)	* (Transect)	*			* (Pb)		Klaminder et al. (2008)
Belgium	Lommel	Zn smelter	Zn (20,476), Pb (2996), Cu (2132), Hg (737) As (312) and other 3 elements		* (Topsoils)	*				Nachtegaal et al. (2005)
Slovakia	Krompachy	Cu smelter	Cu (8087), Zn (2084)	* (Transect)	*			* (Cu, Zn)		Bigalke et al. (2010)
Bulgaria	Kuklen	Pb/Zn smelter	Zn (5231), Pb (4196), Cu (432), Ni (303), Cr (193), Cd (87)		* (Topsoils)			* (Pb)		Bacon and Dinev (2005)
Kosovo	Kosovska Mitrovica	Pb/Zn smelter		* (Topsoils)						Šajn et al. (2013)

(continued on next page)

Country	Locality	Smelting operation	Contaminants (maximum concentrations in mg/kg)	Spatial distribution	Soil profiles	Mineralogy	Extractions	Isotopes	Bioavailability	Reference
			Pb (35,000), Zn (12,000), As (3900), Cu (1600), Sb (1400) and other 31 elements							
Macedonia	Veles	Pb/Zn smelter	Zn (27,000), Pb (15,000), Cu (1700), Cd (600) and other 38 elements	* (Topsoils)						Stafilov et al. (2010)

n.a. — not available.

USEPA (2015) selected screening levels for residential/industrial soils, respectively (in mg/kg): As (0.67/3.0), Cd (7.0/98), Co (2.3/35), Cu (310/4700), Hg (0.94/4), Ni-soluble salts (150/2200), Ni – refinery dust (82/1100), Pb (400/800), Sb (3.1/47), Zn (2300/35,000).

CCME (2007) quality guidelines for agricultural/residential/industrial soils, respectively (in mg/kg): As (12/12/12), Cd (1.4/10/22), Co (40/50/300), Cu (63/63/91), Hg (6.6/6.6/50), Ni (50/50/50), Pb (70/140/600), Sb (20/20/40), Zn (200/200/360).

[a] Estimated from figures.

Table 2. Examples of studies devoted to smelter-affected soils from (sub)tropical or arid areas.

Country	Locality	Smelting operation	Contaminants (maximum concentrations in mg/kg)	Spatial distribution	Soil profiles	Mineralogy	Extractions	Isotopes	Bioavailability	Reference
Zambia	Central-northern Copperbelt	Cu smelters	Cu (41,900), Co (606), Pb (503), Zn (450), As (255) and other 5 elements	* (Topsoils)	*					Kříbek et al. (2010)
			Cu (3067), Zn (280), Co (84.2), Pb (50.0), As (3.81)		* (Topsoils)				* (Cassava)	Kříbek et al. (2014)
	Kitwe (Copperbelt)	Cu smelter	Cu (37,770), Co (676), Zn (549), Pb (419), As (85.2)	* (Topsoils)	*	*	*	* (Pb)		Ettler et al. (2011)
			Cu (27,410), Co (606), Pb (480), Zn (450), As (255)	* (Topsoils)					* (Bioaccessibility)	Ettler et al. (2012a)
			Cu (37,770), Co (676), Zn (549), Pb (419), As (85.2)		*			* (Pb)	* (Tree biomass)	Mihaljevič et al. (2011)
	Mufulira (Copperbelt)	Cu smelter	Cu (8980), Zn (83), Pb (42), Co (46)		*	*	*		* (Worms)	Konečný et al. (2014)
			Cu (8980), Zn (97), Pb (42), Co (46)		*	*	*			Ettler et al. (2014a)
	Kitwe and Mufulira (Copperbelt)	Cu smelter	Hg (0.39) and other 7 elements	* (Topsoils)	*					Podolský et al. (2015)
	Kabwe (Zambia)	Pb smelter	Pb (759), Zn (106), Cu (58.2), Cd (22.3)	* (Topsoil, transects)						Tembo et al. (2006)
DR Congo	Lubumbashi (Katanga, Copperbelt)	Cu smelters	Cu (14,200), Co (6150), Zn (1250), Pb (809) and other 7 elements	*						Narendrula et al. (2012)
			Cu (14,200), Co (6150), Zn (1250), Pb (809) and other 7 elements	*			*			Narendrula et al. (2013)
			Cu (8957), Co (145), Zn (622)	* (Topsoil)			*			Pourret et al. (2015)
			Cu (11,110), Co (152)[a]	* (Topsoils)					* (Metallophytes)	Lange et al. (2014)
			Cu (150) (mean value)	* (Topsoils)					* (Bioaccessibility)	Cheyns et al. (2014)
China	Zhuzhou (Hunan)	Pb/Zn smelter	Zn (3349), Pb (1197), Cu (157), As (93), Cd (41.1), Hg (2.89)	* (Topsoils)	*					Li et al. (2011)
			Ag (3.71), Bi (21.4), Co (30), Cr (199) and other 6 elements	* (Topsoils)	*					Li et al. (2014)
	Hangzhou (Zhejiang)	Cu smelter	Zn (11,840), Cu (716), Cd (8.67)		*		*			Liu et al. (2010)
	Huludao (Lianing)	Zn smelter	Hg (3.80)		* (Topsoils)				* (Crops)	Zheng et al. (2007)
	Xikuangshan (Hunan)	Sb smelter	Sb (5045), As (205)		* (Topsoils)		*		* (Radish)	He (2007)
	Fuyang (Zhejiang)	Cu smelter (secondary)	Hg (15)	* (Topsoils)	*					Yin et al. (2009)
	Magu (Guizhou)	Zn smelter	Pb (37,770), Zn (31,625), Cd (131)		* (Topsoils)	*	*	* (Pb,S)		Yang et al. (2010)
	Nanning (Guangxi)	Pb/Sb smelter	Pb (992), Zn (597), Cu (39), Cd (22) [geometric means]	* (Topsoils)					* (Vegetables)	Cui et al. (2004)
	Zhujiawu (Zhejiang)	Cu/zn smelter	Zn (3219), Cu (658)	* (Topsoils, transect)					* (Microbes)	Wang et al. (2007)
Mexico	Monterrey	Pb smelter	Pb (111,000), Zn (51,500), As (20,400), Cu (594), Cd (573)	* (Topsoils)		*	*			Gutiérrez-Ruiz et al. (2005)
			Pb (31,410), Zn (12,280), As (10,330), Cu (400), Cd (20)	* (Topsoils)			*			Villalobos et al. (2010)
			Pb (31,420), Zn (12,280), As (6043), Cu (726), Sb (270), Cd (88)	* (Topsoils)		*	*			Gutiérrez-Ruiz et al. (2012)
	San Luis Potosí	Cu smelter	Pb (24,615), Cu (15,099), As (9340), Zn (5267), Cd (594)	* (Topsoils)		*	*			Gutiérrez-Ruiz et al. (2005)
			Pb (12,600), As (4424)	* (Topsoils)					* (Bioaccessibility)	Carrizales et al. (2006)
			Pb (2532)		* (Topsoils)				* (Bioaccessibility)	Romero et al. (2008)
			Pb (24,620), Cu (10,620), As (10,330), Zn (5080), Cd (680)	* (Topsoils)			*			Villalobos et al. (2010)
Chile	Puchuncaví	Cu smelter	Cu (3718), Zn (174), Zn (105), Cd (1)	* (Topsoils)	*		*			Ginocchio et al. (2004)
			Cu (636)	* (Topsoils)			*			Neaman et al. (2009)
			Cu (4449), Zn (839), Pb (302) and other 18 elements	* (Topsoils)		*				Parra et al. (2014)
Australia	Port Kembla (New South Wales)	Cu smelter	Cu (1597), Pb (295), Zn (180), As (26)	*						Martley et al. (2004a)
			Cu (1348), Zn (325), Pb (251), Cd (3.8)		*	*	*			Martley et al. (2004b)

(continued on next page)

Namibia	Tsumeb	Cu/Pb smelters	Hg (4.39) and other 7 elements	* (Topsoils)	*					Podolský et al. (2015)
			Zn (9740), Pb (8170), Cu (4970), As (2370), Cd (511)		*			* (Pb)	* (Tree biomass)	Mihaljevič et al. (2015)
Spain	Tharsis, Riotinto and Huelva	Cu smelters	Pb (8306), As (3074), Cu (850), Zn (257)	*		*	*			Chopin and Alloway (2007)
Brazil	Santo Amaro (Bahia)	Pb smelter	Zn (95,940), Pb (37,460), Cu (3196), Cd (771)	* (Topsoils)					* (Earthworms)	Niemeyer et al. (2010)
USA	Hayden (Arizona)	Cu smelter	Pb (340), As (105), Cd (10)[b]		*			* (Pb)		Félix et al. (2015)

USEPA (2015) selected screening levels for residential/industrial soils, respectively (in mg/kg): As (0.67/3.0), Cd (7.0/98), Co (2.3/35), Cu (310/4700), Hg (0.94/4), Ni-soluble salts (150/2200), Ni – refinery dust (82/1100), Pb (400/800), Sb (3.1/47), Zn (2300/35,000).
CCME (2007) quality guidelines for agricultural/residential/industrial soils, respectively (in mg/kg): As (12/12/12), Cd (1.4/10/22), Co (40/50/300), Cu (63/63/91), Hg (6.6/6.6/50), Ni (50/50/50), Pb (70/140/600), Sb (20/20/40), Zn (200/200/360).
[a] Data provided by Olivier Pourret.
[b] Estimated from figures.

The highest levels of metal(loid)s were generally found in the surface soil horizons, but comparisons of the soil horizons with the bottom layers of the soil pits, bedrock, or local-to-regional background have often indicated that the smelter-derived contamination can, to some extent, move downwards in the soil profiles (Derome and Lidroos, 1998,Sterckeman et al., 2000, Kabala and Singh, 2001, Burt et al., 2003, Ettler et al., 2004,Martley et al., 2004b, Ettler et al., 2005b, Ettler et al., 2007a, Neaman et al., 2009, Křibek et al., 2010, Li et al., 2011, Chrastný et al., 2012a, Vaněk et al., 2013 and Podolský et al., 2015). However, contrasting vertical distribution patterns, with higher concentrations of metal(loid)s in the deeper horizons, have also occasionally been observed; particularly at old (medieval) smelting sites, where slag waste dumps were re-vegetated, leading to the formation of a thin soil layer, as recently reported from a site near a former Cu smelting factory in Poland (Kierczak et al., 2013).

Surprisingly consistent data on metal speciation and mobility in Pb/Zn smelter-affected soil profiles were reported from various localities, indicating that Cd and Zn generally seem to be more mobile than Pb; and that the proportion of contaminants in the "labile" fractions of SEP, and subsequent vertical mobility, is generally higher for forest soils than for agricultural soils (e.g., Ettler et al., 2005b, van Oort et al., 2006, Douay et al., 2009 and Chrastný et al., 2012a). The downward movement of contaminants can occur*via* leaching (Dumat et al., 2001 and Fernandez et al., 2007), the movement of metal-bearing colloids (Denaix et al., 2001 and Citeau et al., 2003), mechanical perturbation by ploughing (Rieuwerts and Farago, 1996a and Ettler et al., 2005b), or biological activity (e.g., earthworms; Sterckeman et al., 2000 and Fernandez et al., 2010). Sequential extraction results as well as EXAFS investigations have indicated that the role of organic matter was crucial in the binding of Pb in forest soils; whereas in tilled soils, Pb was predominantly bound to Fe-oxides (Morin et al., 1999, Dumat et al., 2001, Ettler et al., 2005b and van Oort et al., 2006). Despite the fact that Pb is generally less mobile than other contaminants, the calculations based on isotope fingerprinting have indicated that the downward penetration rate of Pb can be remarkably fast, and varies in the range of 0.3–1 cm/year (Ettler et al., 2004, Ettler et al., 2011 and Prapaipong et al., 2008). It has been suggested that this phenomenon may be directly related to the mineralogical composition of the atmospheric fallout. Whereas particulates from primary Pb smelters are composed of relatively insoluble phases ($PbSO_4$, $PbSO_4 \cdot PbO$, PbS) (Sobanska et al., 1999), much higher solubility was observed for secondary Pb smelter particulates, composed of Pb-bearing chlorides (e.g., $PbCl_2$, KPb_2Cl_5 or $Na_3Pb_2(SO_4)_3Cl$), whose higher solubility is probably the main reason for the higher downward leaching of Pb in the soil profiles (Ettler et al., 2005a, Ettler et al., 2005b and Ettler et al., 2012a).

Spherical particles of variable compositions, up to tens of μm in size, were commonly found in the dense fraction separated from topsoils at numerous smelting sites (Adamo et al., 1996, Henderson et al., 1998, Knight and Henderson, 2006, Cabala and Teper, 2007, Ettler et al., 2014a and Konečný et al., 2014). Apart from relatively stable slag-like particles (composed of glass, magnetite, silicates) (Knight and Henderson, 2006 and Chopin and Alloway, 2007), more reactive metal-bearing sulphides, oxides, sulphates, and carbonates were also found. However, the contamination is often mixed, as smelters are commonly located near mining operations and the soil dense fraction not only contains spherical particles formed by condensation at high temperatures during the smelting process, but also wind-dispersed particles from mine tailing sites (Fig. 2). Upon weathering, these particles can release common metallic contaminants (Pb, Cu, Zn, Cd, Ni), and also other associated trace elements (Tl, Co, As, Se), as revealed by laser ablation ICP-MS investigations (Lanteigne et al., 2014) or SEM/EDS analysis (Cabala and Teper, 2007). Lanteigne et al. (2012) recently suggested a weathering sequence for rounded Cu/Ni smelter-derived particulates, exposed for decades in temperate topsoils near Sudbury: (i) sulphide inclusions are exposed to soil pore water through fractures within the surrounding slag-like matrix (glass, Fe-silicates, magnetite); (ii) oxidation and release of metals occurs, followed by their co-precipitation in newly formed Fe-oxides; and (iii) the silicate matrix of rounded inclusions is slowly dissolving for a longer period of time, while magnetite crystals remain unaltered, forming a collapsed skeleton of the original particle. Lanteigne et al. (2012) also found that Cu-bearing spinel ($CuFe_2O_4$) from deposited smelter emissions tended to dissolve more easily than a Ni-bearing one (trevorite, $NiFe_2O_4$), leading to preferential release of Cu and its partitioning into the "labile" fractions in the soil. The most dynamic mineralogical transformations of smelter particulates, which mainly include dissolution and precipitation of the newly formed phase, occur in the organic horizons due to the presence of organic ligands, and also due to generally more acidic conditions (e.g., Ettler et al., 2012a and references therein). Interestingly, Cabala and Teper (2007) found in rhizospheric soils polluted by smelter particulates that plant-root exudation not only enhances the dissolution of primary phases, but also the crystallization of secondary metalliferrous solids (mainly carbonates, sulphosalts, and oxides), which precipitated directly on the plant root epidermis and fungal hyphae. The precipitation of secondary phases formed during the pedogenic processes in the form of Fe-, Mn-, or clay-rich coatings, or spherulites that are often difficult to characterize due to their size or chemical/mineralogical complexity, and which play a key role in the storage of contaminants in smelter-affected agricultural soils from temperate areas (Leguédois et al., 2004 and van Oort et al., 2006). Moreover, the natural precipitation of secondary phases directly in soil can immobilize contaminants, as recently demonstrated by Gutiérrez-Ruiz et al. (2012) in highly polluted soils near a Mexican Pb smelter, where insoluble Pb arsenates formed.

Sampling of soil profiles, from both wooded and non-wooded plots, at the same distance from the smelter indicated that much higher contaminant concentrations were observed in forest soils, especially in organic horizons (Rieuwerts and Farago, 1996a, Ettler et al., 2005b, Douay et al., 2009, Chrastný et al., 2012a and Vaněk et al., 2013). This has been explained by the higher interception of smelter emissions by tree canopies, leading to high levels of contaminants in the litter and in the humus layer (Rieuwerts and Farago, 1996a and Ettler et al., 2005b). It has also been suggested that lower concentrations of contaminants in tilled soils can reflect the effects of ploughing, and subsequent dilution in the topsoil layers (Rieuwerts and Farago, 1996a) or annual crop off-take (Ettler et al., 2005b). Ginocchio et al. (2004) at Cu smelter-affected sites in central Chile found that soils that developed below shrubs exhibited higher concentrations of metals than those that developed in open spaces; however, their bioavailability was much lower due to binding to SOM. Nonetheless, surprisingly contrasting metal distributions have recently been reported by Ettler et al. (2014a) in tropical soil profiles from forested and grassland areas near a Cu smelter in the Zambian Copperbelt. A higher proportion of metal-bearing particles and higher metal concentrations were found in soils from unforested sites. It has been hypothesized that this phenomenon could be related to the more frequent and intense bushfires in the forested areas, leading to mobilization of pollutants contained in biomass-rich topsoils back into the atmosphere (Ettler et al., 2014a). This observation has also been supported by Pb isotope compositions in forest topsoils, which were free of anthropogenic Pb, in contrast to the youngest segments of tree rings from the same area (Mihaljevič et al., 2011). Further research is needed to elucidate this hypothesis.

4. CONTAMINANT BIOAVAILABILITY IN SMELTER SOILS

4.1. Effects on plants and fungi

The accumulation of contaminants in plants grown on smelter-affected soils is either focused on cultivated crops, native species (including trees), or species used for the phytostabilization of smelter sites. Lange et al. (2014) found that Cu and Co accumulation in *Crepidorhopalon tenuis*, which is a metallophyte species growing on Cu- and Co-rich soils, was much lower in Cu smelter-affected areas than naturally enriched soils in Katanga (DR Congo), indicating the relatively lower Cu and Co bioavailability.Nkongolo et al. (2008) studied black spruce (*Picea mariana*) and trees grown on soils near the Sudbury Cu/Ni smelters, and observed low metal contents in the tree biomass, far below the toxic levels for vegetation. In a later study, Narendrula et al. (2013) found that needles of white spruce (*Picea glauca*) trees from the Sudbury region did not exhibit any relationship between genetic variations and soil pollution by metal(loid)s. Tree ring studies at smelter-affected sites indicated that the transfer of contaminants into the wood is mainly determined by the above-ground uptake (dissolution of metal-bearing particles attached to needles, foliage, or bark) and that their transport from soils *via* the root system remains limited (Mihaljevič et al., 2011 and Mihaljevič et al., 2015).

Maize (*Zea mays*) grown in soil artificially contaminated by Pb smelter fly ash exhibited a decrease in biomass yield as a function of the amount of the spike, and showed an increase in the concentration of Cd and Pb in the biomass. If EDTA was added to promote the translocation of contaminants from a highly contaminated spiked soil (7331 mg/kg Pb, 98 mg/kg Cd), the plant showed symptoms of chlorosis and necrosis, and only produced a negligible amount of biomass (Komárek et al., 2009). The translocation of contaminants into the plant (lettuce, *Lactuca sativa*) has been found to be dependent on the size of industrial particles. Uzu et al. (2009) reported 20% greater Pb uptake by plants when soils were spiked with <2.5 μm smelter particulates compared to those spiked with coarser particulates. Nevertheless, grassland species are generally more tolerant to high levels of metals in soils. Bidar et al. (2009) studied seasonal and annual variations of metal uptake and toxicity in two grassland species (clover, *Trifolium repens*; and perennial rye-grass, *Lolium perenne*). Both plants accumulated metals, primarily in their roots, and their translocation to aerial parts of the plants remained limited over time. Both plant species seemed to tolerate high levels of contaminants in soils, and were found suitable for phytostabilization of the polluted site (Bidar et al., 2009).

Douay et al., 2008a, Douay et al., 2008b and Douay et al., 2008c investigated how a smelter closedown affected metal content in wheat (*Triticum aestivum*) cultivated in the area, and found that during the smelter activity up to 0.8 mg/kg Cd and 8 mg/kg Pb were in the grain, and for straw the maximum concentrations of Cd and Pb were 5 mg/kg and 114 mg/kg, respectively. After the smelter's closedown, a large decrease of Pb in the grain (82%) and in the straw (91%) occurred, whereas a smaller decrease was observed for Cd (none for the grain, 39% for the straw). Despite this improvement, 80% of samples remained unacceptable for human consumption (Douay et al., 2008a, Douay et al., 2008b and Douay et al., 2008c). Kříbek et al. (2014) recently compared cassava (*Manihot esculenta*) grown on uncontaminated and contaminated soils in mining and smelting areas of the Zambian Copperbelt, and found that despite higher translocation of contaminants to tubers and leaves from soils in contaminated areas, contaminants in leaves can be attributed to dust from smelting operations (Cu sulphides/sulphates and slag particles) as well as re-suspended dust from contaminated soils and mine tailings, which settle on the surface of leaves. Thus, it is important to stress that contaminated soil dust re-suspension can be responsible for high levels of contaminants in the above-ground parts of the vegetation even after the cessation of industrial activities. Cui et al. (2004) studied soil-to-vegetable transfer of contaminants near a Pb smelter in China, and found highly contrasting bioaccumulations in vegetables, even in those of the same genus. For instance, whereas *Brassica juncea* exhibited high Pb levels, *Brassica campestris* had low Pb accumulation; the authors concluded that by selecting particular vegetables, it would be possible to reduce the risk of human exposure.

Compared to plants, fungi are more effective in the leaching, translocation, and bioconcentration of metal(loid)s in polluted areas due to extensive mycelium growth in the soils. The investigation of arbuscular mycorrhizal fungi (AMF) structure, abundance, and species richness in smelter-impacted soils indicated that some of the species (e.g.*Funneliformis* spp.) were very sensitive to high concentrations of metals and metalloids; whereas richness and diversity of *Glomeraceae* family increased with increasing As, Cd, and Zn concentrations in soils. The authors concluded that such tolerance could be used in plant growth promotion and subsequent phytoremediation process in smelter-affected environments (Krishnamoorthy et al., 2015). Cornejo et al. (2008) found a strong relationship between the amount of glomalin-related soil protein (produced by AMF in soils) and the Cu and Zn contents in copper smelter-affected soils, suggesting that AFM can mitigate metal-related stress in highly polluted soils. Surprisingly, relatively few studies had combined the investigation of contaminant concentrations in edible parts of mushrooms and corresponding soils near non-ferrous metal smelters. For example,Komárek et al. (2007a) studying three different edible species grown in forest soil near a Pb smelter found high levels of Pb (up to 165 mg/kg) and Cd (up to 55 mg/kg) in mushrooms; using Pb isotopic tracing they demonstrated that the "labile" fraction of Pb in the soils was related to contamination from secondary Pb smelting process, and is taken up by the fruiting bodies of mushrooms.

4.2. Effects on soil organisms

At metallurgical sites, it has been found that high concentrations of metals in soils can affect microbial activity (expressed as various parameters such as enzyme activity, biomass carbon, or phospholipid fatty acid profile) as well as viable population size and structure (Kelly and Tate, 1998, Kelly et al., 2003 and Wang et al., 2007). For example, a good correlation was observed between the soil enzyme activity and distance from the smelter (Wang et al., 2007). Metal- or metalloid-tolerant bacteria can be helpful in the mitigation of contaminants in smelter soils. Very recently As-tolerant bacteria were isolated from smelter soils in South Korea and it was evident that these microbes positively affect plant growth during phytoremediation (Shagol et al., 2014).

High levels of contaminants in smelter soils can reduce the overall density of earthworms and other macrofaunal population, but no simple relationship between the total metal concentration and species richness can be resolved, and it is highly site-specific (Kelly and Tate, 1998 and Nahmani and Lavelle, 2002). Thus, some studies also found that earthworms (*Eisenia fetida*) or enchytraeid worms (*Cognettia sphagnetorum*) did not exhibit any response to smelter-contaminated soils (Nahmani et al., 2007 and Salminen and Haimi, 1999); furthermore, increasing distance from the smelting operations did not correlate well with increasing microorganism population, diversity, and dynamics (Anand et al., 2003). In contrast, Tozsa et al. (2010) found along with a pollution gradient in the vicinity of Pb/Zn smelter at Olkusz (southern Poland) that the dominating enchytraeid *C. sphagnetorum* and other species (*Enchytraeus, Fridericia, Henlea*) exhibited a much lower density in the polluted sites. Nevertheless, the authors admitted that the biodiversity of the enchytraeid community was affected by other environmental factors (pH) rather than by metal concentrations. The enchytraeid reproduction test (ERT) with*Enchytraeus crypticus* or *E. albidus* is used within the OECD ecotoxicological testing programme, and is recommended for the toxicity testing of soils. However, only a limited number of studies have been specifically devoted to toxicity testing of smelter soils. For instance, Konečný et al. (2014) studied a set of Cu smelter-affected tropical soils from the Zambian Copperbelt, which exhibited a significant pollution gradient. They found that the number of reproduced enchytraeids negatively correlated with total Cu and Co concentrations and EDTA-extractable Cu; whereas other soil parameters had no effects. No reproduction was possible in soils with Cu levels >5000 mg/kg, and the calculated median effect concentration (EC50) for Cu in soil corresponded to 351 mg/kg. Single chemical extractions are often used as surrogate measures for the determination of the "bioavailable" portion of metal(loid)s from a polluted soil. Conder et al. (2001) compared several chemical extractions and earthworm bioassays, and found that $Ca(NO_3)_2$-extractable metal concentrations were good predictors of metal bioavailability from Zn smelter-affected soils. Bade et al. (2012a) used diffusive gradients in thin films (DGT), an*in situ* method measuring the flux of "labile" contaminants from the soil, plus an *Eisenia foetida* earthworm uptake experiment, and concluded that DGT can be used as a biomimic surrogate for the earthworm uptake of metals (Cu, Pb, Zn) from smelter soils.

Among soil invertebrates, woodlice (*Porcellio scaber*) play an important role in litter decomposition processes, and may be used as a sensitive organism to assess the toxicity of polluted forest soils. For example, Godet et al. (2011) used woodlice for toxicity testing of wooded areas near Pb/Zn smelters in northern France, and found that compared to control sites they exhibited higher mortality, smaller weight gain, and high accumulation of Cd, which was probably related to the Cd/Zn ratio in the food.

5. POTENTIAL REMEDIATION TECHNIQUES

Whereas traditional soil "clean-up" strategies are either based on the simple disposal or isolation of contaminated soil, sustainable soils remediation approaches, based on mineralogical and geochemical knowledge, are needed for the future (Hodson, 2010). While numerous soil remediation studies are available for mining-contaminated areas, there have been relatively few investigations solely on the remediation of smelter soils (Table 3 provides some examples).

Table 3. Selected studies on remediation of smelter-polluted soils.

Country	Locality	Smelting operation	Contaminants	Remediation technique	Reference
Czech Republic	Príbram	Pb smelter	Zn, Cd	Phytoextraction (willow)	Vysloužilova et al. (2003)
			Pb	Phytoextraction (maize, poplar)	Komárek et al. (2007b)
			Pb, Zn, Cd	Phytoextraction (maize, poplar)	Komárek et al. (2007c)
			Pb, Zn, Cd, Cu	Phytoextraction (poplar)	Komárek et al. (2008)
			Pb, Cu, Cd	Amendment (amorphous Mn oxide, maghemite, nano-magnetite)	Michálková et al. (2014)
			Pb, Zn, As, Sb, Cu, Cd	Amendment (amorphous Mn oxide)	Ettler et al. (2015b)
France	Noyelles-Godault/Auby	Pb/Zn smelters	Pb, Cd	Extraction/washing by EDTA	Kedziorek and Bourg (2000)
		Pb/Zn smelters	Pb, Cd	Soil replacement (vegetables)	Douay et al. (2008c)
	Site near Limoges	Cu smelter	Cu, Pb, Zn, Cd	Amendment (amorphous Mn oxide, maghemite, nano-magnetite)	Michálková et al. (2014)
Belgium	Lommel	Zn smelter	Zn, Pb, Cu, As, Hg	Amendment (cyclonic ash, beringite, organic compost), re-vegetation	Vangronsveld et al. (1995, 1996)
			Zn	Amendment (cyclonic ash, beringite, organic compost)	Nachtegaal et al. (2005)
			Zn, Pb, Cu, Cd	Amendment (cyclonic ash, Na silicate, lime, phosphoric acid)	Geebelen et al. (2006)
Austria	Arnoldstein	Pb/Zn smelter	Pb, Cd	15 amendments and accumulation study by barley	Friesl et al. (2006)
			Pb, Zn, Cd	Amendment (lime, red mud, gravel sludge)	Friesl-Hanl et al. (2009)
USA	Joplin (Missouri)	Pb smelter	Pb	Amendment (phosphate)	Yang et al. (2001)
	East Helena? (Montana)	Pb smelter	Pb	Amendment (phosphate)	Ryan et al. (2001)
	Palmerton (Pennsylvania)	Zn smelter	Zn	Amendment (sewage sludge, fly ash, limestone)	Kelly and Tate (1998), Kelly et al. (2003)
	Blackwell (Oklahoma)	Zn smelter	Zn	Amendment (lime-stabilized biosolids, rock phosphate)	Conder et al. (2001)
			Zn, Pb, Cd, As	Amendment (rock phosphate, diammonium phosphate, lime)	McGowen et al. (2001), Basta and McGowen (2004)
	Bartleville, Blackwell, Henryetta (Oklahoma)	Zn/Pb smelters	Pb, Zn, Cd	Amendment (biosolids, rock phosphate)	Basta et al. (2001)
South Korea	Janghang	Cu/Pb/Zn smelter	Pb, Zn, Cu, Cd, As	Amendment (lime)	Bade et al. (2012b)

The simplest means of soil remediation (but often not the cheapest one) is the replacement of contaminated soil, which is dug up, removed from the site, and stored elsewhere (Hodson, 2010), followed by the topsoil layer being substituted by uncontaminated soil. This approach has been applied in areas near Pb/Zn smelters in northern France, where communities cultivated home-grown vegetables (e.g., lettuce, radish, carrots), and had become exposed to high levels of metals *via* their diet (Douay et al., 2008c). A 3-year study indicated a clear improvement in the quality of the vegetables, with 17% of them still exceeding EU limits for foodstuffs; the authors suggested that this was due to foliar contamination by re-suspended dust fallout coming from the closed smelter site and the adjacent polluted soils (Douay et al., 2008c).

Soil decontamination by leaching, with solutions containing EDTA as a chelating agent in laboratory columns, has been found to be relatively efficient in the elution of Pb and Cd (Kedziorek and Bourg, 2000), but it has never been adopted in the field for remediation of soils near smelters.

Assisted (or enhanced) phytoextraction, where chelating agents are also used, has been applied as a cost-effective remediation method at numerous contaminated sites, but the currently achieved contaminant removal rates from soil to plants are not adequate for phytoextraction to be a fast and efficient clean-up technology (Tack and Meers, 2010).Vysloužilová et al. (2003) compared Cd and Zn phytoextraction potential for seven clones of willow (*Salix* sp.) planted on smelter-affected soils with 4.73 mg/kg Cd and 180 mg/kg Zn, and found remediation factors of 20% for Cd, but only 4% for Zn. On soils from the same area containing up to 1360 mg/kg Pb, Komárek et al. (2007b) found that the application of EDTA was more efficient than that of ethylenediaminedisuccinic acid (EDDS), removing up to 60% of the total Pb concentration in the soil. In addition, maize (*Z. mays*) exhibited better results than poplar (*Populus* sp.) when extracting Pb from a more acidic (pH 4.8) smelter-polluted soil than from a mining soil (pH 6.6) (Komárek et al., 2007b). In a subsequent study, Komárek et al. (2008) used poplar for an EDTA- and chloride-enhanced phytoextraction experiment, and found that phytoextraction efficiencies after 2 vegetation periods were comparable, but were relatively low for Cd (<1.27%) and Zn (<0.33%), and extremely low for Pb (<0.04%) and Cu (<0.05%). Moreover, due to low biomass yields, poplars were found to be unsuitable for the phytoextraction of metals from severely contaminated agricultural soils. Many sites are contaminated with more than one element. The distribution of contaminants is highly heterogeneous and phytoextraction targets contaminants, which mainly occur in the plant-available fraction. This bioavailable pool in the soil may decrease over time, and thus the long-term performance of phytoextraction might be limited (Tack and Meers, 2010). Moreover, unwanted leaching of metal-chelant complexes was nevertheless detected, and may lead to groundwater pollution (Komárek et al., 2007b and Tack and Meers, 2010 and references therein).

Various amendments are either used to immobilize contaminants by sorption/ion exchange, encapsulation and/or neoformation of solids, or to improve physicochemical conditions (pH, organic matter content), which are necessary for the stabilization process (O'Day and Vlassopoulos, 2010 and references therein). Phosphoric acid, diammonium phosphate (DAP) or hydroxyapatite have been used with success to immobilize Pb in smelter soil *via* the *in situ* formation of sparingly soluble pyromorphite [$Pb_5(PO_4)_3Cl$] (Basta et al., 2001, McGowen et al., 2001, Ryan et al., 2001, Yang et al., 2001 and Basta and McGowen, 2004). Lime or limestone amendments reduce metal solubility by increasing soil pH, which leads to increased sorption to soil particles and formation of less soluble precipitates; nevertheless, the remediation efficiency is highly variable, with better performance for lime due to cementation reactions such as the formation of calcium-silicate-hydrates (CSH) or calcium-aluminate-hydrates (CAH) (Basta and McGowen, 2004 and Bade et al., 2012b). Interestingly, Basta and McGowen (2004)found that treatment with DAP was highly effective not only for Pb (98.9% reduction compared to control), but also for Cd (94.6%) and Zn (95.8). Limestone, although moderately effective for reducing Cd (55%) and Pb (45.2%) was a poor performer for Zn (21.9%); as a result, co-application of amendments for the remediation of smelter soils was suggested. Iron and manganese oxides with a high potential for adsorption of metal(loid)s are often used as amendments. Recently, Michálková et al. (2014)compared the remediation potential of various Fe- and Mn-(nano)oxides in smelter-contaminated soils (amendments added at a concentration of 1 and 2% [w/w]), and found that treatment with an amorphous Mn oxide (AMO) resulted in >92% decrease in the exchangeable metal fraction, mainly due to the greater sorption capacity of AMO compared to that of nano-magnetite (Fe_3O_4) and nano-maghemite (Fe_2O_3). Similarly, AMO has been added at the concentration level of 2% (w/w) to a Pb smelter-affected agricultural soil, and the remediation potential has been evaluated by a pH-dependent leaching test after 6 months of incubation. The results indicated that the "labile" pool of As, Pb, and Sb significantly decreased in treatments with AMO; leaching of As, Cu, Pb, and Sb decreased down to 20%, 35%, 7%, and 11% of the control, respectively. In contrast, no effects were observed for Cd and Zn (Ettler et al., 2015b).

The efficiency of remediation with amendments (or the combination of various amendments) should be evaluated over longer time perspectives. However, relatively few long-term studies are available in the literature. Near a former Zn smelter in Lommel (Belgium), combinations of various amendments (beringite [modified alumosilicate originating from fluidized bed burning of coalmine slag], cyclonic ash, lime, and phosphoric acid) have been used for highly polluted soils, and the remediation potential was evaluated over time (Vangronsveld et al., 1995, Vangronsveld et al., 1996,Nachtegaal et al., 2005 and Geebelen et al., 2006). The reduction of phytotoxicity by addition of beringite was confirmed 15 months after treatment, and the water-extractable fraction of metals was up to 70 times lower compared to non-treated soils (Vangronsveld et al., 1995 and Vangronsveld et al., 1996). A comparison of different treatments indicated that the metal immobilizing potential of Na-silicate additives was limited (due to reversible sorption of metals), while irreversible fixation at constant pH was the major metal immobilization mechanism for other treatments (cyclonic ash, lime, and phosphoric acid) (Geebelen et al., 2006). Changes in Zn speciation were studied at soils from this site, collected 12 years after being remediated with cyclonic ash and compost. Surprisingly, the EXAFS spectroscopic data coupled to bioavailability tests indicated that Zn had similar solid speciation in both treated and non-treated soils (~8% of Zn was bioavailable); however, the incorporation of Zn into newly formed precipitates (Zn–Al layered double hydroxides and Zn-phyllosilicates) in both soil samples led to natural attenuation of the exchangeable Zn at near neutral pH (Nachtegaal et al., 2005). A similar long-term investigation of smelter soil remediation based on a laboratory batch leaching, a pot study and a pilot-scale testing was carried out near a former Pb/Zn smelter in Arnoldstein (Austria). Based on laboratory batch leaching tests, 5 treatments (combinations of gravel sludge, ferrihydrite, red mud, limestone and biocompost) were selected from 15 potential amendments and subsequently used for a pot experiment with barley (*Hordeum vulgare*). The results indicated that ferrihydrite-bearing amendments, especially in combination with sludge, were the most effective treatments for reducing Cd and Pb uptake, as well as transport in the plant (Friesl et al., 2006). However, a field experiment conducted during one vegetation season did not support the data from the pot trials, and the authors concluded that the application technique of amendments in the field has to be improved by deeper placement (Friesl et al., 2006). The evaluation of the treatments in a long-term perspective (5 years) indicated that the highest efficiency was for an amendment composed of gravel sludge and red mud (easily-extractable Cd, Pb, and Zn were reduced up to 99%); however, it was suggested to use this amendment in combination with metal-excluding cultivars (Friesl-Hanl et al., 2009).

6. CONCLUSIONS AND PERSPECTIVES

Soils near non-ferrous smelters are highly polluted by metal(loid)s, whose concentrations often exceed limit values for residential and industrial soils as stipulated by national environmental agencies. Smelter-related soil contamination generally decreases with distance from the source. It is mostly bound in surface soil layers, follows the prevailing wind direction(s) in the area, and can often be recognized tens of km away from the source. Soils in smelter areas are also commonly connected to mines and/or are developed on mineralized backgrounds, and it is often difficult to differentiate between individual pollution sources by using standard geochemical approaches. To describe and depict the extent and fate of smelter-related contamination in soils, it is very useful to use metal isotope systems, for instance those where fractionation between the feed (ore) and emissions occurs during smelting (Zn, Cd), or those where no isotope fractionation has been reported, and changes in isotope compositions of the emissions reflect the isotope signature of the processed ore (Pb). Due to the fact that some results on isotope fractionation are still contradictory, and that chemical fractionation in the soil seems to complicate the source identification, a multi-isotopic approach is needed. Moreover, source tracing in the smelter-affected environmental systems is further complicated by the fact that the smelter feed used for metal production, over time, is chemically and isotopically variable. For this reason, investigations of the distributions of the so-far less-studied trace elements (Ag, Bi, In, Se, Tl,...) could be another tool for multi-elemental tracing of smelter-derived contamination in soils (for this, there will be a need for new polluted soil CRMs in the future). The combination of traditional geochemical approaches, based on contaminant concentrations and extractabilities in soils, should be combined with detailed mineralogical investigations as well as applications of advanced techniques, such as synchrotron-based XAS, which are useful for the investigation of smelter-derived particulates and their weathering processes, determining the release of contaminants into the soil, and their subsequent binding to individual soil components. Still, relatively little is known about the fate of smelter-derived contamination in soils from tropical and arid areas, despite the fact that they are more vulnerable than are temperate soils. Given the fact that environmental measures are often less efficient in less developed countries where smelting industries are operating, there will be the need for future investigations of smelter-affected soils in the tropics.

ACKNOWLEDGEMENTS

Support from the Czech Science Foundation (Project No. 13-17501S) is acknowledged. The author's team is also supported by the Operational Programme Prague – Competitiveness (Project No. CZ.2.16/ 3.1.00/21516). Peter Lemkin is thanked for revision of the English manuscript. The author is grateful to Mario Villalobos for providing supplementary information on Mexican smelters, and to Olivier Pourret for providing raw concentration data for soils from smelting areas in Katanga (DRC). Also, two anonymous reviewers are thanked for their pertinent comments and valuable suggestions, which have significantly helped to make improvements on the manuscript.

REFERENCES

1. Adamo, P., Dudka, S., Wilson, M.J., McHardy, W.J., 1996. Chemical and mineralogical forms of Cu and Ni in contaminated soils from the Sudbury mining and smelting region, Canada. Environ. Pollut. 91, 11-19.

2. Adamo, P., Dudka, S., Wilson, M.J., McHardy, W.J., 2002. Distribution of trace elements in soils from the Sudbury smelting area (Ontario, Canada). Water Air Soil Pollut. 137, 95-116.

3. Anand, M., Ma, K.M., Okonski, A., Levin, S., McCreath, D., 2003. Characterising biocomplexity and soil microbial dynamics along a smelter-damaged landscape gradient. Sci. Total Environ. 311, 247-259.

4. Bade, R., Oh, S., Shin, W.S., 2012a. Diffusive gradients in thin films (DGT) for the prediction of bioavailability of heavy metals in contaminated soils to earthworms (Eisenia foetida) and oral bioavailable concentrations. Sci. Total Environ. 416, 127-136.

5. Bade, R., Oh, S., Shin, W.S., 2012b. Assessment of metal bioavailability in smeltercontaminated soil before and after lime amendment. Ecotox. Environ. Safe 80, 299-307.

6. Bacon, J.R., Dinev, N.S., 2005. Isotopic characterisation of lead in contaminated soils from the vicinity of a non-ferrous metal smelter near Plovdiv, Bulgaria. Environ. Pollut. 134, 247-255.

7. Banza, C.L.N., Nawrot, T.S., Haufroid, V., Decree, S., De Putter, T., Smolders, E., ⍰ Kabyla, B.I., Luboya, O.N., Ilunga, A.N., Mutombo, A.M., Nemery, B., 2009. High human exposure to cobalt and other metals in Katanga, a mining area of the Democratic Republic of Congo. Environ. Res. 109, 745-752.

8. Baron, S., Carignan, J., Ploquin, A., 2006. Dispersion of heavy metals (metalloids) in soils from 800-year-old pollution (Mont-Lozere, France). Environ. Sci. Technol. 40, 5319-5326.

9. Basta, N.T., Gradwohl, R., Snethen, K.L., Schroder, J.L., 2001. Chemical immobilization of lead, zinc, and cadmium in smelter-contaminated soils using biosolids and rock phosphate. J. Environ. Qual. 30, 1222-1230.

10. Basta, N.T., McGowen, S.L., 2004. Evaluation of chemical immobilization treatments for reducing heavy metal transport in a smelter-contaminated soil. Environ. Pollut. 127, 73e82.

11. Bidar, G., Pruvot, C., Garçon, G., Verdin, A., Shirali, P., Douay, F., 2009. Seasonal and annual variations of metal uptake, bioaccumulation, and toxicity in Trifolium repens and Lolium perenne growing in a heavy metal-contaminated field. Environ. Sci. Pollut. Res. 16, 42-53.

12. Bigalke, M., Weyer, S., Kobza, J., Wilcke, W., 2010. Stable Cu and Zn isotope ratios as tracers of sources and transport of Cu and Zn in contaminated soil. Geochim. Cosmochim. Acta 74, 6801-6813.

13. Birkefeld, A., Schulin, R., Nowack, B., 2006. In situ investigation of dissolution of heavy metal containing mineral particles in an acidic forest soil. Geochim. Cosmochim. Acta 70, 2726-2736.

14. Buatier, M.D., Sobanska, S., Elsass, F., 2001. TEM-EDS investigation on Zn- and Pbcontaminated soils. Appl. Geochem. 16, 1165-1177.

15. Burt, R., Wilson, M.A., Keck, T.J., Dougherty, B.D., Strom, D.E., Lindahl, J.A., 2003. Trace element speciation in selected smelter-

contaminated soils in Anaconda and Deer Lodge Valley, Montana, USA. Adv. Environ. Res. 8, 51-67.

16. Cabala, J., Teper, L., 2007. Metalliferous constituents of rhizosphere soils contaminated by ZnePb mining in southern Poland. Water Air Soil Pollut. 178, 351-362.

17. Carrizales, L., Razo, I., Tellez-Hern � � andez, J.I., Torres-Nerio, R., Torres, A., Batres, L.E., Cubillas, A.C., Díaz-Barriga, F., 2006. Exposure to arsenic and lead of children living near a copper-smelter in San Luis Potosi, Mexico: importance of soil contamination for exposure of children. Environ. Res. 101, 1-10.

18. CCME (Canadian Council of Ministres of the Environment), 2007. Canadian Soil Quality Guidelines for the Protection of Environmental and Human Health. Summary Tables, CCME Ottawa, Canada (accessed May 2015). http://st-ts.ccme. ca/en/index.html.

19. Cecchi, M., Dumat, C., Alric, A., Felix-Faure, B., Predere, P., Guiresse, M., 2008. Multimetal contamination of a calcic cambisol by fallout from a lead-recycling plant. Geoderma 144, 287-298.

20. . Cheyns, K., Banza, C.L.N., Kabamba, L.N., Ngoy, J.A., Haufroid, V., De Putter, T., Nawrot, T., Muleka, C.K., Luboya, O.N., Kabyla, B.I., Nemery, B., Smolders, E., 2014. Pathways of human exposure to cobalt in Katanga, a mining area of D.R. Congo. Sci. Total Environ. 490, 313-321.

21. Chopin, E.I.B., Alloway, B.J., 2007. Trace element partitioning and soil particle characterisation around mining and smelting areas at Tharsis, Ríotinto and Huelva, SW Spain. Sci. Total Environ. 373, 488-500.

22. . Chrastný, V., Vanek, A., Teper, L., Cabala, J., Proch ı azka, J., Pechar, L., Drahota, P., � Peníı zek, V., Kom� arek, M., Nov� ak, M., 2012a. Geochemical position of Pb, Zn and Cd in soils near the Olkusz mine/smelter, South Poland: effect of land use, type of contamination and distance from pollution source. Environ. Monit. Assess. 184, 2517-2536.

23. Chrastný, V., Vanek, A., Kom ı arek, M., Farka � ı s, J., Drabek, O., Vokurkov � a, P., � Nemcov ı a, J., 2012b. Incubation of air-pollution-control residues from secondary � Pb smelter in deciduous and coniferous organic soil horizons: leachability of lead, cadmium and zinc. J. Hazard. Mater. 209e210, 40-47.

24. Chrastný, V., Vanek, A., ı Cadkov ı a, E., R ⍰ ⍰uı ziı ckova, A., Galu ⍰ ı skov⍰ a, I., Faturíkova, D., ⍰ Komarek, M., 2014. Lead migration in smelter-impacted deciduous and conif- ⍰ erous organic soil horizons based on a long-term in-situ implantation and laboratory column experiments. Appl. Geochem. 48, 168-175.

25. Chrastný, V., ı Cadkov⍰ a, E., Vanek, A., Teper, L., Cabala, J., Kom ı arek, M., 2015. Cad- ⍰ mium isotope fractionation within the soil profile complicates source identifi- cation in relation to PbeZn mining and smelting processes. Chem. Geol. 405, 1-9.

26. Citeau, L., Lamy, I., van Oort, F., Elsass, F., 2003. Colloidal facilitated transfer of metals in soils under different land use. Colloids Surf. A Physicochem. Eng. Asp. 271, 11-19.

27. Clemente, R., Dickinson, N.M., Lepp, N.W., 2008. Mobility of metal and metalloids in a multi-element contaminated soil 20 years after cessation of the pollution source activity. Environ. Pollut. 155, 254-261.

28. Cloquet, C., Carignan, J., Libourel, G., Sterckeman, T., Perdrix, E., 2006. Tracing source pollution in soils using cadmium and lead isotopes. Environ. Sci. Technol. 40, 2525-2530.

29. Cloquet, C., Carignan, J., Lehmann, M.F., Vanhaecke, F., 2008. Variation in the isotopic composition of zinc in the natural environment and the use of zinc isotopes in biogeosciences. A review. Anal. Bioanal. Chem. 390, 451-463.

30. Conder, J.M., Lanno, R.P., Basta, N.T., 2001. Assessment of metal availability in smelter soil using earthworms and chemical extractions. J. Environ. Qual. 30, 1231-1237.

31. .Cornejo, P., Meier, S., Borie, G., Rillig, M.C., Borie, F., 2008. Glomalin-related soil protein in a Mediterranean ecosystem affected by a copper smelter and its contribution to Cu and Zn sequestration. Sci. Total Environ. 406, 154e160.

32. Cornu, J.Y., Denaix, L., Schneider, A., Pellerin, S., 2007. Temporal evolution of redox processes and free Cd dynamics in a metal-contaminated soil after rewetting. Chemosphere 70, 306e314.

33. Csavina, J., Landazuli, A., Wonaschütz, A., Rine, K., Rheinheimer, P., Barbaris, B., ⍰ Conant, W., Saez, A.E., Betterton, E.A., 2011. Metal and

metalloid contaminants ⍰ in atmospheric aerosols from mining operations. Water Air Soil Pollut. 221, 145e157.

34. Csavina, J., Field, J., Taylor, M.P., Gao, S., Landazuli, A., Betterton, E.A., S ⍰ ⍰ aez, A.E., 2012. A review on the importance of metals and metalloids in atmospheric dust and aerosol from mining operations. Sci. Total Environ. 433, 58e73.

35. Csavina, J., Taylor, M.P., Felix, O., Rine, K.P., S ⍰ aez, A.E., Betterton, E.A., 2014. Size- ⍰ resolved dust and aerosol contaminants associated with copper and lead smelting emissions: implications for emission management and human health. Sci. Total Environ. 493, 750e756.

36. Cui, Y.J., Zhu, Y.G., Zhai, R.H., Chen, D.Y., Huang, Y.Z., Qiu, Y., Liang, J.Z., 2004. Transfer of metals from soil to vegetables in an area near a smelter in Nanning, China. Environ. Int. 30, 785e791.

37. Davis, A., Ruby, M.V., Bloom, M., Schoof, R., Freeman, G., Bergstrom, P.D., 1996. Mineralogic constraints on the bioavailability of arsenic in smelter-impacted soils. Environ. Sci. Technol. 30, 392e399.

38. Denaix, L., Semlali, R.M., Douay, F., 2001. Dissolved and colloidal transport of Cd, Pb, and Zn in a silt loam soil affected by atmospheric industrial deposition. Environ. Pollut. 113, 29e38.

39. Derome, J., Lidroos, A.J., 1998. Copper and nickel mobility in podzolic forest soils subjected to heavy metal and sulphur deposition in western Finland. Chemosphere 36, 1131e1136.

40. Douay, F., Roussel, H., Fourrier, H., Heyman, C., Chateau, G., 2007. Investigation on heavy metal concentrations on urban soils, dust and vegetables nearby a former smelter site in Mortagne du Nord, Northern France. J. Soils Sedim. 7, 143e146.

41. Douay, F., Pruvot, C., Roussel, H., Ciesielski, H., Fourrier, H., Proix, N., Waterlot, C., 2008a. Contamination of urban soils un an area of northern France polluted by dust emissions of two smelters. Water Air Soil Pollut. 188, 247e260.

42. Douay, F., Roussel, H., Pruvot, C., Waterlot, C., 2008b. Impact of a smelter closedown on metal content of wheat cultivated in the neighbourhood. Environ. Sci. Pollut. Res. 15, 162e169.

43. Douay, F., Roussel, H., Pruvot, C., Loriette, A., Fourrier, H., 2008c. Assessment of a remediation technique using the replacement of contaminated soils in kitchen gardens nearby a former lead smelter in Northern France. Sci. Total Environ. 401, 29e38.

44. Douay, F., Pruvot, C., Waterlot, C., Fritsch, C., Fourrier, H., Loriette, A., Bidar, G., Grand, C., de Vaufleury, A., Scheifler, R., 2009. Contamination of woody habitat soils around a former lead smelter in the North of France. Sci. Total Environ. 407, 5564e5577.

45. Dumat, C., Chiquet, A., Goody, D., Aubry, E., Morin, G., Juillot, F., Benedetti, M.F., 2001. Metal ion geochemistry in smelter impacted soils and soil solutions. Bull. Soc. Geol. Fr. 172, 539 ? e548.

46. Eckel, W.P., Rabinowitz, M.B., Foster, G.D., 2002. Investigation of unrecognized former secondary lead smelting sites: confirmation by historical sources and elemental ratios in soils. Environ. Pollut. 117, 273e279.

47. . Ettler, V., Mihaljevii c, M., Kom? arek, M., 2004. ICP-MS measurements of lead isotopic ratios in soils heavily contaminated by lead smelting: tracing the sources of pollution. Anal. Bioanal. Chem. 378, 311e317.

48. Ettler, V., Johan, Z., Barronnet, A., Jankovský, F., Gilles, C., Mihaljevii c, M., ı Sebek, O., Strnad, L., Bezdii cka, P., 2005a. Mineralogy of air-pollution-control residues from a secondary lead smelter: environmental implications. Environ. Sci. Technol. 39, 9309e9316.

49. Ettler, V., Vanek, A., Mihaljevi ı ı c, M., Bezdii cka, P., 2005b. Contrasting lead speciation in forest and tilled soils heavily polluted by lead metallurgy. Chemosphere 58, 1449e1459.

50. . Ettler, V., Rohovec, J., Navr? atil, T., Mihaljevii c, M., 2007a. Mercury distribution in soil profiles polluted by lead smelting. Bull. Environ. Contam. Toxicol. 78, 12e16.

51. Ettler, V., Mihaljevii c, M., ı Sebek, O., Nechutný, Z., 2007b. Antimony availability in highly polluted soils and sediments e a comparison of single extractions. Chemosphere 68, 455e463.

52. Ettler, V., Mihaljevii c, M., ı Sebek, O., Grygar, T., 2007c. Assessment of single extractions for the determination of mobile forms of metals

in highly polluted soils and sediments e analytical and thermodynamic approaches. Anal. Chim. Acta 602, 131e140.

53. Ettler, V., Tejnecký, V., Mihaljevi c, M., ı Sebek, O., Zuna, M., Vanek, A., 2010. Antimony ı mobility in lead smelter-polluted soils. Geoderma 155, 409e418.

54. Ettler, V., Mihaljevi c, M., Kı ríbek, B., Majer, V., ı Sebek, O., 2011. Tracing the spatial distribution and mobility of metal/metalloid contaminants in oxisols in the vicinity of the Nkana copper smelter, Copperbelt province, Zambia. Geoderma 164, 73-84.

55. Ettler, V., Kı ríbek, B., Majer, V., Knesl, I., Mihaljevi ı c, M., 2012a. Differences in the bioaccessibility of metals/metalloids in soils from mining and smelting areas (Copperbelt, Zambia). J. Geochem. Explor. 113, 68-75.

56. Ettler, V., Mihaljevi c, M., ı Sebek, O., Matys Grygar, T., Klementov a, M., 2012b. Experimental in situ transformation of Pb smelter fly ash in acidic soils. Environ. Sci. Technol. 46, 10539-10548.

57. . Ettler, V., Mihaljevi c, M., ı Sebek, O., Valigurov a, R., Klementova, M., 2012c. Differ- ences in antimony and arsenic releases from lead smelter fly ash in soils. Chem. Erde Geochem. 72 (S4), 15-22.

58. Ettler, V., Kone cný, L., Kova rov a, L., Mihaljevi c, M., ı Sebek, O., Kı ríbek, B., Majer, V., Veselovský, F., Pení zek, V., Vanek, A., Nyambe, I., 2014a. Surprisingly contrasting ı metal distribution and fractionation patterns in copper smelter-affected tropical soils in forested and grassland areas (Mufulira, Zambian Copperbelt). Sci. Total Environ. 473e474, 117-124.

59. Ettler, V., Vítkov a, M., Mihaljevi c, M., ı Sebek, O., Klementov a, M., Veselovský, F., Vybíral, P., Kı ríbek, B., 2014b. Dust from Zambian smelters: mineralogy and contaminant bioaccessibility. Environ. Geochem. Health 36, 919e933.

60. Ettler, V., Petra nov ı a, V., Vítkov a, M., Mihaljevi ı c, M., ı Sebek, O., Kı ríbek, B., 2015a. Reactivity of fly ash from copper smelters in an oxisol: implications for smelterpolluted soil systems in the tropics. J. Soils Sedim.

61. Ettler, V., Toma ı sova, Z., Kom arek, M., Mihaljevi c, M., ı Sebek, O., Michalkov a, Z., 2015b. The pH-dependent long-term stability of an amorphous manganese oxide in smelter-polluted

soils: implication for chemical stabilization of metals and metalloids. J. Hazard. Mater. 286, 386-394.

62. Felix, O.I., Csavina, J., Field, J., Rine, K.P., S ? ? aez, A.E., Betterton, E.A., 2015. Use of lead isotopes to identify sources of metal and metalloid contaminants in atmospheric aerosol from mining operations. Chemosphere 122, 219-226.

63. Fernandez, C., Labanowski, J., Cambier, P., Jongmans, A.G., van Oort, F., 2007. Fate of airborne metal pollution in soils as related to agricultural management. 1. Zn and Pb distributions in soil profiles. Eur. J. Soil Sci. 58, 547-559.

64. Fernandez, C., Monna, F., Labanowski, J., Loubet, M., van Oort, F., 2008. Anthropogenic lead distribution in soils under arable land and permanent grassland estimated by Pb isotopic compositions. Environ. Pollut. 156, 1083-1091.

65. Fernandez, C., Labanowski, J., Jongmans, T., Bermond, A., Cambier, P., Lamy, I., 2010. Fate of airborne metal pollution in soils as related to agricultural management. 2. Assessing the role of biological activity in micro-scale Zn and Pb distributions in A, B and C horizons. Eur. J. Soil Sci. 61, 514-524.

66. Friesl, W., Friedl, J., Platzer, K., Horak, O., Gerzabek, M.H., 2006. Remediation of contaminated agricultural soils near a former Pb/Zn smelter in Austria: batch, pot and field experiments. Environ. Pollut. 144, 40-50.

67. Friesl-Hanl, W., Platzer, K., Horak, O., Gerzabek, M.H., 2009. Immobilizing of Cd, Pb, and Zn contaminated arable soils close to a former Pb/Zn smelter: a field study in Austria over 5 years. Environ. Geochem. Health 31, 581-594.

68. Geebelen, W., Sappin-Didier, V., Ruttens, A., Cerleen, R., Yperman, J., Bongue-? Boma, K., Mench, M., ven der Lelie, N., Vangronsveld, J., 2006. Evaluation of cyclonic ash, commercial Na-silicates, lime and phosporic acid for metal immobilisation purposes in contaminated soils in Flanders (Belgium). Environ. Pollut. 144, 32-39.

69. Ginocchio, R., Carvallo, G., Toro, I., Bustamante, E., Silva, Y., Sepúlveda, N., 2004. Micro-spatial variation of soil metal pollution and plant recruitment near a copper smelter in Central Chile. Environ. Pollut. 127, 343-352.

70. Godet, J.P., Demuynck, S., Waterlot, C., Lemiere, S., Souty-Grosset, C., Schei fler, R., Douay, F., Lepretre, A., Pruvot, C., 2011. Growth and metal accumulation in ^ Porcellio scaber exposed to poplar litter from Cd-, Pb-, and Zn-contaminated sites. Ecotox. Environ. Safe 74, 451-458.

71. Gregurek, D., Melcher, F., Pavlov, V.A., Reimann, C., Stumpfl, E.F., 1999. Mineralogy and mineral chemistry of snow filter residues in the vicinity of the nickelcopper processing industry, Kola Peninsula, NW Russia. Miner. Petrol. 65, 87-111.

72. Gutierrez-Ruiz, M.E., Villalobos, M., Romero, F., Fern ? andez-Lomelín, P., 2005. Natural ? attenuation of arsenic in semiardi soils contaminated by oxidized arsenic wastes. In: O'Day, P.A., Vlassopoulos, D., Meng, X., Benning, L.G. (Eds.), Advances in Arsenic Research: Integration of Experimental and Observational Studies and Implications for Mitigation, Amm. Chem. Soc. Symp. Series, vol. 915, pp. 235-352 (Chapter 17).

73. Gutierrez-Ruiz, M.E., Ceniceros-G ? omez, A.E., Villalobos, M., Romero, F., Santiago, P., ? 2012. Natural arsenic attenuation via metal arsenate precipitation in soils contaminated with metallurgical wastes: II. Cumulative evidence and identifi- cation of minor processes. Appl. Geochem. 27, 2204-2214.

74. He, M., 2007. Distribution and phytoavailability of antimony at an antimony mining and smelting area, Hunan, China. Environ. Geochem. Health 29, 209-219.

75. Henderson, P.J., McMartin, I., Hall, G.E., Percival, J.B., Walker, D.A., 1998. The chemical and physical characteristics of heavy metals in humus and till in the vicinity of the base metal smelter at Flin Flon, Manitoba, Canada. Environ. Geol. 34, 39-58.

76. Hodson, M.E., 2010. The need for sustainable soil remediation. Elements 6, 363e368. http://dx.doi.org/10.2113/gelements.6.6.363. Hou, X., Parent, M., Savard, M.M., Tasse, N., B ? egin, C., Marion, J., 2006. Lead con- ? centrations and isotope ratios in the exchangeable fraction: tracing soil contamination near a copper smelter. Geochem. Explor. Environ. Anal. 6, 229-236.

77. IUSS Working Group WRB, 2014. World Reference Base for Soil Resources 2014. International Soil Classification System for Naming

Soils and Creating Legends for Soil Maps. World Soil Resources Reports No. 106. FAO, Rome.

78. Juillot, F., Morin, G., Ildefonse, P., Trainor, T.P., Benedetti, M., Galoisy, L., Calas, G., Brown Jr., G.E., 2003. Occurrence of Zn/Al hydrotalcite in smelter-impacted soils from northern France: evidence from EXAFS spectroscopy and chemical extractions. Am. Mineral. 88, 509-526.

79. Kabala, C., Singh, B.R., 2001. Fractionation and mobility of copper, lead and zinc in soil profiles in the vicinity of a copper smelter. J. Environ. Qual. 30, 485-492.

80. Karczewska, A., 1996. Metal species distribution in top-and sub-soil in an area affected by copper smelter emissions. Appl. Geochem. 11, 35-42.

81. Kedziorek, M.A.M., Bourg, A.C.M., 2000. Solubilization of lead and cadmium during the percolation of EDTA through a soil polluted by smelting activities. J. Contam. Hydrol. 40, 381-392.

82. Kelly, J.J., Tate, R.L., 1998. Effects of heavy metal contamination and remediation on soil microbial communities in the vicinity of a zinc smelter. J. Environ. Qual. 27, 609e617. Kelly, J.J., H€ aggblom, M.M., Tate, R.L., 2003. Effects of heavy metal contamination and remediation on soil microbial communities in the vicinity of a zinc smelter as indicated by analysis of microbial community phospholipid fatty acid pro- files. Biol. Fertil. Soils 38, 65e71.

83. Ketterer, M.E., Lowry, J.H., Simon Jr., J., Humphries, K., Novotnak, M.P., 2001. Lead isotopic and chalcophile element compositions in the environment near a zinc smelting-secondary zinc recovery facility, Palmerton, Pennsylvania, USA. Appl. Geochem. 16, 207-229.

84. Kierczak, J., Neel, C., Puziewicz, J., Bril, H., 2009. The mineralogy and weathering of ▯ slag produced by the smelting of lateritic Ni ores, Szklary, southwestern Poland. Can. Mineral. 47, 557-572.

85. Kierczak, J., Potysz, A., Pietranik, A., Tyszka, R., Modelska, M., Neel, C., Ettler, V., ▯ Mihaljevıı c, M., 2013. Environmental impact of the historical Cu smelting in the Rudawy Janowickie Mountains (south-western Poland). J. Geochem. Explor. 124, 183-194.

86. . Klaminder, J., Bindler, R., Rydberg, J., Renberg, I., 2008. Is there a chronological record of atmospheric mercury and lead deposition

preserved in the mor layer (O-horizon) of boreal forest soils? Geochim. Cosmochim. Acta 72, 703-712.

87. Kom arek, M., Chrastný, V., Ettler, V., Tlustoı s, P., 2006. Evaluation of extraction/ digestion techniques used to determine lead isotopic composition in forest soil. Anal. Bioanal. Chem. 385, 1109-1115.

88. . Kom arek, M., Chrastný, V., ı Stíchov a, J., 2007a. Metal/metalloid contamination and isotopic composition of lead in edible mushrooms and forest soils originating from a smelting area. Environ. Int. 33, 677-684.

89. Kom arek, M., Tlustoı s, P., Szakov a, J., Chrastný, V., Ettler, V., 2007b. The use of maize and poplar in chelant-enhanced phytoextraction of lead from contaminated agricultural soils. Chemosphere 67, 640e651.

90. Kom arek, M., Tlustoı s, P., Szakov a, J., Chrastný, V., 2007c. The role of chloride salts in chemically enhanced phytoextraction of heavy metals from a contaminated agricultural soil. Bull. Environ. Contam. Toxicol. 78, 176e180.

91. Kom arek, M., Tlustoı s, P., Sz akova, J., Chrastný, V., 2008. The use of poplar during a two-year induced phytoextraction of metals from contaminated agricultural soils. Environ. Pollut. 151, 27e38.

92. Kom arek, M., Ettler, V., Sz akov a, J., ı Sebek, O., Tlustoı s, P., 2009. Bioavailability of lead and cadmium in soils artificially contaminated with smelter fly ash. Bull. Environ. Contam. Toxicol. 83, 286e290.

93. Koneı cný, L., Ettler, V., Munch Kristiansen, S., Amorim, M.J.B., Kı ríbek, B., Mihaljeviı c, M., ı Sebek, O., Nyambe, I., Scott-Fordsmand, J.J., 2014. Response of Enchytraeus crypticus worms to high metal levels in tropical soils polluted by copper smelting. J. Geochem. Explor. 144, 427e432.

94. Knight, R.D., Henderson, P.J., 2006. Smelter dust in humus around Rouyn-Noranda, Quebec. Geochem. Explor. Environ. Anal. 6, 203 e214.

95. Kı ríbek, B., Majer, V., Veselovský, F., Nyambe, I., 2010. Discrimination of lithogenic and anthropogenic sources of metals and sulphur in soils in the centralnorthern part of the Zambian

Copperbelt Mining District: a topsoil vs. subsurface soil concept. J. Geochem. Explor. 104, 69e85.

96. Kı ríbek, B., Majer, V., Knesl, I., Nyambe, I., Mihaljevi ı c, M., Ettler, V., Sracek, O., 2014. Concentrations of arsenic, copper, cobalt, lead and zinc in cassava (Manihot esculenta Crantz) growing on uncontaminated and contaminated soils of the Zambian Copperbelt. J. Afr. Earth Sci. 99, 713e723.

97. Krishnamoorthy, R., Kim, C.G., Subramanian, P., Kim, K.Y., Selvakumar, G., Sa, T.M., June 2, 2015. Arbuscular mycorrhizal fungi community structure, abundance and species richness changes in soil by different levels of heavy metal and metalloid concentration. PLoS One 1e15. http://dx.doi.org/10.1371/ journal.pone.0128784. Lange, B., Faucon, M.P., Meerts, P., Shutcha, M., Mahy, G., Pourret, O., 2014. Prediction of the edaphic factor influence upon the copper and cobalt accumulation in two metallophytes using copper and cobalt speciation in soils. Plant Soil 379, 257-287.

98. Lanteigne, S., Schindler, M., McDonald, A.M., Skeries, K., Abdu, Y., Mantha, N.M., Murayama, M., Hawthorne, F.C., Hochella Jr., M.F., 2012. Mineralogy and weathering of smelter-derived spherical particles in soils: implications for the mobility of Ni and Cu in the surficial environment. Water Air Soil Pollut. 223, 3619-3641.

99. Lanteigne, S., Schindler, M., McDonald, A., 2014. Distribution of metals and metalloids in smelter-derived particulate matter in soils and mineralogical insights into their retention and release in a low-T environment. Can. Mineral. 52, 453-471.

100. Leguedois, S., van Oort, F., Jongmans, T., Chevallier, P., 2004. Morphology, chemistry and distribution of neoformed spherulites in agricultural land affected by metallurgical point-source pollution. Environ. Pollut. 130, 135-148.

101. Li, Z., Feng, X., Li, G., Bi, X., Sun, G., Zhu, J., Qin, H., Wang, J., 2011. Mercury and other metal and metalloid soil contamination near a Pb/Zn smelter in east Hunan province, China. Appl. Geochem. 26, 160-166.

102. Li, Z., Feng, X., Bi, X., Li, G., Lin, Y., Sun, G., 2014. Probing the distribution and contamination levels of 10 trace metal/metalloids in soils near a Pb/Zn smelter in Middle China. Environ. Sci. Pollut. Res. 21, 4149-4162.

103. Liu, L., Wu, L., Luo, Y., Zhang, C., Jiang, Y., Qiu, X., 2010. The impact of a copper smelter on adjacent soil zinc and cadmium fractions and soil organic carbon. J. Soils Sedim. 10, 808e817.

104. Mattieli, N., Petit, J.C.J., Deboudt, K., Flament, P., Perdrix, E., Taillez, A., RimetzPlanchon, J., Weis, D., 2009. Zn isotope study of atmospheric emissions and dry depositions within a 5 km radius of a PbeZn refinery. Atmos. Environ. 43, 1265-1272.

105. Martley, E., Gulson, B.L., Pfeifer, H.R., 2004a. Metal concentrations in soils around the copper smelter and surrounding industrial complex of Port Kembla, NSW, Australia. Sci. Total Environ. 325, 113-127.

106. Martley, E., Gulson, B., Louie, H., Wu, M., Di, P., 2004b. Metal partitioning in soil profiles in the vicinity of an industrial complex, New South Wales, Australia. Geochem. Explor. Environ. Anal. 4, 171-179.

107. McGowen, S.L., Basta, N.T., Brown, G.O., 2001. Use of diammonium phosphate to reduce heavy metal solubility and transport in smelter-contaminated soil. J. Environ. Qual. 30, 493-500.

108. Michalkov □ □ a, Z., Kom□ arek, M., ı Sillerov□ a, H., Della Puppa, L., Joussein, E., Bordas, F., Vanek, A., Van ı ek, O., Ettler, V., 2014. Evaluating the potential of three Fe- and ı Mn-(nano)oxides for the stabilization of Cd, Cu and Pb in contaminated soils. J. Environ. Manag. 146, 226-234.

109. Mihaljeviı c, M., Ettler, V., ı Sebek, O., Sracek, O., Kı ríbek, B., Kyncl, T., Majer, V., Veselovský, F., 2011. Lead isotopic and metallic pollution record in tree rings from the Copperbelt mining-smelting area, Zambia. Water Air Soil Pollut. 216, 657-668.

110. Mihaljeviı c, M., Ettler, V., Vanek, A., Pení ı ı zek, V., Svoboda, M., Kı ríbek, B., Sracek, O., Mapani, B.S., Kamona, A.F., 2015. Trace elements and the lead isotopic record in marula (Sclerocarya birrea) tree rings and soils near the Tsumeb smelter, Namibia. Water Air Soil Pollut. 226, 177.

111. Morin, G., Ostergren, J.D., Juillot, F., Ildefonse, P., Calas, G., Brown Jr., G.E., 1999. XAFS determination of the chemical form of lead in smelter-contaminated soils and mine tailings: importance of adsorption processes. Am. Mineral. 84, 420-434.

112. Nachtegaal, M., Marcus, M.A., Sonke, J.E., Vabgronsveld, J., Livi, K.J.T., van der Lelie, D., Sparks, D.L., 2005. Effects of in situ remediation on the speciation and bioavailability of zinc in a smelter contaminated soil. Geochim. Cosmochim. Acta 69, 4649-4664.

113. Nahmani, J., Lavelle, P., 2002. Effects of heavy metal pollution on soil macrofauna in a grassland of Northern France. Eur. J. Soil Biol. 38, 297-300.

114. Nahmani, J., Hodson, M.E., Black, S., 2007. Effect of metals on life cycle parameters of the earthworm Eisenia fetida exposed to field-contaminated, metal-polluted soils. Environ. Pollut. 149, 44-58.

115. Narendrula, R., Nkongolo, K.K., Beckett, P., 2012. Comparative soil metal analyses in Sudbury (Ontario, Canada) and Lubumbashi (Latanga, DR-Congo). Bull. Environ. Contam. Toxicol. 88, 187-192.

116. Narendrula, R., Nkongolo, K.K., Beckett, P., Spiers, G., 2013. Total and bioavailable metals in two contrasting mining regions (Sudbury in Canada and Lubumbashi in DR-Congo): relation to genetic variation in plant populations. Chem. Ecol. 29, 111-127.

117. Neaman, A., Reyes, L., Trolard, F., Bourrie, G., Sauv ⍰ e, S., 2009. Copper mobility in ⍰ contaminated soils of the Puchuncaví valley, central Chile. Geoderma 150, 359e366.

118. Niemeyer, J.C., Moreira-Santos, M., Nogueira, M.A., Carvalho, G.M., Ribeiro, R., Da Silva, E.M., Sousa, J.P., 2010. Environmental risk assessment of a metal contaminated area in the tropics. Tier I: screening phase. J. Soils Sedim. 10, 1557-1571.

119. Nkongolo, K.K., Vaillancourt, A., Dobrzeniecka, S., Mehes, M., Beckett, P., 2008. Metal content in soil and black spruce (Picea mariana) trees in the Sudbury region (Ontario, Canada): low concentration of arsenic, cadmium, and nickel detected near smelters. Bull. Environ. Contam. Toxicol. 80, 107-111.

120. Nowack, B., Schulin, R., Luster, J., 2010. Metal fractionation in a contaminated soil after reforestation: temporal changes versus spatial variability. Environ. Pollut. 158, 3272-3278

121. Nriagu, J.O., 1996. A history of global metal pollution. Science 272, 223-224..

122. Nriagu, J.O., Pacyna, J.M., 1988. Quantitative assessment of worldwide contamination of air, water and soils by trace metals. Nature 333, 134-139.

123. O'Day, P.A., Vlassopoulos, D., 2010. Mineral-based amendments for remediation. Elements 6, 375-381.

124. Pacyna, J.M., Pacyna, E.G., Aas, W., 2009. Changes of emissions and atmospheric deposition of mercury, lead, and cadmium. Atmos. Environ. 43, 117-127.

125. Parra, S., Bravo, M.A., Quiroz, W., Moreno, T., Karanasiou, A., Font, O., Vidal, V., Cereceda, F., 2014. Distribution of trace elements in particle size fractions for contaminated soils by a copper smelter from different zones of the Puchuncaví Valley (Chile). Chemosphere 111, 513-521.

126. Pelfrene, A., Waterlot, C., Mazzuca, M., Nisse, C., Bidar, G., Douay, F., 2011. Assessing ^ Cd, Pb, Zn human bioaccessibility in smelter-contaminated agricultural topsoils (northern France). Environ. Geochem. Health 33, 477-493.

127. Pelfrene, A., D ^ etrich ▯ e, S., Douay, F., 2015. Combining spatial distribution with oral ▯ bioaccessibility of metals in smelter-impacted soils: implications for human health risk assessment. Environ. Geochem. Health 37, 49-62.

128. Pirrone, N., Cinnirella, S., Feng, X., Finkelman, R.B., Friedli, H.R., Leaner, J., Mason, R., Mukherjee, A.B., Stracher, G.B., Streets, D.G., Telmer, K., 2010. Global mercury emissions to the atmosphere from anthropogenic and natural sources. Atmos. Chem. Phys. 10, 5951-5964.

129. . Podolský, F., Ettler, V., ı Sebek, O., Jeı zek, J., Mihaljevıı c, M., Kı ríbek, B., Sracek, O., Vanek, A., Pení ı ı zek, V., Majer, V., Mapani, B., Kamona, F., Nyambe, I., 2015. Mercury in soil profiles from metl mining and smelting areas in Namibia and Zambia: distribution and potential sources. J. Soils Sedim. 15, 648-658.

130. Popescu, I., Biasioli, M., Ajmone-Marsan, F., Stanescu, R., 2013. Lability of potentially toxic elements in soils affected by smelting activities. Chemosphere 90, 820-826.

131. Pourret, O., Lange, B., Houben, D., Colinet, G., Shutcha, M., Faucon, M.P., 2015. Modeling of cobalt and copper speciation in

metalliferous soils from Katanga (Democratic Republic of Congo). J. Geochem. Explor. 149, 87-96.

132. Prapaipong, P., Enssle, C.W., Morris, J.D., Shock, E.I., Lindvall, R.E., 2008. Rapid transport of anthropogenic lead through soils in southeast Missouri. Appl. Geochem. 23, 2156-2170.

133. . Qasim, B., Motelica-Heino, M., Jousseil, E., Soubrand, M., Gauthier, A., 2015. Potentially toxic element phytoavailability assessment in technosols from former smelting and mining area. Environ. Sci. Pollut. Res. 22, 5961-5974.

134. Rabinowitz, M.B., 2005. Lead isotopes in soils near five historic American lead smelters and refineries. Sci. Total Environ. 346, 138-148.

135. .Rauret, G., Lopez-Sanchez, J.F., Sahuquillo, A., Rubio, R., Favidson, C., Ure, A., ? Quevauviller, P., 1999. Improvement of the BCR three step sequential extraction procedure prior to the certification of new sediment and soil reference materials. J. Environ. Monit. 1, 57-61.

136. Rieuwerts, J.S., Farago, M.E., 1995. Lead contamination in smelting and mining environments and variations in chemical forms and bioavailability. Chem. Spec. Bioavail. 7, 113-123.

137. Rieuwerts, J.S., Farago, M., 1996a. Heavy metal pollution in the vicinity of a secondary lead smelter in the Czech Republic. Appl. Geochem. 11, 17-23.

138. Rieuwerts, J.S., Farago, M., 1996b. Mercury concentrations in a historic lead mining and smelting town in the Czech Republic: a pilot study. Sci. Total. Environ. 188, 167-171.

139. Rieuwerts, J.S., Farago, M., Cikrt, M., Bencko, V., 1999. Heavy metal concentrations in and around households near a secondary lead smelter. Environ. Monit. Assess. 58, 317-335.

140. Rieuwerts, J.S., Farago, M., Cikrt, M., Bencko, V., 2000. Differences in lead bioavailability between a smelting and a mining area. Water Air Soil Pollut. 122, 203-229.

141. Rieuwerts, J.S., 2007. The mobility and bioavailability of trace metals in tropical soils: a review. Chem. Spec. Availab. 19, 75-85.

142. Roberts, D.R., Scheinost, A.C., Sparks, D.L., 2002. Zinc speciation in a smeltercontaminated soil profile using bulk and microspectrometric techniques. Environ. Sci. Technol. 36, 1742-1750.

143. Romero, F.M., Villalobos, M., Aguirre, R., Gutierrez, M.E., 2008. Solid-phase control ⍰ on lead bioaccessibility in smelter-impacted soils. Arch. Environ. Contam. Toxicol. 55, 566-575.

144. Roussel, H., Waterlot, C., Pelfrene, A., Pruvot, C., Mazzuca, M., Douay, F., 2010. Cd, Pb ^ and Zn oral bioaccessibility of urban soils contaminated in the past by atmospheric emissions from two lead and zinc smelters. Arch. Environ. Contam. Toxicol. 58, 945-954.

145. Ryan, J.A., Zhang, P., Hesterberg, D., Chou, J., Sayers, D.E., 2001. Formation of chloropyromorphite in a lead-contaminated soil amended with hydroxyapatite. Environ. Sci. Technol. 35, 3798-3803.

146. Sajn, R., Aliu, M., Stafilov, T., Alijagic, J., 2013. Heavy metal contamination of topsoil around a lead and zinc smelter in Kosovska Mitrovica/Mitrovice, Kosovo/ € Kosove. J. Geochem. Explor. 134, 1 € -16.

147. Salminen, J., Haimi, J., 1999. Horizontal distribution of copper, nickel and enchytraeid worms in polluted soils. Environ. Pollut. 104, 351-358.

148. . Shagol, C.C., Krishnamoorthy, R., Kim, K., Sundaram, S., Sa, T., 2014. Arsenic-tolerant plant-growth-promoting bacteria isolated from arsenic-polluted soils in South Korea. Environ. Sci. Pollut. Res. 21, 9356-9365.

149. Shiel, A.E., Weis, D., Orians, K.J., 2010. Evaluation of zinc, cadmium and lead isotope fractionation during smelting and refining. Sci. Total Environ. 408, 2357-2368.

150. Sobanska, S., Ricq, N., Laboudigne, A., Bremard, C., Laureyns, J., Merlin, J.C., ⍰ Wignacourt, J.P., 1999. Microchemical investigations of dust emitted by a lead smelter. Environ. Sci. Technol. 33, 1334-1339.

151. Sobanska, S., Ledesert, B., Deneele, D., Laboudigne, A., 2000. Alteration in soils of ⍰ slag particles resulting from lead smelting. C. R. Acad. Sci. Ser. II A 331, 271-278.

152. Soil Survey Staff, 2014. Keys to Soil Taxonomy, twelfth ed. USDA-Natural Resources Conservation Service, Washington, DC, USA. Sonke, J.E., Sivry, Y., Viers, J., Freydier, R., Dejonghe, L., Andre, L., Aggarwal, J.K., ⍰ Fontan, F., Dupre, B., 2008. Historical variations in the isotopic composition of ⍰ atmospheric zinc deposition from a zinc smelter. Chem. Geol. 252, 145-157.

153. . Sorooshian, A., Csavina, J., Shingler, T., Dey, S., Brechtel, F.J., Saez, A.E., Betterton, E.A., ⍰ 2012. Hygroscopic and chemical properties of aerosols collected near a copper smelter: implications for public and environmental health. Environ. Sci. Technol. 46, 9473-9480.

154. Stafilov, T., ı Sajn, R., Penı cevski, Z., Boev, B., Frontasyeva, M.V., Strelkova, L.P., 2010. Heavy metal contamination of topsoils around a lead and zinc smelter in the Republic of Macedonia. J. Hazard. Mater. 175, 896-914.

155. Sterckeman, T., Douay, F., Proix, N., Fourrier, H., 2000. Vertical distribution of Cd, Pb and Zn in soils near smelters in the North of France. Environ. Pollut. 107, 377-389.

156. Sterckeman, T., Douay, F., Proix, N., Fourrier, H., Perdrix, E., 2002. Assessment of the contamination of cultivated soils by eighteen trace elements around smelters in the North of France. Water Air Soil Pollut. 135, 173e194.

157. Suchara, I., Sucharov⍰ a, J., 2004. Distribution of 36 elements deposition rates in a historic mining and smelting area as determined through fine-scale biomonitoring technique. Part II: relative long-term accumulated atmospheric deposition levels. Water Air Soil Pollut. 153, 229-252.

158. Tack, F.M.G., Meers, E., 2010. Assisted phytoextraction: helping plants to help us. Elements 6, 383-388.

159. Takaoka, M., Fukutani, S., Yamamoto, T., Horiuchi, M., Satta, N., Takeda, N., Oshita, K., Yoneda, M., Morisawa, S., Tanaka, T., 2005. Determination of chemical form of antimony in contaminated soil around a smelter using X-ray absorption fine structure. Anal. Sci. 27, 769-773.

160. Tembo, B.D., Sichilongo, K., Cernak, J., 2006. Distribution of copper, lead, cadmium and zinc concentrations in soils around Kabwe town in Zambia. Chemosphere 63, 497-501.

161. Tozsa, E., Dumnicka, E., Niklinska, M., Rozen, A., 2010. Enchytraeid and earthworm communities along a pollution gradient near Olkusz (southern Poland). Eur. J. Soil Biol. 46, 218-224.

162. Tyszka, R., Pietranik, A., Kierczak, J., Ettler, V., Mihaljevıı c, M., Weber, J., 2012. Anthropogenic and lithogenic sources of lead in Lower Silesia (Southwest Poland): an isotope study of soils, basement rocks and anthropogenic materials. Appl. Geochem. 27, 1089e1100.

163. U.S. Environmental Protection Agency (EPA), 2015. Regional Screening Levels (accessed May 2015). http://www.epa.gov/region9/superfund/prg/index.html.

164. Uzu, G., Sobanska, S., Aliouane, Y., Pradere, P., Dumat, C., 2009. Study of lead phy- toavailability for atmospheric industrial micronic and sub-micronic particles in relation with lead speciation. Environ. Pollut. 157, 1178-1185.

165. Uzu, G., Sobanska, S., Sarret, G., Sauvain, J.J., Pradere, P., Dumat, C., 2011. Charac- terization of lead-recycling facility emissions at various workplaces. Major insights for sanitary risk assessment. J. Hazard. Mater. 186, 1018-1027.

166. Vanek, A., Chrastný, V., Kom ı arek, M., Pení ▯ ı zek, V., Teper, L., Cabala, J., Dr▯ abek, O., 2013. Geochemical position of thallium in soils from a smelter-impacted area. J. Geochem. Explor. 124, 176-182.

167. Vangronsveld, J., Van Assche, F., Clijsters, H., 1995. Reclamation of a bare industrial area contaminated by non-ferrous metals: in situ metal immobilization and revegetation. Environ. Pollut. 84, 51-59.

168. Vangronsveld, J., Colpaert, J.V., Van Tichelen, K.K., 1996. Reclamation of a bare industrial area contaminated by non-ferrous metals: physico-chemical and biological evaluation of the durability of soil treatment and revegetation. Environ. Pollut. 94, 131-140.

169. van Oort, F., Jongmans, A.G., Citeau, L., Lamy, I., Chevallier, P., 2006. Microscale Zn and Pb distribution patterns in subsurface soil horizons: an indication for metal transport dynamics. Eur. J. Soil Sci. 57, 154-166.

170. Verner, J.F., Ramsey, M.H., Helios-Rybicka, E., Jedrzejczyk, B., 1996. Heavy metal contamination of soils around a Pb-Zn smelter in Bukowno, Poland. Appl. Geochem. 11, 11-16.

171. Vespa, M., Lanson, M., Manceau, A., 2010. Natural attenuation of zinc pollution in smelter-affected soil. Environ. Sci. Technol. 44, 7814-7820.

172. Villalobos, M., García-Payne, D.G., Lopez-Zepeda, J.L., Ceniceros-G ▯ omez, A.E., ▯ Gutierrez-Ruiz, M.E., 2010. Natural arsenic attenuation via metal arsenate ▯ precipitation in soils contaminated with metallurgical wastes: I. Wet chemical and thermodynamic evidences. Aquat. Geochem. 16, 225-250.

173. Vyslouı zilova, M., Tlusto ▯ ı s, P., Sz▯ akov▯ a, J., 2003. Cadmium and zinc phytoextraction potential of seven clones of Salix spp.planted on heavy metal contaminated soils. Plant Soil Environ. 49, 542-547.

174. Wang, Y.P., Shi, J.Y., Wang, H., Lin, Q., Chen, X.C., Chen, Y.X., 2007. The influence of soil heavy metals pollution on soil microbial biomass, enzyme activity, and community composition near a copper smelter. Ecotox. Environ. Safe 67, 75-81.

175. Wen, H., Zhang, Y., Cloquet, C., Zhu, C., Fan, H., Luo, C., 2015. Tracing sources of pollution in soils from the Jinding PbeZn mining district in China using cadmium and lead isotopes. Appl. Geochem. 52, 147-154.

176. Wu, Q., Wang, S., Wang, L., Liu, F., Che-Jen, L., Zhang, L., Wang, F., 2014. Spatial distribution and accumulation of Hg in soil surrounding a Zn/Pb smelter. Sci. Total Environ. 496, 668-677.

177. Yang, J., Mosby, D.E., Casteel, S.W., Blanchar, R.W., 2001. Lead immobilization using phosphoric acid in a smelter-contaminated urban soil. Environ. Sci. Technol. 35, 3553-3559.

178. Yang, Y., Li, S., Bi, X., Wu, P., Liu, T., Li, F., Liu, C., 2010. Lead, Zn, and Cd in slags, stream sediments, and soils in an abandoned Zn smelting region, southwest of China, and Pb and S isotopes as source tracers. J. Soil Sedim. 10, 1527-1539.

179. Yin, X., Yao, C., Song, J., Li, Z., Zhang, C., Qian, W., Bi, D., Li, C., Teng, Y., Wu, L., Wan, H., Luo, Y., 2009. Mercury contamination in vicinity of secondary copper smelters in Fuyang, Zhejiang Province,

China: levels and contamination in topsoils. Environ. Pollut. 157, 1787-1793.

180. Zhang, X., Yang, L., Li, Y., Li, H., Wang, W., Ge, Q., 2011. Estimation of lead and zinc emissions from mineral exploitation based on characteristics of lead/zinc deposits in China. Trans. Nonferrous Met. Soc. China 21, 2513-2519.

181. Zheng, N., Wang, Q., Zheng, D., 2007. Mercury contamination and health risk to crops around the zinc smelting plant in Huludao City, northeastern China. Environ. Geochem. Health 29, 385-393.

3

SATELLITE IMAGERY FOR STRUCTURAL GEOLOGICAL INTERPRETATION IN WESTERN SIRT BASIN, LIBYA: IMPLICATION FOR PETROLEUM EXPLORATION

Abdunaser K. M

Libyan Petroleum Institute, Libya

ABSTRACT

Interpretation and analysis of Landsat imagery, integrated with published geological maps, has provided the basis for a re-examination of the structural geology of the area, and has enabled a new interpretation of the tectonic development of the western part of the Sirt Basin to be made. In addition to detecting normal faults, and strike-slip fault systems, a number of important fold structures were mapped and analysed using the imagery, and the resulting maps represent an improvement on existing publications. Tectonic features, particularly folds and faults near the basin margin, may provide traps for hydrocarbon accumulations. Direction statistics and rose diagrams for all the interpreted linear features acquired in the study area were prepared and calculated which given as a total distribution and then the totals are broken down by the major basin elements of the area. Hundreds of lineaments were recognized. Their lengths, range between a hundred meters up to hundreds of kilometers long and the dominant trends are between N25W and N55W which coincides with Sirt Basin structures. The correlation between surface linear features and subsurface oil and gas traps was assessed, and the implication of using surface lineament and fracture analysis for delineating hydrocarbon reservoirs in the region discussed. It was observed that the surface linear features were extremely consistent in orientation with the basement faults in the Sirt Basin region suggests that the systematic fault systems at the basement in the study area have probably been reactivated many times and have propagated upward all the way to the surface. They may have acted as the loci for the development of other geological structures, including oil and gas traps. Indication of the relative timing of structures and movement along faults has been determined where possible. The results presented in this paper prove the validity and indicate the significance of using surface linear features for inferring subsurface oil and gas reservoirs in the study area. Any new potential oil and gas reservoirs in the study area, if they exist, will be likely associated with the northeast- and northwest-trending surface lineaments and fracture traces in the region.

KEYWORDS

Sirt Basin, Rose diagrams, Surface lineament, Normal faults, Relay ramp

1. INTRODUCTION

Geological mapping is evolving from the conventional field survey to the use of remote sensing technologies. The remote sensing technology were useful and has shown a great promise for large-scale geological mapping to identified of structural features, such as faults and folds. In addition the integrating of remote sensing data and geographical information systems (GIS) technique has further improvised the geological mapping techniques.

In other hand remote sensing provides a relatively inexpensive and quick method for assessing the exploration potential of large areas. Satellite sensors and data processing techniques can now provide enhanced resolution data, both spatially and spectrally, as well as three-dimensional models of an area.

In this study a regional-scale structures have been mapped using remote sensing techniques, particularly lineament analysis, and the digitization of pre-existing geological maps. Fault structures at this scale were interpreted from Landsat enhanced Thematic Mapper (ETM) and Google Earth data.

So, this paper, contributes new findings based on an integrated approach utilizing data from the various disciplines of remote sensing, and existing geological maps in the NW portion of the Sirt Basin. The aim of the study was to critically review the existing body of knowledge on the structural geology of the western part of the Sirt Basin and to build a new understanding based on my own interpretations of used data. These new insights provide a basis for a new exploration strategy that could stimulate future petroleum exploration activities in the study area.

1.1. Location of the Study Area

The Sirt Basin in north-central Libya (Figure 1) covers an approximately 600,000 km^2 area that is bordered to the east by the block-faulted region of the Cyrenaican Platform, including Al Jabal al Akhdar, and extend as far as the Dakhla Basin in Egypt. It is bounded to the south by the Tibisti Mountains and the Al Kufrah Basin to the southeast, to the west by the Nafusah, Al Qargaf Arch, the Ghadamis-As Sawda and Murzuq Basin, and the Mediterranean Sea to the north. This study is focusing mainly on north western Sirt Basin, Libya between 1500' E to 1830' E and 2800' N to 3130' N (Figure 1).

1.2. Database and Methodology

A large database containing information about structural features such as faults, folds and fractures was produced in the form of GIS thematic layers enabling interpretation and analysis. This information was then integrated using GIS software and used to construct a range of thematic maps.

Geographical Information Systems (GIS) have proved to be ideal for that purpose [1]. A GIS can be used for data capture, storage, analyses, prognostication, presentation, and follow-up and constitutes an excellent tool for modelling work.

It is important to start building the GIS database at an early stage of any project, enabling data interpretation and prognostication as soon as enough data have been entered into the system.

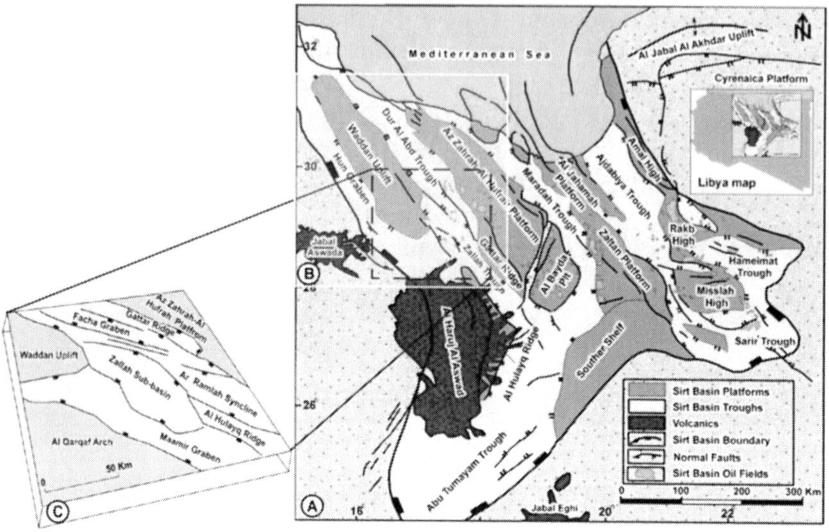

Figure 1. A) Generalized tectonic map of the Sirt Basin as located in the north central Libya map (Modified after Abadi, 2002 [27]) showing the major structural elements and oil and gas fields, B) Location of the study area, C) Dashed box marks the Zallah Trough area with its structural domains referred to in the text. Inset illustrates the Libya map

In specific the main purpose of conducting this remote sensing interpretation is to identify faults and fractures as well as major stratigraphic boundaries.

Imagery data from Google Earth™ and MrSid (Multi-resolution seamless image database) covering the entire study area were used for the remote sensing component of this study. The Google Earth™ were used for this study because it provided the highest resolution data and is available free.

The structural interpretation was performed digitally on-screen using ArcGIS software and then summarized as major faults in the region. The satellite imagery provided effective images on which to map the surface geological structure and geomorphology of the region and therefore enables validation of existing geological mapping.

In addition to the satellite imagery, 1:250,000 scale published geological maps and accompanying reports were also used during the interpretation. The maps were scanned and geo-referenced to the satellite imagery.

An orientation analysis was performed, and correlations between surface structures and subsurface oil and gas traps were investigated. The significance of using surface faults and fracture analysis for delineating additional traps in the region was assessed.

The other satellite date used was MrSid imagery (Figure 2) with a Landsat TM (Thematic Mapper) band combination 7,4, and 2 as red, green and blue, which combines spectral information from the near- to mid-infrared part of the electromagnetic spectrum and therefore highlights features that would otherwise be invisible to the naked eye. This type of band combination enables interpretation of geological structure as well as spectral discrimination of surface lithologies [2, 3].

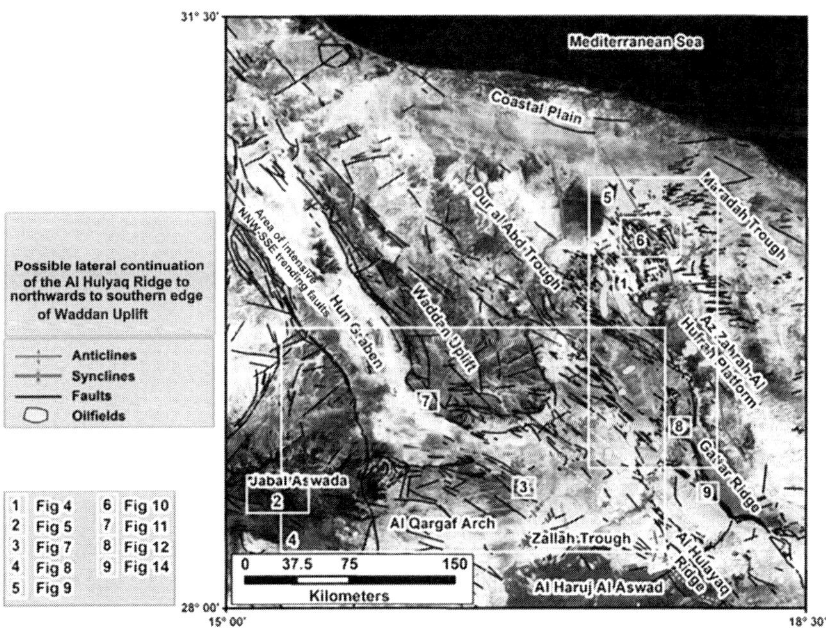

Figure 2. MrSid imagery that covers the study area showing the interpreted surface faults and folds, the location of the structural domains and other key structural features with figures locations

1.3. Geological Setting of Sirt Basin

The Sirt Basin is a wide Meso-Cenozoic rift basin (Figure 1) composed of several troughs separated by intrabasin highs deepening towards the east, played as an aborted intracontinental rift of Late Cretaceous age considered to be typical of a Tethyan extensional rift system, began in the Early Cretaceous and peaked in the Late Cretaceous [4, 5]; [6], with active rifting initiated possibly in the Early Cretaceous [7], which might reactivating pre-existing Paleozoic tectonic lineaments [8, 9], possibly triggered by Cretaceous lithosphere-plume interaction [10], even though, the [11] opine that the major subsidence is recorded only later in Late Cretaceous–Eocene, over the entire basin.

In the Alpine-related tectonic pulses during Middle–Late Eocene the Sirt Basin underwent compression resulted in northward tilting of the basin, causing abrupt subsidence in the north and uplift on the basin southern shoulders, possibly driving the latest stage of regional minor subsidence [12].

The complex tectonic and sedimentary history of the Sirt Basin has resulted in multiple reservoirs and conditions that favoured hydrocarbon generation, migration, and accumulation. Most oil pools discovered are located on horst structures at relatively shallow depths. Production in the Sirt Basin comes from more than 20 intervals ranging in age from Precambrian to Oligocene, but the most important finds have been in the Lower Cretaceous Nubian sandstone, although Palaeocene reefs are also important reservoirs.

2. SURFACE STRUCTURES ANALYSIS

The major faults systems were defined very clearly based on the satellite image mapping and the published geological maps (Figures 2 and 3). Most of the lineaments measured on remote sensing imagery were linear segments of stream valleys or other geomorphological alignments. These alignments are assumed to have formed as a result of differential erosion along lines of weakness (fracture zones) in the crust, produced by numerous episodes of tectonism. The main NW to NNW trending faults trend are up to tens of kilometers long. Viewed at different scales, the faults are offset by NNE trending 'hard-linked', oblique-slip transfer faults. The combined effect is a characteristic rhomboidal or zigzag fault pattern.

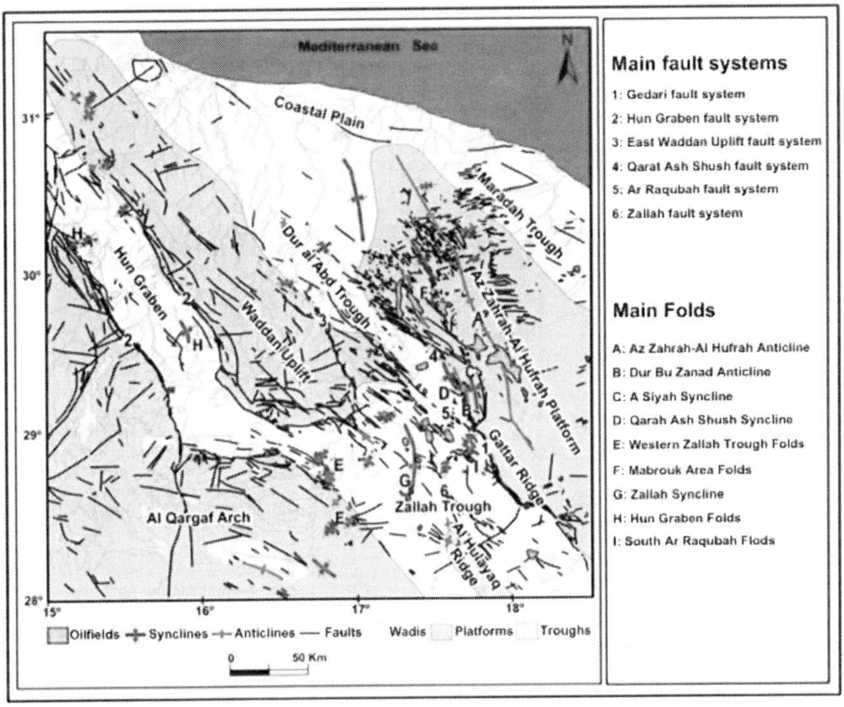

Figure 3. The location of the main interpreted fault systems and folds overlaid on the structural map which has been interpreted from satellite imagery

In the hangingwalls to the major faults, beds are gently inclined and warped forming synclines that trend parallel to sub-parallel to the main faults (Figure 4). Many folds within the study area are located sub-parallel to and bounded by the offset fault zones and occur as trains of hangingwall synclines and anticlines trending NNW to NW as well (Figure 4). The beds are folded about NNE trending axes and limb dips rarely exceed 10° and the folds plunge both to the NW and to the SE.

Many eruptions of basalt lavas are found in the southern part of the study area and some of them (Figure 5) form linear trending NW-SE shapes consistent with the main fault systems of the study area. The distribution of these basalt lava eruptions may indicate the existence of faults that cannot be seen due to the younger sediments and the extensive lava flows (Figure 6).

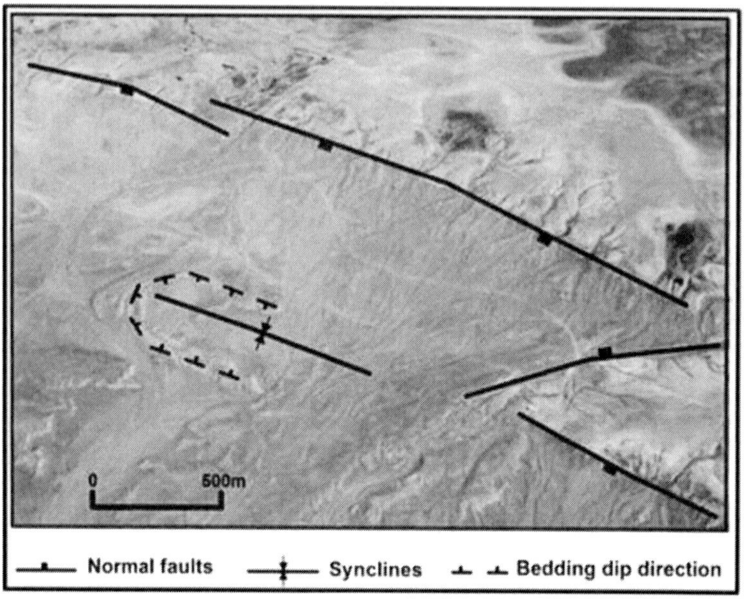

Figure 4. Example of fault related folds as they appear on satellite imagery. They are located sub-parallel to and bounded by the offset fault zones as well. For location see fig. 2. (adapted from Google Earth)

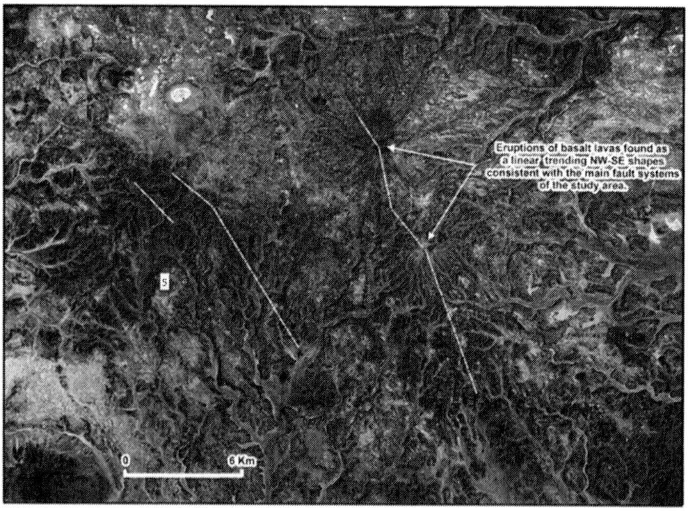

Figure 5. Example of basalt lavas eruptions which found in the southern part of the study area form linear trending NW-SE shapes consistent with the main fault systems of the study area. For location see fig. 2

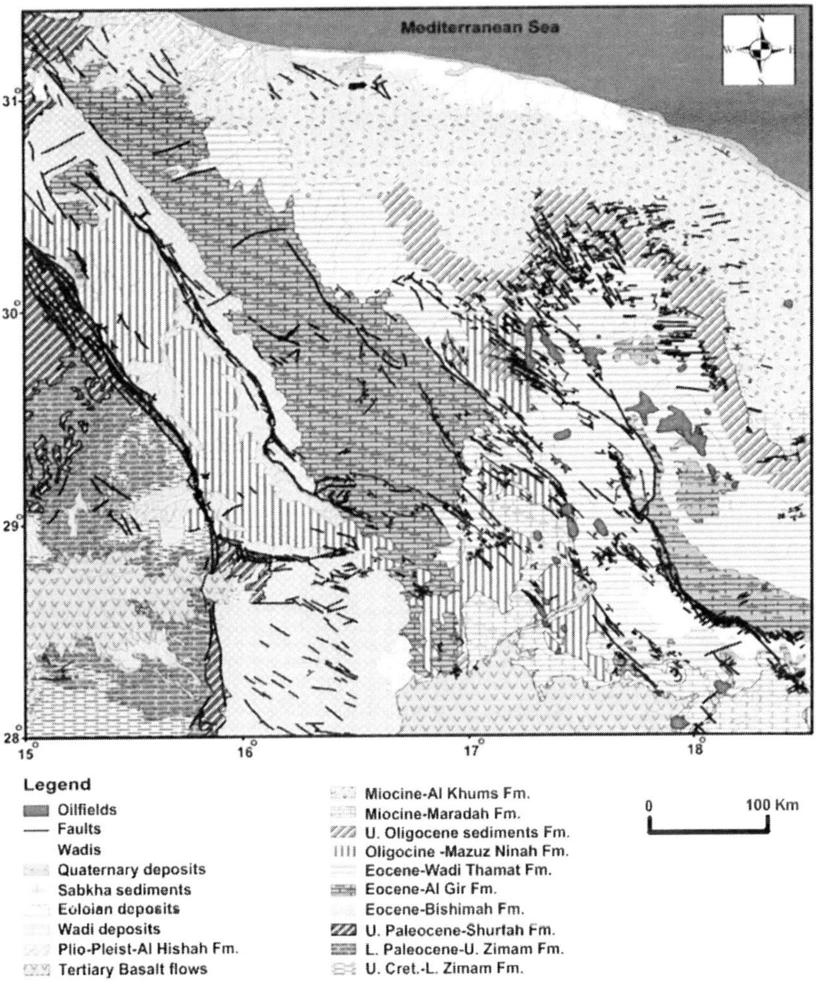

Figure 6. Simplified surface geologic map of study area shows the exposed rock units, and oilfields compiled after Vesely, (1985) [13], Jurak, (1985) [14] and Anketell & Kumati, (1991) [21]

These structures suggest that the area remained tectonically active even during post-Eocene times.

The main basin (trough) boundaries of the study area (Figure 3) are :- 1) the Gedari fault system located along the eastern boundary separating the Az Zahrah-Al Hufrah Platform from the Gattar Ridge; 2) the western boundary represented by Hun Graben fault system; 3) the

East Waddan Uplift fault system along the eastern margin of the Waddan Uplift; 4) the Qarat Ash Shush fault system; 5) the Ar Raquba fault system, and 6) the Zallah fault system, which together divide the main trough into a number of sub-blocks.

3. BASIN ELEMENTS AND GEOTECTONICS OF THE AREA

The satellite image interpretation integrated with the surface geological maps shows that the study area contains several basinal and geotectonic elements; namely, the Zallah, Dur al Abd, Maradah, Hun Graben, the Waddan Uplift, Gattar Ridge, the Az Zahrah-Al Hufrah Platforms, the Al Qargaf Arch, and the intrabasinal Al Hulayq Ridge (Figures 2 and 3).

The study area can be divided into a number of tectono-stratigraphic domains based on the following criteria:

- *Structural grain, i.e., the dominant trend of faults and associated folds.*
- *Homogeneity and orientation of bedding.*
- *Relative intensity of faults and fractures.*
- *Style of deformation.*
- *Surface stratigraphy.*
- *Structural geomorphology including drainage pattern and morphology.*

Using these criteria the following structural elements were identified from the data used. The structural domains generally show good internal correlation and are consistent with the known regional tectonic elements of the Sirt Basin.

3.1. Zallah Trough

The Zallah Trough (Figures 2 and 3) is situated in the southern part of the study area and is an asymmetrical faulted graben which connects northwards with the Dur al Abd Trough and southwards with the Abu Tumayam Trough where they appear more or less as a continuous low NW-SE trending embayment. It is bounded by the Waddan Uplift to the northwest, the Al Qargaf Arch to the southwest and the Az Zahrah-Al Hufrah Platform to the east.

On the basis of satellite image mapping and the published geological maps [13, 14] the Zallah Trough is divided into a number of sub-basins (Figure 1): the Facha Graben, Ar Ramlah Syncline, Zallah Sub-basin and Ma'amir Graben [15, 16]. The Al Hulayq Ridge (Figures 2 and 3), projects into the area from the south, and forms a spur that divides the Ar Ramlah Syncline, to the east and the Ma'amir Graben to the west, and then passes beneath the Al Haruj Al Aswad volcanic centre to the south.

Due to down-faulting, the deepest part of the trough lies on the NE side of the axial Al Hulayq Ridge and the southwestern Az Zahrah-Al Hufrah Platform, with the Gattar Ridge recognized.

This interpretation reveals that the Zallah limits are reasonably well defined at surface and from satellite images (Figures 6 and 2). The Zallah Trough is connected with the Hun Graben to the northwest by a complicated zone of wrench tectonics whose extension beneath the Al Haruj al Aswad volcanics remains uncertain. The Zallah Trough is an extremely complex area in terms of structural relationships in the western Sirt Basin.

Towards the northwest the Zallah Trough is bounded by several east trending, south downthrowing faults that are part of a normal fault zone that defines the southeastern margin of Waddan Uplift.

The southwestern margin is bounded by a north and east downthrowing, normal fault zone of the Al Qargaf Arch. Some large triangular bedding facets or 'flat-irons' of the exposed sediment along the western side of the trough indicate eastward dipping strata into the trough along the eastern end of the Al Qargaf Arch (Figure 7).

A partial discontinuity is also represented by faults striking NW-SE in the southwestern corner of the study area, which belong to the periphery of the Hun Graben and are mostly covered with younger sediments and is manifest on the surface only by indications of a straight lineament (Figure 8) indicated by a change of a series of wadis (dry valleys) direction and a lithology (spectral reflectance) which seems to be differ from the pattern of adjacent features.

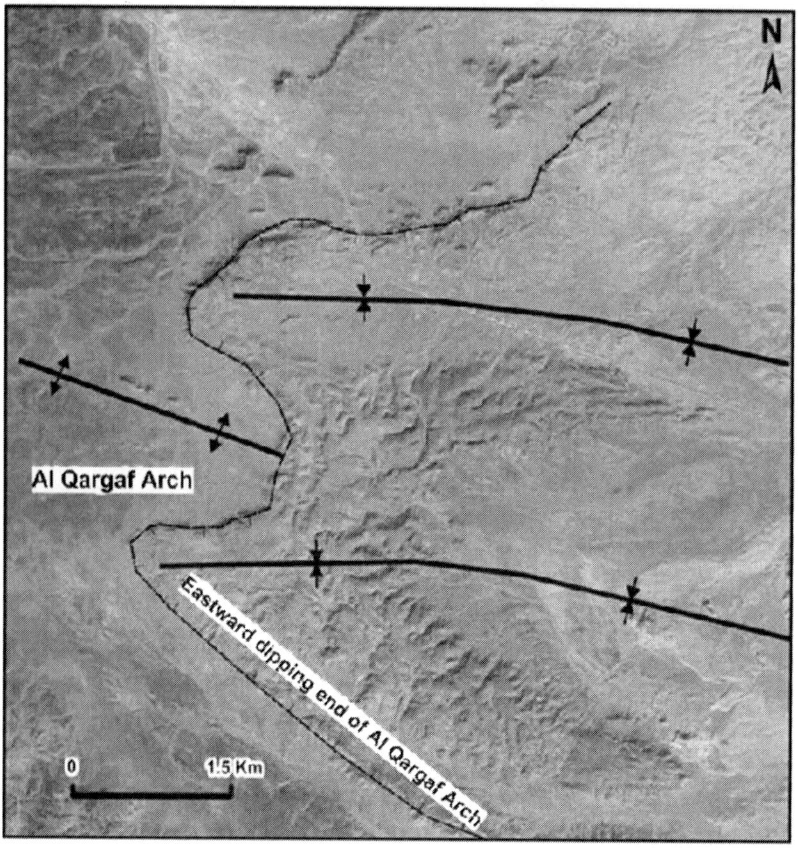

Figure 7. Some large triangular bedding facets or 'flat-irons' in exposed sediment along the western side of the Zallah Trough indicate eastward dipping strata into the trough along the eastern end of the Al Qargaf Arch. (For location see fig. 2)

They are probably the southernmost spurs of this structure and the amplitude of the movement along the faults is clearly smaller than in the northerly continuation of the Hun Graben.

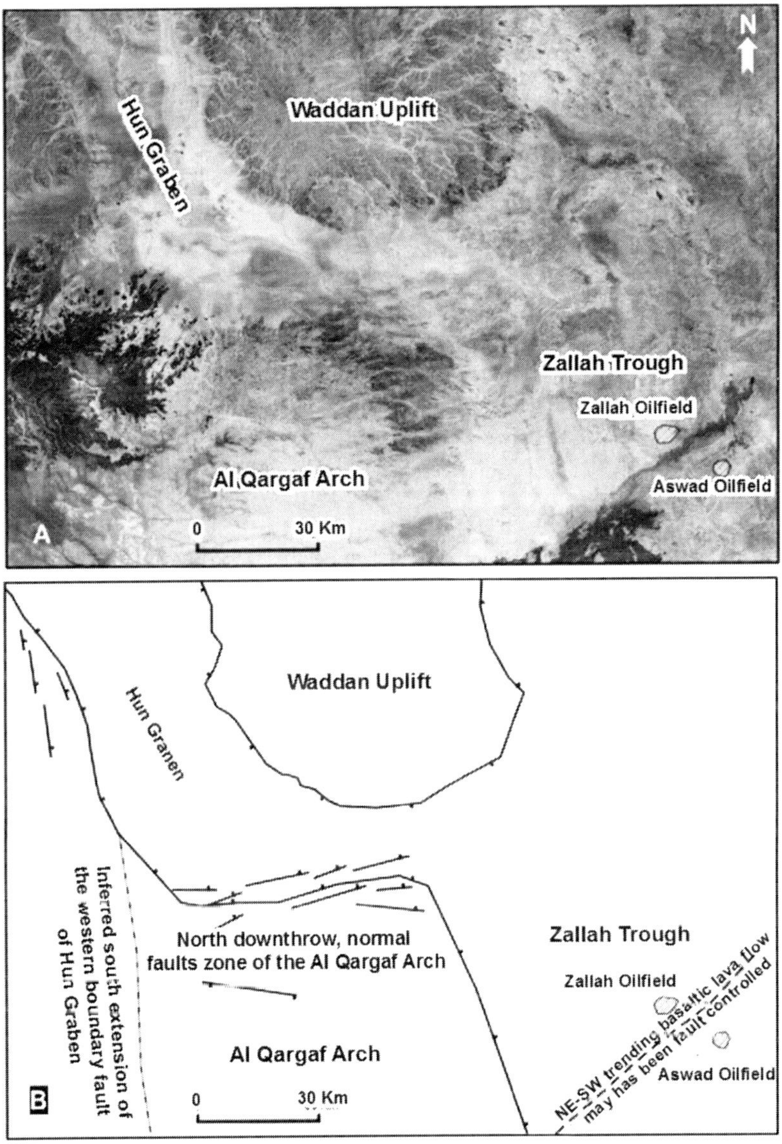

Figure 8. (A) The southern portion of Hun Graben as appears on satellite imagery. (B) Satellite imagery interpretation of the southern portion of the Hun Graben suggests that it extends beyond the E-W cross-cutting fault and may link to a major NNE-SSW trending fault system to the south that is covered by lava flows of Jabal Al Aswda. For location see fig. 2

Compared with the western margin, the eastern side of the Zallah Trough is more defined and marked by a series of major fault scarps which coincide with a normal fault zone defining the western edge of the Gattar Ridge as SW fringe of Az Zahrah-Al Hufrah Platform.

The fault zone extends for approximately 180km along strike in the study area and has been described as the Gedari fault.

The interpretation also reveals that faults up to tens of kilometers long are the dominant structural element in the study area where the main NW-NNW structural trend coincides with the orientation of tectonic elements of the Sirt Basin. Most of these faults are associated with normal displacement whilst others show no visible displacement at the resolution of the imagery. NE-SW trending faults are also developed, and in most cases, appear to post-date and off-set the NNW-SSE trending faults.

Many fault-related folds (drag-folds) formed adjacent to the fault structures and are typically narrow asymmetric monoclinal structures with beds tilted generally less than 5, although dips up to 20-25 are estimated adjacent to faults.

Furthermore there are numerous elongate asymmetric synclines such as the Zallah syncline (Figure 3) located in the central part of Zallah Trough associated with shallow west limbs and steeper dipping east limbs that are developed in the hangingwall and adjacent to west downthrowing normal faults on the eastern margin [13]. These hangingwall basins contain Oligocene Ma'zul Ninah, Miocene Al Khums and Late Tertiary-Quaternary sediments (Figure 6). The period of folding is connected with subsidence movements during the filling of the Sirt Basin [13]. Some of the fold structures may be due to compaction of sediments in troughs, which supplement the tectonic folding [13] where the sedimentary sequences are mildly undulated into synclines and anticlines, the axes of which are parallel to the principal tectonic direction, i.e. NW-SE.

According to satellite interpretation most of the subtle low amplitude periclinal structures which are exposed at surface coincide with producing oil fields. The Zallah Trough is a well-known oil-producing area with over 20 significant oil discoveries, (Figure 3) and production from various oil fields [17]. The Zallah Trough oil fields are situated mainly on the updip edges of interpreted folds.

3.2. Dur al Abd Trough

The Dur al Abd Trough (Figures 2 and 3) represents a more or less continuous elongated low of three troughs, including Dur al Abd, Zallah and Abu Tumayam Troughs from north to south respectively which runs from the Sirt embayment coast in the north to the southern (shelf) boundary of the Sirt Basin. Some authors believe that the Zallah Trough narrows northwards and becomes shallower to form the Dur al Abd Trough [18].

The Dur al Abd Trough separates the Az Zahrah-Al Hufrah Platform from the Waddan Uplift in the area of the Mabruk field and it is difficult to follow north of the Mabruk field where it loses its identity in the faulted northeast flank of the Waddan Uplift. The Dur al Abd Trough is a 50km wide NW-SE trending, fault bounded graben in the Mabruk area but broadens out further south and has greatest fault displacement in the east. It extends approximately 175km northwards from the Facha field. The Dur al Abd, and Zallah Troughs seem to be in the first instance a continuous trough that follows the NNW-SSE trending structural grain of the region, however the data reveal a structure that acts as a bridge that separates them from each other possibly extending from northern Gattar Ridge to the southeastern part of the Waddan Uplift (Ar Raqubah fault system, (Figure 3).

The Dur al Abd Trough is extensively faulted and characterized mainly by two fault trends, one that is parallel to the NNW-SSE dominant fault trend of the Sirt Basin and has an associated subsidiary later ENE-WSW transverse fault set and a second N-S trending fault set that are evident on the northeastern dipping flank of the Waddan Uplift and extend into the Dur al Abd Trough. Displacement across the graben bounding faults is approximately 250m [19, 18].

Based on this surface interpretation it is difficult to split the northern extension of Dur al Abd Trough from the adjacent Az Zahrah-Al Hufrah Platform to the east.

The Mabruk field is the only field that lies within the Dur al Abd Trough with the Facha field on its faulted western flank on the Az Zahrah-A Hufrah Platform.

3.3. Az Zahrah-Al Hufrah Platform

The Az Zahrah-Al Hufrah Platform (Figures 2 and 3) is located in the eastern part of the study area between the Dur al Abd and Zallah Troughs to the west and the Maradah Trough to the east and is bounded on the south by the NNE-SSW trending Kotlath Graben which separates it from Al Bayda Platform (Figure 1). On its southwest margin, the Gattar Ridge is formed by downfaulting to the WSW. The platform occupies a 40,000 km^2 area and dips towards the northeast with a northern boundary that is poorly defined and may be affected by wrench-faulting close to the present coastline [20]. The western platform boundary fault in the SW region is characterised by a complex series of en echelon faults indicating a sinistral strike-slip movement, with associated riedel shears forming small-scale horst and graben structures during the Eocene [21].

The western platform boundary fault (Gedari fault) lies close to the assumed junction between the basement west and east African plates, which were active during the Cretaceous and reactivated as sinistral wrench faults in the Eocene time [20, 22] as shown in Figure 1.

This western boundary fault zone of the Az Zahrah-Al Hufrah Platform is defined by: NNE-SSW trending monoclinal flexures in the bedding (Figure 4) which provide a sense of relative downthrow across the fault zone. The northeasterly tilt of the platform occurred during the post-Eocene and is associated with extensive faulting and fracturing with identification of the main faults trending NNE-SSW being considered as a reactivation of pre-rift faults, NW-SE Cretaceous syn-sedimentary faults and NNW-SSE Late Eocene (post-rift) faults [19, 23, 15].

Faults within the platform have generally less than 60m displacement whereas the platform bounding faults have 800-1000m displacement [24, 19, 23, 15, 18]. From a surface perspective, the northern end of the platform is expressed at surface by a homoclinal shallow northeast dipping panel predominantly of U. Eocene Wadi Thamat Formation. An elongate erosional window occurs along the western margin of the platform exposing the M. Eocene Al Gatta Member.

Large SW dipping triangular bedding facets of the lower member of the Oligocene sediments help to define the limits of the platform at surface forming elongated synclines parallel to the western platform boundary faults (Figure 4). The Az Zahrah-Al Hufrah Platform at the northern end (Figure 9) is marked by a two north-northwest plunging nose of a large low-amplitude gentle flexures anticlines which are expressed by erosional features and surface geomorphology. An anticlinal monoclinal fold (Figure 3) defines the western margin of the dominant fault trend within the platform is NNW-SSE with subsidiary E-W to ENE-WSW trending faults (Figure 4) consistent with the Sirt Basin structures. Combination of the NNW-SSE and ENE-WSW trending faults impart a grid-like pattern that resembles crosscut fault structures (Figure 10). The Az Zahrah Al Hufrah Platform comprises carbonates of the Lutetian to Priabonian (Eocene) Wadi Thamat Formation (Figure 6) and is outcrop equivalents to major deeper hydrocarbon reservoirs specifically in the eastern portion of Sirt Basin.

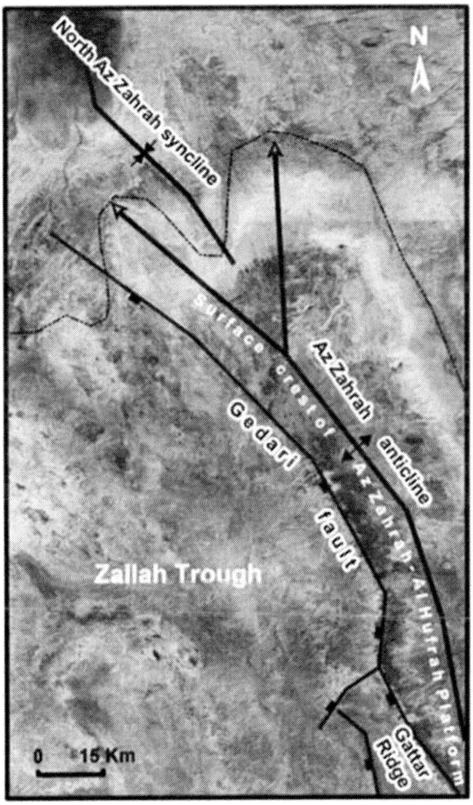

Figure 9. Satellite (MrSid) imagery showing the northern end of Az Zahrah-Al Hufrah Platform which is marked by a two north-northwest plunging nose of large low-amplitude gentle flexures anticlines which are expressed by erosional features and surface geomorphology. For location see fig. 2

These beds are cut by a NW-trending system of fractures, characterized by swarms of vertical extension joints accompanied by crosscut sets (Figure 10) of hybrid and shear joints, and normal faults. Many of the NE to E-W interpreted lineaments exhibit fractures, and are particularly conspicuous in the north of the platform where they are possibly due to the influence of additional fracture systems related to major flexures trending obliquely to the NW-SE direction. As any crosscutting feature must be younger than the rock or sediment it cuts across, these major E-W trending flexures and associated fracture systems were formed during the post-Eocene phase of deformation, and are cross-cut obliquely by fractures of the NW-trending system.

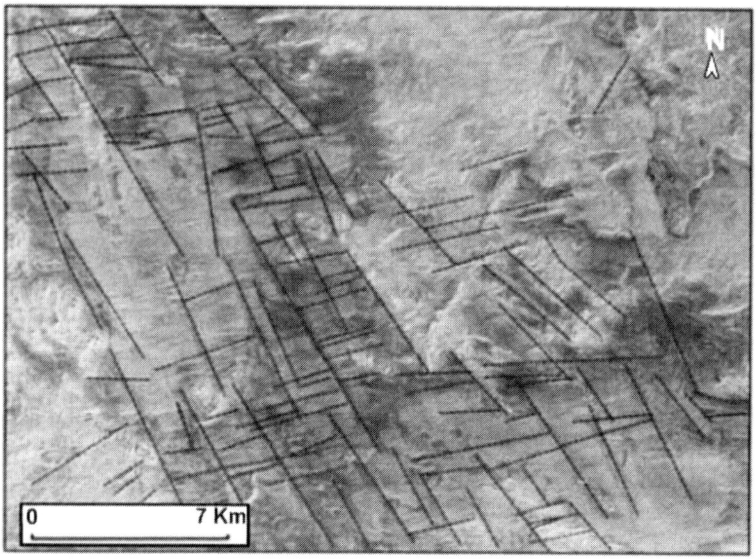

Figure 10. Satellite (MrSid) imagery showing a combination of the NNW-SSE and ENE-WSW trending faults that impart a grid-like pattern that resembles crosscut fault structures at the northern end of Az Zahrah-Al Hufrah Platform. For location see fig. 2

3.4. Hun Graben

The Hun Graben (Figures 2 and 3) is an elongated structure feature measuring 300km long and averaging 40km in width, has a distinct surface expression and forms a present day fault bounded depression indicating that it is one of the youngest tectonic elements in the region. The development of the Hun Graben, however, is controversial. [25] indicated that the graben was downthrown 500-800m relative to the Tripoli-Tibesti Uplift (Figure 1) and also suggested that the graben was initiated prior to the Late Cretaceous and that the western margin was Paleocene in age. [26] estimated a vertical throw of 100-120m and an Oligocene age for the development of the graben whilst [27], suggested the major faulting occurring in post-Oligocene time.

The graben narrows at the southeast near the town of Hun (Figures 2 and 3) and its connection with Zallah Trough is complicated by strike-slip faulting. The structure underlying the Al Haruj al Aswad volcanics is uncertain, but [25] inferred the presence of a remnant of the Palaeozoic Kalanshiyu Trough below the volcanics. Based on the drainage pattern analysis (Figure 6) the Hun Graben appears to cross-cut existing southwest draining wadi channels.

Along the eastern margin of the basin, (Figure 11) wadis are generally narrow and become incised as they approach the margin of the graben indicating that they have been rejuvenated by a relative drop in the surface elevation within the graben.

This is evidenced by a relatively linear major northeast draining channel present on both flanks of the Hun Graben. The drainage channel is deflected as it enters the Hun Graben due to normal faults and related gentle folding of strata adjacent to the faults. These relationships imply that the palaeo-drainage (at least pre-Oligocene age) was to the northeast and was subsequently modified by development of the Hun Graben in the Oligocene.

Drainage within the graben is characterized by large arcuate drainage channels or Wadis (dry valleys) in the Hun Graben. The shape of these wadis reflects the deflection of the drainage channels along the bedding contact between relatively hard and soft lithological units, which have been deformed into a subtle synclinal warp (Figure 11). A northward draining NNW-SSE trending axial drainage system is located mainly along the eastern margin of the graben.

Most drainage channels entering the western side of the graben converge towards a common point or drainage sink towards the northern end of the Hun Graben.

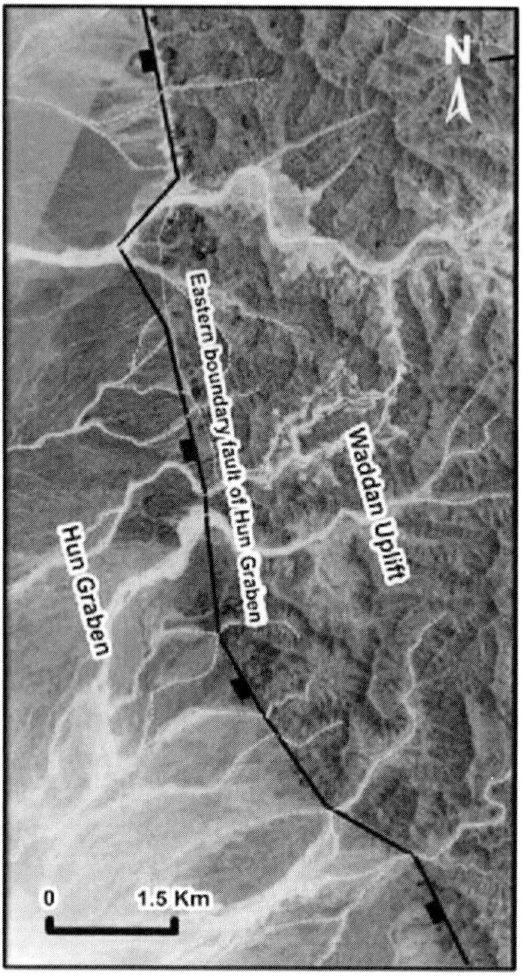

Figure 11. Satellite (Google Earth) imagery showing wadis along the eastern margin of the Hun Graben are generally narrow and become incised as they approach the margin of the graben indicating that they have been rejuvenated by a relative drop in the surface elevation within the graben. For location see fig. 2

A number of NNW-SSE open folds (Figure 3) have been interpreted within Ma´zul Ninah Formation strata along the margins of the graben. The folds have probably formed in response to drag in the hangingwall to the basin bounding faults or possibly due to roll-over on non-planar faults in the sub-surface.

3.5. Waddan Uplift

The Waddan Uplift (Figures 2 and 3) is a gentle north-northeast tilted block, approximately 250km long with a width of 105km in the south, narrowing to 75km in the north before it disappears below the coastal plain. The uplift terminates in the south against the northeastern extension of the Al Qargaf Arch (Figure 8).

The dominant fault trend is NNW-SSE with a subsidiary orthogonal set trending ENE-WSW (Figure 2). The northeastern margin of the Waddan Uplift is extensively faulted due to dextral shear along a WSW-ESE basement fault [20].

The southern end of the uplift has a convex morphology to the south and is characterised by a discrete fault bounded margin with downthrow to the southwest, south and southeast. The western boundary is marked by the eastern boundary fault of the Hun Graben (Figure 8). The topographically highest point of the Waddan Uplift is located towards the southern end. This area characterised is by radial drainage although the crest of the structure is marked by NNW-SSE trending drainage divide, located some 20km east of the western faulted margin of the Waddan Uplift.

The internal structure of the Waddan Uplift is comparatively simple with a NNW-SSE trending trellised drainage pattern controlled by NNW-SSE trending, south-southwest downthrowing normal faults (antithetic to the northeast to north-northeast drainage direction).

The northern half of the platform (Figure 3) is characterised by a north and north-northeast-trending dendritic drainage pattern down the sub-horizontal dip slope of the structure. A more organized trellis drainage pattern is developed towards the northern end of the structure parallel to the NNW-SSE and ENE-WSW trending fracture system.

Elongate erosional escarpments of more resistant lithologies and/or fault scarps may also control linear NNW-SSE trending Wadi channels. The southeastern part of the Waddan Uplift appears to be associated with a downthrown fault block which is defined by a broad low amplitude anticlinal structure developed in Wadi Thamat Formation.

3.6. Gattar Ridge

The Gattar Ridge (Figures 2 and 3) is coincident with a 15km wide, 120km long NNW-SSE trending northeast tilted panel of Al Jir Formation at surface. It appears as a terrace adjacent to the south western boundary fault of Az Zahrah-Al Hufrah Platform which projects into the eastern side of the Zallah Trough.

The western edge of the fault block is bounded by a fault zone up to 1km wide, consisting of overlapping parallel normal faults consistently downthrowing to the west. It is also characterized by segmented normal faults with relay ramp structures associated with horsetail splay faults that may indicate a component of sinistral strike-slip (Figure 12).

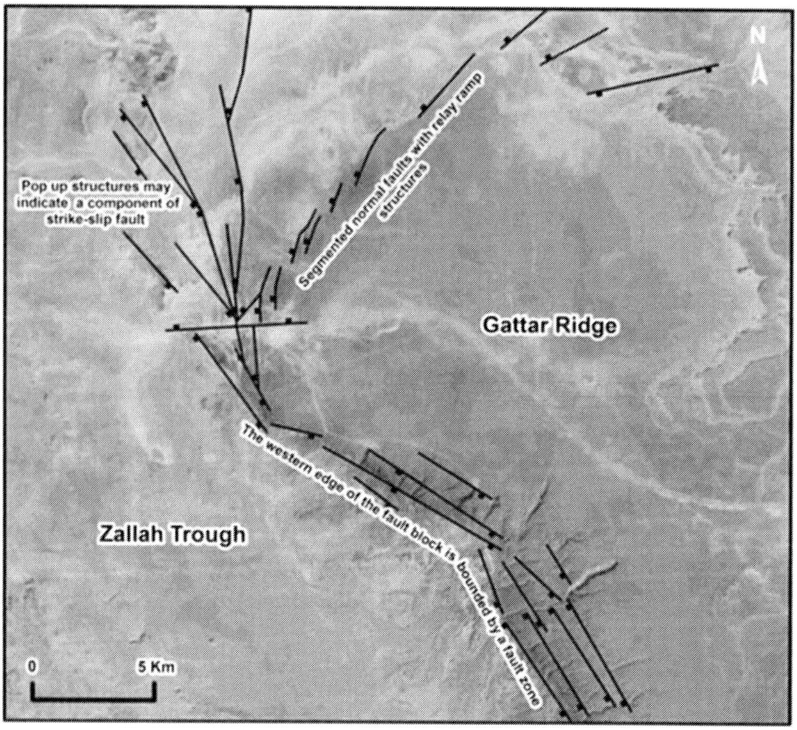

Figure 12. Segmented normal faults with relay ramp structures associated with horsetail splay faults that may indicate a component of sinistral strike-slip as interpreted from satellite (Google Earth) imagery. For location see fig. 2

The panel is bounded to the north-northwest by an ENE-WSW trending transverse fault. To the north-northwest the Gattar Ridge dies out into an area of close spaced faults with a NW-SE trend. The faults cross-cut Eocene-aged Wadi Thamat Formation, Oligocene-aged Ma´zul Ninah Formation and Late Tertiary-Quaternary deposits. The dominant fault trend is NNW-SSE and subsidiary E-W and NE-SW faults sets are also developed.

3.7. Maradah Trough

The Maradah Trough (Figures 1, 2 and 3) is located in the northeastern portion of the study area and is a deep, narrow, NW-SE-trending basin bounding the Az Zahrah-Al Hufrah Platform to the west and swings to a NNE trend around the southern termination of the Al Baydah Platform and then shallows rapidly to the southeast onto the Southern Shelf. In the north, the Maradah Trough narrows and shallows rapidly between the Az Zahrah-Al Hufrah Platform and the Shammar and An Nawfalfyah Highs. North of this it widens and deepens again towards the Sirt Gulf [20].

The dominant fault trend bounding the trough is NNW-SSE with subsidiary NNE-SSW trending consistent with the Kotlath Graben and the basin southern shelf towards the south (Figure 1).

3.8. Al Hulayq Ridge

The Al Hulayq Ridge is a NNW-SSE trending narrow basement ridge or horst block which protrudes into the southern part of the Zallah Trough (Figures 2 and 3). The western margin of the high is structurally complex and associated with a series of drag folds developed within Paleocene-Eocene strata. The folds create hydrocarbon traps and may represent part of a flower structure which formed in response to dextral shear during the Oligocene [28].

The surface structure appears to suggest that there is a lateral continuation of this feature northwards to the southern edge of the Waddan Uplift (Figure 2). The Hakim and Fidda oil fields are on strike with the Al Hulayq Ridge, while two broad, low amplitude periclines at surface may be related to the reactivation of faults bounding the Al Hulayq Ridge. The fields are expressed at surface by periclines developed mainly in Eocene Wadi Thamat Formation. The Zallah fault system bounds the Al Hulayq Ridge on both sides from the Ar Ramlah Syncline to the east and the Ma'amir Graben to the west (Figure 1). This fault system forms a series of NNW trending fault segments arranged as overlapping normal faults downthrowing to the SW, changing to NE throw when the fault bounds the Al Hulayq Ridge on the east side.

3.9. Al Qargaf Arch

The Al Qargaf Arch (Figures 2 and 3) is an E-W trending Hercynian structure located on the southwestern edge of the study area and exposes Palaeozoic strata (outside of study area) onlapped by Oligocene cover rocks. Regionally the structure forms a WSW-ENE topographic high which separates the Murzuq and Ghadamis Basins.

The core of the structure is formed by Jabal Hasawnah which is elevated 800m a.s.l. The structural grain of the Al Qargaf Arch is dominated by ENE-WSW trending faults and resistant ridges (possibly silicified relict fault planes and/or igneous dykes). Subsidiary WNW-ESE to NW-SE and N-S trending faults possibly also occur.

4. ORIENTATION ANALYSIS OF THE STRUCTURAL FEATURES

Orientation is one of the most revealing characteristics of linear features, such as surface lineaments, surface fracture zones, and surface fracture traces. Rose diagrams provide one of the most informative ways of representing orientation data. Rose diagrams for all the linear features acquired in the study area were generated and interpreted which provide one of the most informative ways of representing orientation data [2, 3].

Direction statistics for all the interpreted linear features acquired in the study area were calculated and given as a total distribution and then the totals are broken down by the major basin elements of the area. Hundreds of lineaments were recognized.

Their lengths, range between a hundred meters up to hundreds of kilometers long and the dominant trends are between N25W and N55W (Figure 13) which coincides with Sirt Basin structures.

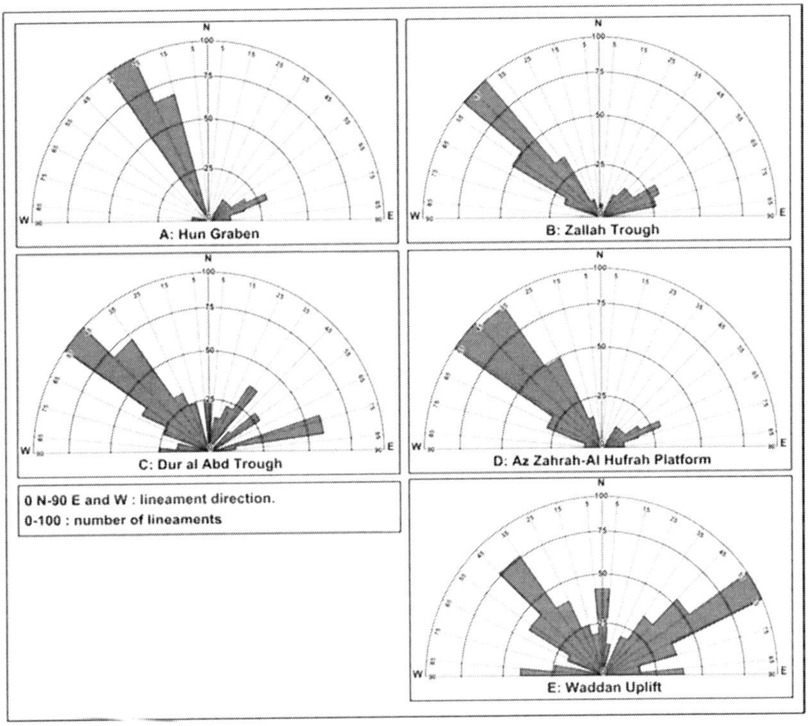

Figure 13a-e. Rose diagrams of the trend of surface lineaments, major fracture zones, and/or fracture traces in various parts of the study area

The rose diagrams were generated based mainly on total frequency. A majority of the surface linear features in the region from these rose diagrams have five preferred orientations: N25-35W in the Hun Graben, N40-50W in the Zallah Trough, N45-55W in the Dur al Abd Trough, N35-55W in the Az Zahrah-Al Hufrah Platform, and in contrast in the Waddan Uplift a N55-65E trend. In general the NW to NNW trends form a series of an elongate fault segments arranged as overlapping normal faults dipping to the SW, changing to NW dip where the faults are arranged in a closely spaced en echelon pattern with NW-SE trend. These can measure up to 300km long in the Hun Graben. This trend forms the dominant structural grain in the region and may reflect the regional structural styles of the study area and is in good agreement with the orientation of major tectonic elements of the Sirt Basin structures.

On the contrary, those surface linear features oriented northeast-southwest appear to be related to local geological structures which are less prevalent than those oriented northwest to southeast. In addition, ENE-WSW trending sets of faults are developed on both shoulders of the Hun Graben, orthogonal to the main structural grain. The faults clearly off-set NNW-SSE trending bounding faults and appears to post-date and off-set the NNW-SSE trending faults.

The Gedari fault system is one of which composed of two different trends where it changes from NNW-SSE trending to ENE-WSW trending, at 17°50', 29°05' (Figure 3) before swinging directly north to follow the trend of an escarpment until 17°50' and 29°15'. Then it can be followed to 17°20', 29°45' where it branches to the northwest to link up with the northernmost faults of the Qarat Ash Shush fault system and then continuing beyond the Mabruk oilfield.

A similar combination of NNW-SSE and ENE-WSW trending faults particularly in the northern part of Az may reflect conjugate fault structures. Zahrah-Al Hufrah Platform imparts a grid-like pattern and may reflect conjugate fault structures.

It should be mentioned that some lineaments, mapped from the remote sensing data were not present in the geologic maps, which suggest that this image lineament interpretation provides new geologic perspectives even in an area which has been well mapped because of petroleum exploration. Some of the newly recognized lineament may be faults that were not previously mapped. It seems to be that the longest linear features in the study area were found on the western and eastern margins of the Hun Graben (Figures 3, and 14).

5. RELATIVE TIMING AND STYLES OF THE AREA STRUCTURES FEATURES

The dominant structural features in the study area are composed of NNW-SSE trending extensional faults and associated monoclinal folds which form the dominant structural grain of the area.

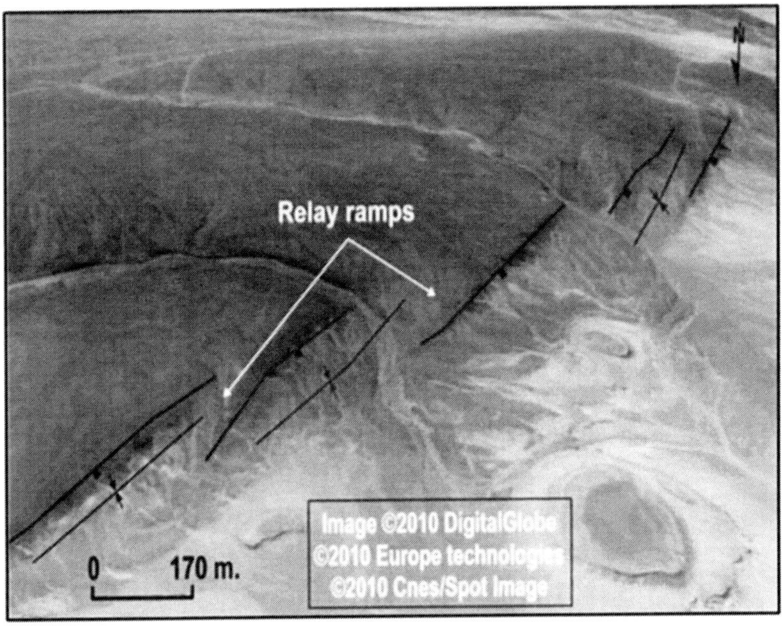

Figure 14. Example of relay ramp structures with adjacent elongated synclines as appear on satellite imageries. It shows a fault displacement folds and relay ramp associated with stepping normal faults. (adapted from Google Earth). For location see fig. 2

These structures are related to pre-Upper Cretaceous NE-SW extension possibly associated with local reactivation of existing Pan-African structures. In addition the NNW-SSE to NW-SE trending broad, low amplitude, periclinal folds and intervening anastomosing synclines are likely to be related to differential compaction above and adjacent to Cretaceous age structures.

On the contrary, those surface linear features oriented extensional faults of the Hun Graben which associated with present day topographic and geomorphological expressions are at least Oligocene in age. The western boundary fault of the Hun Graben appears to cross-cut the E-W northern boundary fault of the Al Qargaf Arch. Some ENE-WSW trending faults appear to off-set the bounding faults to the Hun Graben suggesting they developed during the Oligocene or during Alpine deformation.

In the other hand the ENE-WSW trending normal fault zones and associated monoclinal flexures clearly offset all Cretaceous age structures and may be related to Alpine deformation. They could also be influenced by reactivation of deeper Hercynian structures. Whereas the E-W trending normal faults were probably formed by Miocene deformation and reactivation of older Hercynian structures especially around the Al Qargaf Arch.

All of these trends are associated with similar trending minor faults with no apparent sense of offset. All faults appear to be steep dipping and generally associated with normal displacement. Most normal faults show soft-domino linkage and hard linkage is not common. The western boundary fault of the Gattar Ridge may be associated with transtensional sinistral strike-slip, possibly related to reactivation of an originally Cretaceous age fault [21].

6. THE SIGNIFICANCE OF SURFACE STRUCTURE FEATURES FOR PETROLEUM EXPLORATION IN THE STUDY AREA

It has been observed in this study on the basis of remote sensing interpretation that at least in some parts of the study area, oil and gas traps line up well with the trends of surface major structural features such as folds, lineaments, fracture zones, and/or fracture traces. As a result, surface structural features have been used to examine any relationship with the existence of explored oilfields in the study area.

Some of the existing oil and gas fields in the central part of Zallah Trough apparently are structures oriented in a northwest-southeast direction consistent with the trends of the Zallah fault system (Figure 3) which is characterized by the existence of major oil fields (Figures 2 and 15) that are located parallel to this fault system, e.g., the Hakim and Fidda oilfields, which are associated with a narrow NNW-SSE trending surface anticline expressed in Maradah Formation and Late Tertiary-Quaternary sediments (Figure 6). The Hakim and Fidda structure is approximately 36.5 km long and approximately 4km wide exposed in the erosional core of the structure along strike with the Al Hulayq Ridge, while two broad, low amplitude periclines at surface may be related to the reactivation of faults bounding the Al Hulayq Ridge. The Ghani and Zenab fields are also located in the middle portion of this fault system in the vicinity of a NNW-SSE to NW-SE trending arcuate fault zone associated with normal downthrow to the west.

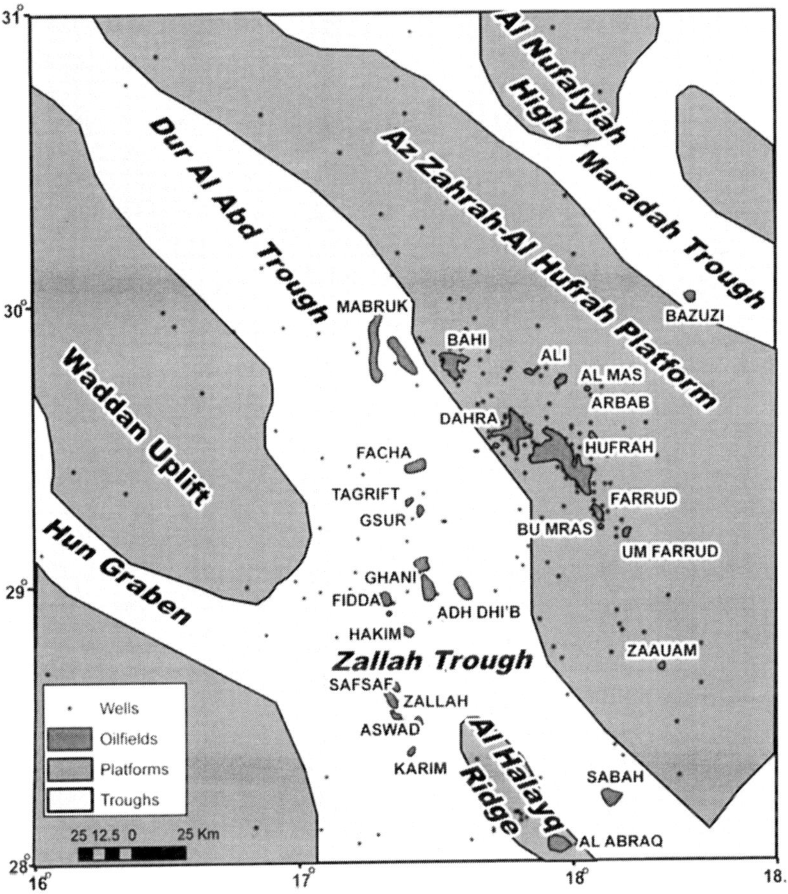

Figure 15. Generalized tectonic map of the western Sirt Basin shows the location of the main oil and gas fields

The Karim field is one of the largest in the study area. It is a NNW-SSE trending periclinal structure expressed at surface in Oligocene Ma´zul Ninah and Miocene Maradah Formations (Figure 6). The structure has gentle plunge to the SSE and may be structurally closed at its northern end by NE-SW trending transverse faults associated with normal displacement. NW-SE trending normal faults cross the structure with downthrow dominantly to the southwest. The main axis of the structure is 38km long and its width is 24.5km at surface. A narrow NE-SW trending basaltic lava flow (Figure 8) separates Aswad from the Zallah oilfields to the north and appears to occupy a palaeo-valley which is now topographically inverted. The NE-SW linear trend of the preserved flow may have been fault controlled.

The Aswad and Zallah fields (Figures 15, and 3) are associated with a large subtle domal structure, 27km long by 22.5 km wide expressed at surface within Ma´zul Ninah and Maradah Formations. The Ma´zul Ninah Formation is exposed in the eroded core of the structure (Figures 6, and 3). The fold structure is bounded by a NNW-SSE trending fault on its eastern side with downthrow to the west. Similar trending normal faults occur near the crest of the structure.

The Mellugh, Ghani and Zenab fields (Figures 15, and 3) are located in the vicinity of a NNW-SSE to NW-SE trending arcuate fault zone associated with normal downthrow to the west. The Daba, Gsur, Tagrifet and Facha fields are situated towards the northern end of the Zallah Trough on a NNW-SSE trend on the hangingwall side of the boundary fault of the Gattar Ridge. The area is also situated to the north of a synclinal area. Faults in the vicinity of the oil fields are downthrow to the northeast along the strike continuation of the eastern boundary faults of the Waddan Uplift.

The Mabruk field is considered to be the only field that lies within the Dur al Abd Trough with the Facha field on its faulted western flank on the Az Zahrah-A Hufrah Platform.

Additionally many major oil fields (Figure 15) are located along the western crest of the Az Zahrah-Al Hufrah Platform such as the Bahi field which is situated in an erosional window developed above the structural crest of the Az Zahrah-Al Hufrah Platform, 4.5km east of the fault bounded western margin of the platform. The Az Zahrah oil field is situated within a NNW-SSE trending open syncline east of the monoclinal margin of the Az Zahrah-Al Hufrah Platform.

A number of NE-SW trending transverse faults occur in the area separating the Az Zahrah oil field from the Bahi field. The Al Hufrah oilfield is located to the east of the monoclinal margin of the Az Zahrah-Al Hufrah Platform in a similar structural position to the Az Zahrah field. A significant NE-SW trending, possibly fracture controlled, erosional window is present to the north of the field, and the Ali and Almas fields are located on the eastern edge of the study area. They are located on the gentle northeast dipping limb of the Az Zahrah-Al Hufrah Platform.

Figures 15 and 3 show the distribution of the oil and gas fields in the study area and one can see from this interpreted satellite data that both the geometry of individual oilfields and the relative locations of most oilfields clearly follow the trends of the surface lineaments in the area. The consistency in orientation and the association in position between the surface structural features and the oilfields make it very attractive to use surface structural features as a guide for delineating new oil and gas traps in the study area.

Based on the above observations, it is reasonable to conclude that any potential new oil and gas reservoirs in the remaining parts of the study area, if they exist, will be likely associated with the surface structural features and fracture traces in the region. Therefore, surface structural features and fracture analysis should be employed as a cost effective tool for delineating preliminary locations for more sophisticated and more expensive geophysical and/or geochemical surveys in the study area.

7. Discussion

The surface geology of this area (Figure 6) is composed primarily of Late Upper Cretaceous, Tertiary and Quaternary surficial deposits. The study area is highly faulted, with a complex geological pattern, bounded to the west by a portion of the Hun Graben (Figures 2 and 3) and containing prominent geological features of the Sirt Basin. The faulting style exhibited in this region is that of normal and strike-slip faulting with a general NW to NNW trend.

Figures 2 and 3 display the interpreted segmented fault pattern composed of NW-SE and N-S to NNE-SSW striking fault systems that have formed at both the rift borders and within the basin. As a general rule, the NW-SE faults that are parallel to the major structural trend of the basin controlled the rate of rifting and subsidence whereas the NE-SW structure modifies the pattern formed by previous system to form the block structures.

The NE-SW faults are limited in length and not continuous. Based on their characteristics, the NE-SW fault system can be classified as transfer faults.

Both sets of faults combine to control several structural and geotectonic elements; namely, the Zallah, Dur al Abd, Maradah Troughs, Hun Graben, the Waddan Uplift, Gattar Ridge, Az Zahrah-Al Hufrah Platforms, Al Qargaf Arch, and the intrabasinal Al Hulayq Ridge (Figures 2 and 3).

These blocks are bounded on one side by a major NW trending border fault system and a predominant stratal dip direction toward the border fault system. The Al Hulayq Ridge (Figures 2 and 3), projects into the area from the south, and forms a spur that divides the Ar Ramlah Syncline, to the east and the Ma'amir Graben to the west, and then passes beneath the Al Haruj Al Aswad volcanic centre to the south.

8. CONCLUSIONS

The study area is highly faulted, with a complex geological pattern, bounded to the west by a portion of the Hun Graben and containing prominent geological features of the Sirt Basin. The faulting style exhibited in this region is that of normal and strike-slip faulting with a general NW to NNW trend.It is also characterized by segmented normal faults with relay ramp structures associated with horsetail splay faults that may indicate a component of sinistral strike-slip as well. In general the NW to NNW trends form a series of an elongate fault segments arranged as overlapping normal faults dipping to the SW, changing to NW dip where the faults are arranged in a closely spaced en echelon pattern with NW-SE trend.

In the hangingwalls to the major faults, beds are gently inclined and warped forming synclines that trend parallel to sub-parallel to the main faults. These numerous elongate asymmetric synclines associated with shallow west limbs and steeper dipping east limbs are developed in the hangingwall adjacent to west downthrowing normal faults predominantly along the eastern fault bounded margin of the Gedari fault system and the Gattar Ridge.

Segmentation of the Gedari fault system has resulted in the development of relay ramps between overlapping fault tips at the southeast part of the study area. These relay ramps are arranged as right lateral stepping faults dipping to the NW in the extreme south and to NNW when they swing to ENE-WSW.

The satellite image interpretation integrated with the surface geological maps shows that the study area contains several basinal and geotectonic elements; namely, the Zallah, Dur al Abd, Maradah, Hun Graben, the Waddan Uplift, Gattar Ridge, the Az Zahrah-Al Hufrah Platforms, the Al Qargaf Arch, and the intrabasinal Al Hulayq Ridge. Moreover the Zallah Trough is divided into a number of sub-basins: the

Facha Graben, Ar Ramlah Syncline, Zallah Sub-basin and Ma'amir Graben.

The produced rose diagrams reveal that the majority of the surface linear features in the region have five preferred orientations: N25-35W in the Hun Graben, N40-50W in the Zallah Trough, N45-55W in the Dur al Abd Trough, N35-55W in the Az Zahrah-Al Hufrah Platform, and in contrast in the Waddan Uplift a N55-65E trend.

According to satellite interpretation most of the subtle low amplitude periclinal structures which are exposed at surface coincide with producing oil fields. It has been observed in this study on the basis of remote sensing interpretation that at least in some parts of the study area, oil and gas traps line up well with the trends of surface major structural features such as folds, lineaments, fracture zones, and/or fracture traces.

It should be mentioned that some lineaments, mapped from the remote sensing data were not present in the geologic maps, which suggest that this image lineament interpretation provides new geologic perspectives even in an area which has been well mapped because of petroleum exploration.

ACKNOWLEDGEMENTS

The author is indebted to the Management of Libyan Petroleum Institute and National Oil Corporation, Libya for their support throughout the study and funding, and the permission to publish this paper.

Also I would like to express my sincere thanks to all colleagues who have offered many good ideas during the preparation of this article and for their continuous encouragement together with providing other varied assistance.

REFERENCES

1. Gustafsson, P. 1993. High resolution satellite imagery and GIS as a dynamic tool in groundwater exploration in a semi-arid area.

HydroGIS93: Application of Geographic Information Systems In Hydrology and Water Resources (proceedings of the Vienna Conference, April 1993). IAHS Publ. no. 211, 1993.

2. Drury, S., 2001. Image Interpretation in Geology, Blackwell Science, Nelson Thornes, UK, pp. 290.

3. Rajesh, H.M., 2004. Application of remote sensing and GIS in mineral resource mapping - An overview: Journal of Mineralogical and Petrological Sciences, 99, p. 83-103.

4. Harding, T. 1984. Graben hydrocarbon occurrences and structural style. Bulletin of American Assiciation of Petroleum Geologists, 68, 333-362.

5. Gras, R., and Thusu, B., 1998. Trap architecture of the Early Cretaceous Sarir Sandstone in the eastern Sirt Basin, Libya, in MacGregor, D.S., Moody, R.T.J., and Clark-Lowes, D.D., eds., Petroleum geology of North Africa: Geological Society, Special Publication 132, p. 317–334.

6. Ambrose, G., 2000. The geology and hydrocarbon habitat of the Sarir Sandstone, SE Sirt Basin, Libya: Journal of Petroleum Geology, 23, 165-191.

7. Goudarzi, G.H. 1980. Structure-Libya. In The Geology of Libya Volume III, Salem, M.J., Busrewil (eds), Academic Press, London p. 879-891.

8. Burke, K., and J. Dewey, 1974. Two plates in Africa during Cretaceous: Nature, 249, 313-316.

9. Conant, L.C. and Goudarzi, G.H. 1967. Stratigraphic and tectonic framework of Libya. Bull Amer. Assoc. Pet. Geolvol. 51, p. 719-730.

10. Van Houten, F.B., 1983. Sirte Basin, north-central Libya: Cretaceous rifting above a fixed mantle hotspot? Geology, 11, 115-118.

11. Gumati, Y.D., and Nairn, A.E.N., 1991. Tectonic subsidence of the Sirte Basin, Libya: Jour. Petrol. Geol., 14, 93-102.

12. Van der Meer, F. and Cloetingh, S., 1993. Intraplate stresses and the subsidence history of the Sirte basin (Libya). Tectonophysics, 226, 37–58.

13. Vesely, J., 1985. Sheet Zallah (NH 33-16), Geological Map of Libya, scale 1:250,000, Explanatory Booklet, Industrial Research Centre, Tripoli.

14. Jurak, L., 1985. Sheet Wadi Bu Ash Shaykh (NH 33-12), Geological Map of Libya, scale 1:250,000, Explanatory Booklet, Industrial Research Centre, Tripoli.

15. Schroter, T. 1996. Tectonic and sedimentary development of the central Zallah Trough Western Sirt Basin, Libya. In: The Geology of Sirt Basin III, Salem, M.J., Busrewil, M.T., Misallati, A.A. and M.A. Sola (eds), Elsevier, Amsterdam 155-166.

16. Johnson, B.A. and Nicoud, D.A. 1996. Integrated exploration for Beda reservoirs in the southern Zallah Trough (West Sirt Basin, Libya). In The Geology of Sirt Basin Volume II, Salem, M.J., El-Hawat, A.S. and A.M. Sbeta (eds), Elsevier, Amsterdam, 211-222.

17. Hallett, D. and El Ghoul, A. 1996. Oil and gas potential of the deep trough areas in the Sirt Basin. In The Geology of Sirt Basin Volume II, Salem, M.J., El-Hawat, A.S. and A.M. Sbeta (eds), Elsevier, Amsterdam 455-482.

18. Hallett, D. 2002. Petroleum Geology of Libya. Elsevier, Amsterdam. 503p

19. Roohi, M., 1996. Geological history and hydrocarbon migration pattern of the central Az Zahrah–Al Hufrah platform, in M. J. Salem, A. S. El-Hawat, and A. M. Sbeta, eds., Geology of the Sirt Basin: Amsterdam, Elsevier, v. 2, p. 435–454.

20. Anketel, J.M. 1996. Structural History of the Sirt Basin and its relationships to the Sabratah Basin and Cyrenaican Platform, Northern Libya. In The Geology of Sirt Basin Volume II, Salem, M.J., El-Hawat, A.S. and A.M. Sbeta (eds), Elsevier, Amsterdam. 57-87.

21. Anketell, J.M. and Kumati, S.M. 1991. Structure of Al Hufrah Region – Western Sirt Basin, G.S.P.L.A.J. In The Geology of Libya Volume V, Salem, M.J. and M.N. Belaid (eds), 2353-2370.

22. Jerzykiewicz, T., Ghummed, M.A., Abugares, M.M. and Tshakreen S.O., 2002. Evolution of the Western Margin of the Sirt Basin of Libya in Late Cretaceous Time. Abstract, Extended Abstract and Talk at CSPG Convention in Calgary, Compact Disc and Abstract Volume, p. 179.

23. Roohi, M., 1996. A geological view of source-reservoir relationships in the western Sirt Basin, in M. J. Salem, A. S. El-Hawat, and A. M. Sbeta, eds., Geology of the Sirt Basin: Amsterdam, Elsevier, v. 2, p. 323–336.

24. Wennekers, J.H.N., Wallace, F.K. and Abugares, Y.I. 1996. The geology and hydrocarbons of the Sirt Basin: a synopsis. First Symposium on the Sedimentary Basins of Libya, Geology of the Sirt Basin, vol. 1. (eds. M.J. Salem, A.J. Mouzughi and O.S. Hammuda), Elsevier, Amsterdam, p. 3-56.

25. Klitzsch, E., 1970. Die Structurgeschichte der Zentralsahara: Neue Erkenntnissezum Bau und der Palaeogeographieeines Tafellandes. Geologische Rundschau, 59 (2), p.459-527.

26. Cepek, P. 1979. Sheet Al Qaryat ash Sharqiyah (NH 33-6), Geological Map of Libya, scale 1:250,000, Explanatory Booklet, Industrial Research Centre, Tripoli-Libya.

27. Abadi, A.M., 2002. Tectonics of the Sirt basin. Inferences from tectonic subsidence analysis, stress inversion and gravity modelling. PhD thesis, Vrije University, Amsterdam, the Netherlands, 187 pp.

28. Knytl, J., Brydle, G. and Greenwood, D. 1996. Tectonic history and structural development of the Kaf-Themar trend of the western Sirt Basin. First Symposium on the Sedimentary Basins of Libya, Geology of the Sirt Basin, vol. 3. (eds. M.J. Salem, M.T. Busrewil, A.A. Misallati and M.J. Sola), Elsevier, Amsterdam, p. 167-200.

4

UNDERSTANDING THE CONTROLS ON SEDIMENT-P INTERACTIONS AND DYNAMICS ALONG A NON-TIDAL RIVER SYSTEM IN A RURAL–URBAN CATCHMENT: THE RIVER NENE

A.M. Tye [a,*], B.G. Rawlins [a], J.C. Rushton [a], R. Price [b]

[a] British Geological Survey, Keyworth, Nottingham, NG12 5GG, UK

[b] Environment Agency, Nene House, Pytchley Lodge Industrial Estate, Kettering, NN15 6JQ, UK

ABSTRACT

The release of Phosphorus (P) from river sediments has been identified as a contributing factor to waters failing the criteria for 'Good Ecological Status' under the EU Water Framework Directive (WFD). To identify the contribution of sediment-P to river systems, an understanding of the factors that influence its distribution within the entire non-tidal system is required. Thus the aims of this work were to examine the (i) total (P_{Total}) and labile (P_{Labile}) concentrations in sediment, (ii) the sequestration processes and (iii) the interactions between sediment P and the river water in the six non-tidal water bodies of the River Nene, U.K. Collection of sediments followed a long period of flooding and high stream flow. In each water body, five cores were extracted and homogenised for analysis with an additional core being taken and sampled by depth increments. Comparing the distribution of sediment particle size and P_{Total} data with soil catchment geochemical survey data, large increases in P_{Total} were identified in sediments from water body 4–

6, where median concentrations of P_{Total} in the sediment (3603 mg kg^{-1}) were up to double those of the catchment soils. A large proportion of this increase may be related to in-stream sorption of P, particularly from sewage treatment facilities where the catchment becomes more urbanised after water body 3. A linear correlation (r = 0.8) between soluble reactive phosphate (SRP) and Boron in the sampled river waters was found suggesting increased STW input in water bodies 4–6.

P_{Labile} concentrations in homogenised cores were up to 100 mg kg^{-1} PO_4–P (generally < 2% of P_{Total}) and showed a general increase with distance from the headwaters. A general increase in Equilibrium Phosphate Concentrations (EPC_0) from an average of 0.9–~1.7 µm L^{-1} was found between water bodies 1–3 and 4–6. Fixation within oxalate extractable phases (Al, Fe and Mn) accounted for ~90% of P binding in water bodies 4–6, but only between 31 and 74% in water bodies 1–3. Statistical models predicting P_{Total} (R^2 = 0.78), oxalate extractable P (R^2 = 0.78) and Olsen P (R^2 = 0.73) concentrations in river sediments identified Mn oxy-hydroxides (MnO*x*) as a strong predictive variable along with the location within the river system. It is suggested that MnO_x within model predictions is identifying a pool of mixed Fe–Mn oxy-hydroxides (MnO*x*–FeOOH) or Fe oxy-hydroxide (FeOOH) from the wider $Fe_{Oxalate}$ pool that are particularly effective at sorbing and fixing P. The findings demonstrate how sediment and P may accumulate along a 100 km non-tidal river system, the extent to which a range of processes can fix P within mineral phases and how natural flooding processes may flush sediment from the river channel. The processes identified in this study are likely to be applicable to similar river systems over their non-tidal water bodies in eastern England.

KEYWORDS

Phosphorus; Sediment; P enrichment; Labile P; EPC_0; Nene

1. INTRODUCTION

A principal aim of the EU Water Framework Directive (Directive EC, 2000/60/EC) is to prevent further deterioration and to improve the quality of surface waters (rivers and lakes), groundwater and coastal waters and to promote 'good ecological status' (GES) with respect to biodiversity in rivers (Johnes et al., 2007 and EA, 2014). Phosphate is the major nutrient in rivers that is typically in shortest supply, relative to the stoichiometric requirements of plants, and therefore has the greatest potential to limit river productivity (Mainstone and Parr, 2002). Thus, soluble reactive P (SRP) concentrations exceeding target concentrations (Mainstone, 2010) in river water is a common reason why GES is often not achieved. Major inputs of phosphate in river waters are from point sources such as sewage treatment plants (Jarvie et al., 2006 and Neal et al., 2010) or diffuse sources such as agricultural land where phosphorus (P) enters the river channel primarily attached to soil particles (Bilotta et al., 2010 and Quinton et al., 2010). Common pathways for agriculturally derived sediment bound P are either via soil erosion (Haygarth et al., 2006 and Quinton et al., 2010) or through under field land drainage systems (Bilotta et al., 2008 and Reid et al., 2012). In addition, dissolved P from field drains is increasingly recognised as a source (King et al., 2015). Detrimental outcomes for rivers include (i) shifts to more eutrophic communities, and (ii) potential future desorption of phosphate from the sediment to the water body (McDowell et al., 2003 and Jarvie et al., 2005).

In-channel cycling of P involves a complex set of interactions between water, sediment and aquatic biota. It involves various species of P including soluble reactive phosphorus (SRP), particulate P (PP) and organic forms of phosphorus (Mainstone et al., 2000). Despite significant efforts by regulatory bodies, water companies and land managers to decrease SRP concentrations in river waters, these efforts have often failed to improve SRP concentrations to the desired extent. The problem of 'legacy P' (the P that has accumulated along the land-water continuum from past management practices) have been cited (Sharpley et al., 2013 and Jarvie et al., 2013) with sediment P being identified as a potential source (House and Denison, 1997, Collins et al., 2007 and Bilotta et al., 2010).

River sediment can act as both a sink and a source for SRP, depending on the prevailing geochemical conditions. These interactions are highly dependent on sediment architecture, water velocity, sediment and water chemistry (House and Warwick, 1999). Within a catchment, variation in geology and land-use will change (i) the geochemistry and mineralogy of the sediment and (ii) the amount of P entering the channel. Within the length of the river channel the hydromorphology, vegetation and water velocity will determine the size distribution of and residence time of the sediment (Naiman et al., 2000, Fisher et al., 2004 and Heppell et al., 2009). In addition, redox status and the dissolution and re-precipitation of metal oxides are considered essential for the sorption and fixation of SRP (Reddy et al., 1995 and House and Denison, 2000). The nature of the catchment geology will also determine river water chemistry which may influence in-channel mineral precipitation (e.g. Ca-apatite) contributing to the removal of SRP from the river system (House and Denison, 1997 and House and Denison, 2002).

Three parameters interact to determine the role of sediment in controlling SRP concentrations in the water column; (i) the size of the labile pool of P (P_{Labile}) in the sediment, (ii) the equilibrium phosphate concentration (EPC_0) and (iii) the rate of fixation or release that P may undergo. Where P_{Total} includes all the inorganic and organic P species, the P_{Labile} pool represents that pool of P that is potentially able to respond to changes in the solid-solution equilibria. The concentration of P_{Labile} can increase rapidly through sorption with increases in SRP inputs as has been demonstrated downstream of sewage works (Jarvie et al., 2006). The most easily desorbable component of P_{Labile} is likely to buffer the solid $\leftrightarrow$ solution equilibria through the EPC_0, which is defined as the solution concentration of SRP at which there is no net release or uptake of SRP by the sediment and is sometimes referred to as the instantaneous labile pool (Hartikainen et al., 2010). The instantaneous labile pool and EPC_0 are often related through Quantity/Intensity relations using either Langmuir or Freundlich isotherms (Hartikainen, 1991, Jarvie et al., 2005 and Withers et al., 2009). After SRP is sorbed onto the sediment, it has the potential to move to the non-labile pool through fixation onto predominantly metal oxides, with the amorphous or non-crystalline pool of Fe, Al and Mn oxides considered important sorptive phases (Hartikainen et al., 2010).

This paper reports on changes in the sediment-P status in the six main Water Framework Directive (WFD) water bodies of the non-tidal River Nene (Tye et al., 2013). Within the River Nene, sediment and P are often interlinked with plant growth and the siltation of the river course. Therefore, to understand the contribution that sediment-P may make to SRP concentrations the spatial distribution and dynamics of sediment-P within the whole or substantial parts of the river system needs to be understood. Thus the aims of this work are (i) to examine the how P_{Total} varies in sediments and the degree to which the sediment has been enriched or depleted in P relative to catchment soils, (ii) to understand the processes through which P is fixed into non-labile pools with distance from the headwaters and (iii) to understand how P_{Labile} concentrations might effect SRP concentrations. Whilst previous studies have examined sediment P relationships in selected river locations in the UK, few studies have examined sediment P interactions along complete stretches of the non-tidal river systems (e.g. ~100 km). In particular, this paper focuses on the possible effects of the rural–urban catchment transition and how sediment geochemistry controls the interactions between sediment P and water.

2. MATERIALS AND METHODS

2.1. The Nene Catchment

The river Nene is the 10th longest river in the UK and rises in Northamptonshire, flowing in a north-easterly direction out to the sea via the town of Northampton and City of Peterborough (Fig. 1). It is 161 km long and has a total catchment area of 2270 km^2. The river is navigable from the sea as far as Northampton where it connects with the Grand Union Canal. It is a slow flowing river that has suffered from major siltation and macrophyte growth in the past. For example, in 1930 the River Nene Catchment Board was established to undertake extensive dredging and improve sediment management (Meadows, 2007).

The catchment for the upper six water bodies of interest in this study has an area of 1590 km^2 (Fig. 1). The floodplain is relatively wide (from a few hundred metres to ~2 km) and the channel frequently bifurcates and re-joins, as well as being heavily engineered in sections (Williams and Fawthrop, 1988 and Meadows, 2007). The river course falls from ~160 m above ordnance datum (AOD) at source to ~6 m AOD at Peterborough. The majority of this fall occurs in the first 9.5 km, with the channel lying at 80 m AOD in water body 1 (Meadows, 2007). Whereas several urban centres are based on the Nene, particularly Northampton, the major land use is agriculture, mainly arable (cereals and oilseed rape) and grazing. Buffer strips are commonly present on arable land adjacent to the river. A major source of sediment input from arable land is via under field drainage which is extensive as the soils developed on the mudstone geology that dominates the catchment are slow draining. This was identified as a major source during sampling where drains were discharging straight into the river (A. Tye, *pers. comm.*) The concentration of 0.12 mg L^{-1} (3.87 µm L^{-1}) SRP is proposed for Good Ecological Status for the Nene (River Type 4n; Altitude >80 m; Alkalinity >50 mg L^{-1} $CaCO_3$) (Mainstone, 2010 and UKTAG, 2012). Current concentrations of SRP in the Nene often exceed the WFD limits (Atkins, 2014).

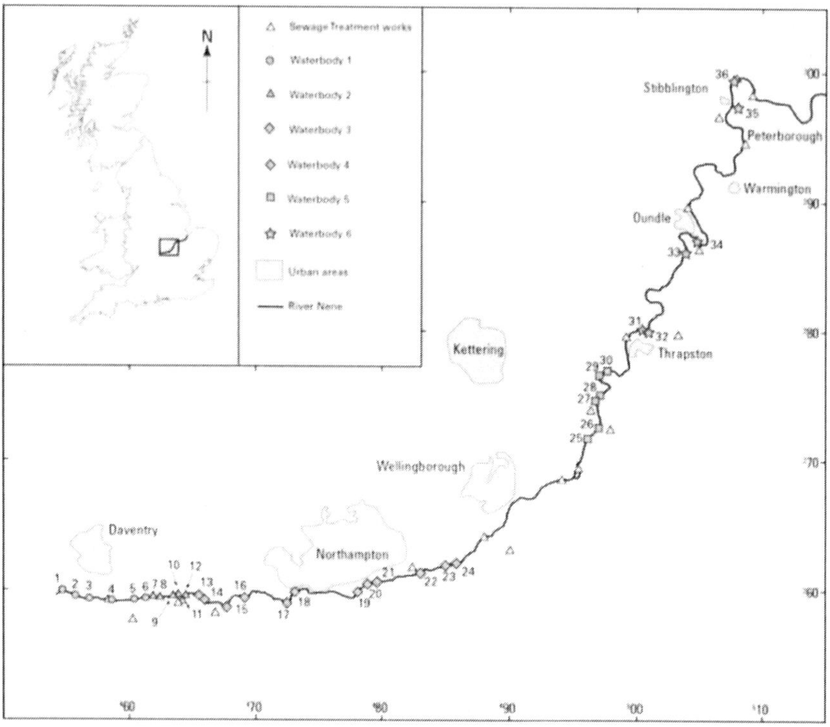

Figure 1. Map of the sampling locations in each of the six water bodies of the River Nene referred to in this study. Six samples were taken from each of the six water bodies marked by different symbols (see key). Sewage Treatments works that discharge to the Nene are also shown. Coordinates show kilometres of the British National Grid.

2.2. Catchment geology

The bedrock geology and soil parent materials of the River Nene catchment comprises rocks of the Lias group and are predominately the Dyrham Formation (interbedded siltstones and mudstones) for water bodies 1–3, and the Whitby Mudstone formation (water bodies 4–6) (Cox et al., 1999). There is also a small area of Charnmouth mudstone within water bodies 1–3, but this is very much a minor component. The Nene Catchment is beyond the southern ice limit of the Devensian stadial but some glacial outwash is found in the catchment. These glaciogenic sediments are underlain by sands and gravels (Milton Sand) that represent earlier trunk rivers (Brown et al., 1994). Significant extraction of this sand and gravel has occurred along some reaches of the river, leaving a range of adjacent wetland environments. On the modern floodplains alluvial soils and deposits are found.

2.3. Sampling strategy and site selection

The River Nene was sampled from the source waters down to the west of Peterborough, this being the non-tidal section, and divided into six WFD water bodies by the Environment Agency (see Fig. 1). The sampling strategy was designed to obtain robust estimates of sediment P_{Total} and P_{Labile} concentrations in each of the six water bodies with a reasonable estimate of variability. Therefore five sediment cores were collected from each water body to be homogenised prior to analysis, whilst a sixth core ($Core_D$) was extracted from each water body to allow analysis of P_{Total} and P_{Labile} variations with depth. A desk-top study was undertaken to locate sampling sites. In the upper water bodies these were as regularly spaced as possible. For water bodies 4, 5 and 6 a boat was used for sampling, so positions were partially determined by the available boat launch sites and travelling distance upstream and downstream. Sampling was undertaken in the winter so little information regarding potential habitats of sediment sampled was available. Thus the sampling aimed just to take representative samples of sediment at each site. Fig. 1 shows the location of the cores extracted. The $Core_D$samples were extracted from positions 2, 7, 15, 20, 28 and 35 (Fig. 1).

2.4. Collection of sediments

Sampling followed the very wet autumn/winter of 2012/13 in January and February 2013, a time of unusually long periods of flooding and high river flows. River flow was still above base-flow when the samples were collected. Cores were taken by pushing a length of polycarbonate tube (diameter = 58 mm internal diameter) into the sediment and retention was through the use of a core catcher, at the base of the tube. Due to the long period of flooding and high river flow (6–8 weeks) prior to sampling insufficient sediment was found at 9 sites in the upper 3 water bodies to collect a core sample. At these sites a sample of the sediment (predominantly sand and gravel) was taken from shallow water close to the edge of the river. The sample size collected from these sites was >3 kg so that a representative sample could be obtained. All samples were kept <4 °C after sampling. A water sample was collected at each of the $Core_D$ sample sites to determine SRP and hydrochemistry. The water sample was filtered immediately after collection through a 0.45 µm syringe filter and kept in the dark at <4 °C until SRP analysis, within 5 days of collection (Jarvie et al., 2002). The total depth of sediment present at each sampling location was examined by probing at each site. Measurements of sediment depths were collected (n = 10) within a 20 m distance up and down from the sampling site.

2.5. Sediment sample characterisation

Due to the size and number of samples the five homogenised core samples taken from each water body were air-dried (<30 °C) before being sieved to <2 mm. The depth of core was recorded and the mass of air-dry sediment from each of these cores in each water body was weighed. The core designated for depth analysis within each water body ($Core_D$) was divided into 5 segments, with the top section always being 0–5 cm depth from the sediment surface, this being the standard depth on which EPC_0 and sorption/desorption kinetic analysis are undertaken (Jarvie et al., 2005). The top 0–5 cm was wet-sieved <2 mm and stored <4 °C. After measurements of EPC_0 and P uptake kinetics were undertaken, the remainder of these samples was air-dried for other analyses. The rest of $Core_D$ was divided into four equal sections and air-dried immediately, before being sieved to <2 mm.

Estimates of the quantity of organic matter in the sediment samples were undertaken by loss on ignition; 1 g of sample was ignited at 475 °C for 4 h. P_{Total} (all species of inorganic and organic P) analysis was undertaken using a UKAS accredited method based onSulcek and Povandra (1992). Subsamples of the dried sediment (~30 g) were ground in an agate mill to <150 µm. Milled samples were digested by weighing 0.25 g of soil into a Savillex™ vial and adding HF, HNO_3 and $HClO_4$ concentrated and analytical grade acids, with a subsequent stepped heating program up to 170 °C overnight, to digest silicate and oxide phases. The dry residue was re-constituted after warming with Milli-Q water, HNO_3 and H_2O_2, to 25 ml of 5% v/v HNO_3 and stored in HDPE bottles. Reference materials (NIST SRM2710, SRM2711, GSS-6, BGS102 and BCR-2), duplicated samples and blanks were all prepared in a similar manner to check accuracy of the analytical and digestion method. In addition to the quantification of P_{Total}, a range of elements associated with P mineralogy and phosphate sorption were measured including Al, Fe, Mn and Ca by ICP-MS. P_{Labile} was determined using an equilibrating solution of 0.5 M $NaHCO_3$ adjusted to pH 8.5 on the air-dry samples from each of the homogenised cores. Triplicate samples were shaken in a 1:20 sediment:solution ratio for 30 min before centrifuging at 1327 RCF for 15 min and filtering (0.45 μm). Analysis was carried using the molybdenum blue method (Murphy and Riley, 1962) and the results expressed on an oven dry weight basis. Estimates of amorphous and poorly crystalline oxyhydroxides of Fe (FeOOH), Mn (MnO_x) and Al (AlOOH) in sediment were determined using 0.2 M ammonium oxalate and 0.125 M oxalic acid extractions. Samples were shaken in darkness for 2 h (McKeague and Day, 1966) (see Supplementary Information 1 for additional information). Particle size distribution (PSD) of sediment samples (0.01–2000 μm) was determined using a Beckman Coulter LS13 320 laser diffraction particle size analyser. Prior to analysis, organic matter was removed using H_2O_2. Mineralogy of the sediments was examined using Scanning Electron Microscopy (SEM) and Scanning Transmission Electron Microscopy (STEM) (see Supplementary Information 2 for methodologies and results).

2.6. Solution analysis

Phosphate concentrations in (a) river water samples (b) Olsen P extracts, (c) isotherm analysis and (d) kinetic analysis were undertaken using the molybdenum blue method (Murphy and Riley, 1962). Calibration curves and analysis were undertaken using a Perkin Elmer Lambda 35 uv/vis spectrometer. Limits of Detection (LOD) were calculated for each of the different matrices and were 0.22 μg P for water, 0.37 μg P for 2 mMol $CaCl_2$ and 0.59 μg P for the 0.5 M $NaHCO_3$ Olsen P extractions. Total elements in water samples collected were analysed by ICP-MS.

2.7. Methodologies for EPC_0 and rates of desorption or sorption

When considering the dynamics of the interactions of SRP with river sediment, it is important to know whether the sediment has the potential to act as a source or a sink. The EPC_0 represents the equilibrium concentration of SRP in solution (i.e. when there is no net sorption or desorption in a laboratory simulation undertaken over a 24 h period). Thus, when SRP concentrations in the overlying water are greater than EPC_0, the sediment has the potential to sorb SRP from the water column and act as a sink. By contrast when SRP concentrations are less than the EPC_0, the sediment has the potential to release SRP to the water column and act as a source. EPC_0 analysis was undertaken on wet sediment (0–5 cm) within seven days of sampling to minimize sample deterioration as suggested by Jarvie et al. (2005). A measured mass of sediment (equivalent to 0.5 g dry weight) was placed in 6 bottles with 200 ml of a synthetic water composition roughly matching the major element chemistry of the River Nene (2 mmol $CaCl_2$). The bottles were spiked with different concentrations of KH_2PO_4 (the amounts of P added were equivalent to 0, 2.5, 5, 10, 15 and 20 μmol l^{-1}), and placed in an orbital shaker in the dark at 10 °C for 24 h. Samples were then centrifuged and their SRP concentration determined (Murphy and Riley, 1962). The isotherms produced were based on the Freundlich model (Equation (2)) which was fitted using the least squares method. The Freundlich model takes the form

$$\Delta N_a = K_f C_i^n \qquad (1)$$

where ΔNa = change in adsorbed P (μmol g^{-1}), K_f is the Freundlich constant, C_i is the concentration of SRP in solution and n is a constant. Using the fitted isotherms the Equilibrium Phosphorus Concentrations (EPC_0) was calculated for each sample (seeHouse et al., 1995; Jarvie et al., 2005). The EPC_0 and SRP measurements were compared using an 'EPC_0 percentage saturation' term (EPC_{0sat}) calculation, which describes the increase or decrease in SRP compared to the EPC_0. This is defined as

$$EPC_{0Sat}(\%) = 100*(EPC_0 - SRP)/EPC_0 \qquad (2)$$

2.8. Calculation of rate constants for SRP sorption/desorption from sediments

To calculate the rate of release or sorption of SRP from/to sediments, the kinetic methodology described by Jarvie et al. (2005) was used. For the SRP release experiments (where EPC_0 > dissolved SRP), a measured mass of wet sediment (equivalent to 0.5 g dry sediment) was placed in polypropylene bottles with 200 mL of $CaCl_2$ solution, pre-chilled to 10 °C. For SRP uptake experiments (where SRP > EPC_0), the synthetic river solution in each bottle was spiked with KH_2PO_4 to an appropriate SRP concentration (greater than the EPC_0 and close to the measured SRP concentration in the water column at the time of sediment sampling). For water body 1 & 2, a 2 μM P solution was used and for the other 4 water bodies a 4 μM P equilibrating solution was used, these concentrations being slightly greater than those measured in the river waters at the time of collection. Bottles were placed in an orbital incubator in the dark and shaken at 150 rpm at 10 °C and then one bottle was sampled after specific time intervals (5 min, 15 min, 30 min, 1 h, 3 h, 6 h, 15 h, 24 h), centrifuged and analysed for their SRP concentrations. The sorption/desorption rates of SRP (House and Warwick, 1999 and Jarvie et al., 2005) to or from the sediments were calculated based on the equation:

$$\boldsymbol{R} = \boldsymbol{K_r}(\boldsymbol{C_t} - \boldsymbol{C}_0)^n \tag{3}$$

R is the change in amount of orthophosphate sorbed (μmol g^{-1} h^{-1}), *Ct* is the orthophosphate concentration (μmol l^{-1}) in the overlying water, *C0* is the orthophosphate concentration in the overlying water after 24 h (μmol l^{-1}), *Kr* is a rate constant (μmol^{1-n} l^n g^{-1} h^{-1}) and *n* is a power term. The Nelder–Mead algorithm as implemented in Matlab (Lagarias et al., 1998) was used to determine the parameter values by minimizing the squared difference between the observed concentrations and those predicted by the rate curve.

2.9. Statistical analysis

Multiple linear regression by ordinary least squares was used to examine the geochemical and environmental controls on P_{Total} and oxalate extractable P in the river sediments. Based on previous research, there are geochemical associations (e.g. concentrations of Mn, Fe and Al or oxalate extractable components of them) which may account for the variation in total sediment P or its reactive components (e.g. $P_{Oxalate}$). The relative importance of these associations accounting for P_{Total} and $P_{Oxalate}$ were investigated by including the concentrations of specific elements or their extractable components as predictors in regression models and exploring the interactions between the geochemical predictors using the linear model (*lm*) function in the R statistical environment. After initial analysis of the data, and to improve the model, each sample point was allocated to one of two parts of the river system (water bodies 1–3 or water bodies 4–6) and this was included within the regression model. This allocation represented whether water bodies were in a rural (water bodies 1–3) or more urbanised (water bodies 4–6) part of the catchment. The reasons why this allocation was undertaken is explained in Section 3.3. In each case we checked that the residuals from the models were approximately normally distributed.

3. RESULTS AND DISCUSSION

3.1. Sediment distribution and depth in the River Nene

Few detailed descriptions of sediment depth and distribution for relatively deep (~2 m) rivers in the UK exist, although some have been published for shallow chalk bed rivers (e.g. Cotton et al., 2006). The generalised patterns of sediment distribution, found in each water body of the Nene at the time of sampling and after the period of flooding and high river flow (Section 2.4) were as follows. For water bodies 1, 2, and the top part of water body 3, sediment deposition at the sampling sites generally consisted of gravel and sand, although an occasional silt deposit was found in areas of slower flow, particularly on sharp bends or where field drainage entered the channel. For water bodies 3–6, most sediment deposits located extended <1 m into the water channel and were largely confined to small inlet areas of the river channel where the current was slower. No fine grained sediment deposits were found in the central areas of the main channel, with probing only revealing a compact armoured bed. It is not clear to what extent the absence of sediment along much of the river channel was a function of the extended period of flooding and high river flow that was prevalent throughout the winter period of 2012/13 that preceded sampling. However, large areas of macrophyte vegetation had been removed by the flooding throughout the length of the river, allowing greater sediment removal (Cotton et al., 2006). Previous work has recorded that extensive scouring of sediment can occur in the River Nene as a consequence of long periods of flooding and high water flow with the resultant removal of significant macrophyte communities (Brierley et al., 1989). A visit to sampling sites in the spring of 2013 showed the areas from which sediment was sampled for this work was usually the substrate for macrophyte (e,g, *Iris pseudacorus*) growth and it is likely that the root systems prevented erosion of this sediment. Typically in water bodies 1 & 2 the mean depth of the sand and gravel lag was <10 cm to the clay river bed. Water bodies 3–6 had sediment deposits with a mean of ~60 cm. However, large standard deviations demonstrate the variability of mean sediment depth within each water body (see graph 1 in Supplementary Information 3).

3.2. Sediment properties

Particle size distribution (PSD < 2 mm) was analysed in the homogenised samples from each water body and plotted on a ternary diagram, along with estimates of particle size from the soils derived from the major soil parent materials (Dyrham mudstone, Charnmouth mudstone, Whitby mudstone and the alluvial floodplain soils) in the catchment (Fig. 2). Particle size varied between 6 and 91% for sand, 4–35% for silt and 2–66% for clay for the homogenised cores taken from the six water bodies. The percentage of stones by weight (>2 mm) in the samples were from 0 to 30% with a median value of 2.98%. A higher percentage weight of stones was found in the samples taken from sand and gravel sediments in water bodies 1–3 where coring was not possible, with the range being 0–30% with a median of 8.6%. In water bodies 4–6 the range for the weight of stones in the sediment cores was 0–22% with the median being 0.5%. Sediments in water bodies 1–3 were generally coarser than the soils, suggesting transport of finer particles downstream. A general, but with occasional exceptions, fining in PSD occurred in each water body as distance from the headwaters increased but the variability within each water body demonstrates that the major influences on the particle size distributions is the depositional environment (e.g. position in relation to current, density of plants that can trap sediment, sediment input). However, the PSD of the cores generally did not become finer than the alluvial soil within the catchment, but 7 out of the 30 sediment samples became finer to varying degrees than the soil generated directly from the mudstone parent materials.

Organic matter content (LOI) varied between 3.18 and 12.06% in the homogenised cores taken from the six water bodies. Fig. 3 shows that % LOI tends to be higher after water body 2. The results are generally associated with the % clay in the samples. A linear regression between LOI and % clay was found (% LOI = 3.05 + 0.145% Clay; df = 29; r^2 = 0.65, $P < 0.001$). This is possibly a result of slower water currents enabling greater quantities of fine silt and clay particles to settle; these along with FeOOH and AlOOH being the particles that organic matter tends to sorb in soils and sediments. Weaker regression relationships were found between $Fe_{Oxalate}$ (% LOI = 2.88 + 0.0002 $Fe_{Oxalate}$; r^2 = 0.42; df = 29; $P < 0.001$) or $Al_{Oxalate}$ (% LOI = 1.712 + 0.0034 $Al_{Oxalate}$, r^2 = 0.46 df = 29; $P < 0.001$) and (% LOI where oxalate concentrations are in mg kg^{-1}).

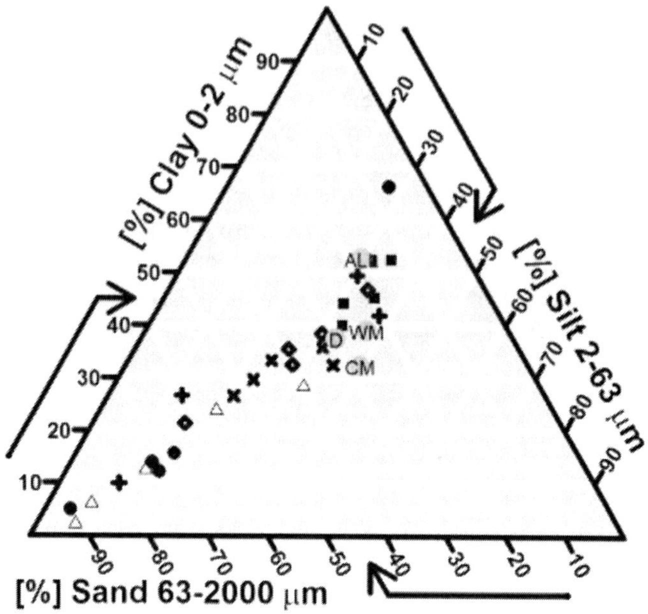

Figure 2. Particle size distributions of homogenised samples (n = 30) from the river Nene channel and soils derived from major soil parent materials from the catchment. Sediments are marked as (•) for Water Body 1, (△) for Water Body 2, (+) for Water Body 3, (×) for Water Body 4, (■) for Water Body 5 and (◊) for Water Body 6. Particle size of the soils derived from the parent materials that make up the Nene catchment are marked as ALL = Alluvial soils, D = Dyrham Formation, CM = Charnmouth Mudstone, and WM = Whitby Mudstone formation.

3.3. P_{Total} in homogenised sediment cores

Concentrations of P_{Total} was analysed in the homogenised sediment cores and results are shown in Fig. 3. An increase in P_{Total} was found after core 18, the end of water body 3. The whole P_{Total} dataset was assessed along with particle size, organic matter and elements (Ca, Mg, Fe and Mn) commonly associated with P bearing minerals or sorptive surfaces. Initial correlation analysis suggested no relationships existed between $P_{Total}$and these parameters. However, when the data was plotted there were indications that separate relationships existed between the upper water bodies (1–3) and lower water bodies (4–6). In particular, relationships were found between P_{Total} and Fe_{Total} and Mn_{Total} (Fig. 4).

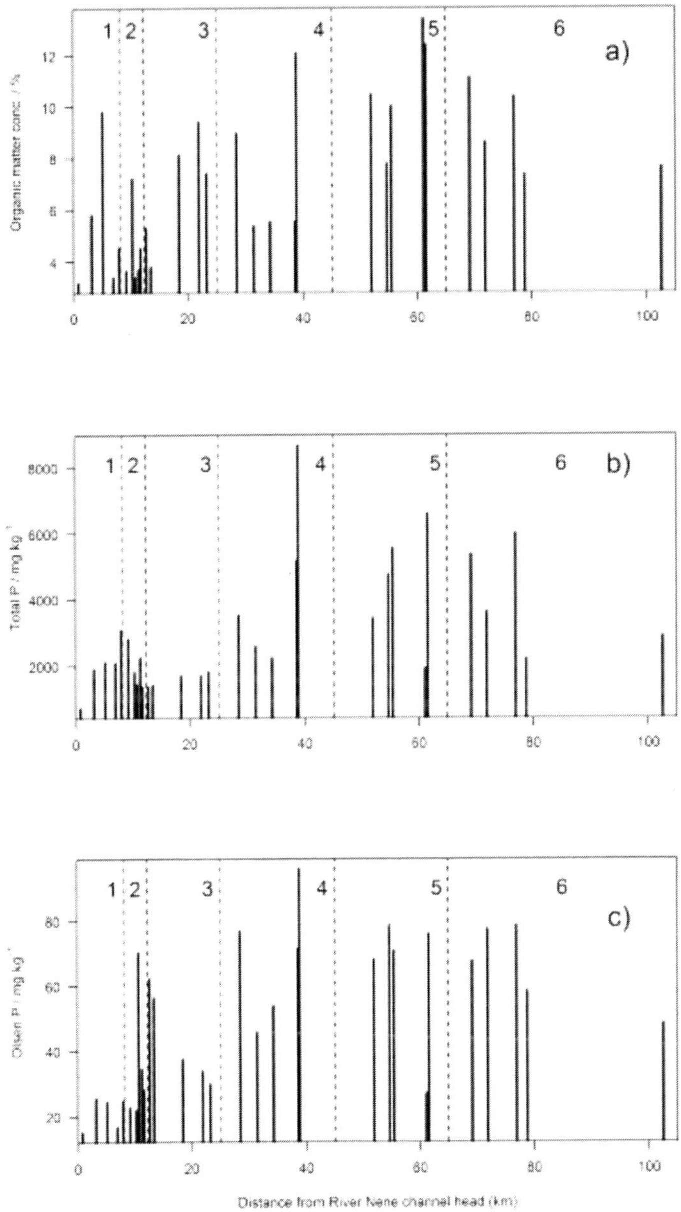

Figure 3. Changes in (a) organic matter as determined by loss on ignition (%), (b) Total sediment P (mg kg^{-1}) and (c) Olsen extractable P (mg kg^{-1}) for the 5 homogenised samples collected from the six water bodies of the River Nene. The end of each water body is marked by the broken vertical line and the sampling points are measured as distance from the channel head.

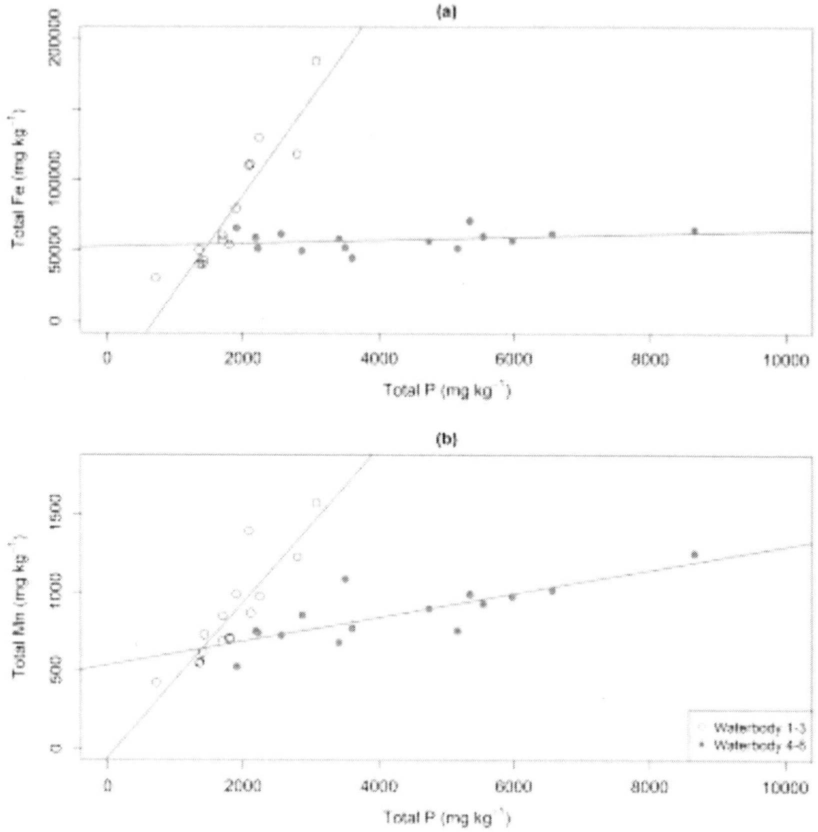

Figure 4. Relationships between P_{Total} (mg kg^{-1}) and (a) Total Fe (mg kg^{-1}) and (b) Mn_{Total} (mg kg^{-1}) in the 5 homogenised sediment samples taken from each of the six water bodies of the River Nene.

Four possible explanations were considered for the relationships found between $P_{Total}$and Fe_{Total} or Mn_{Total} for the two sets of grouped water bodies, these being; (i) a change in geology with associated changes in the geochemistry of sediment inputs (soil) to the river, (ii) a change in agricultural practice, (iii) changes caused by the fining of the sediment as previously described and (iv) a change from a rural catchment to a more urban catchment, with associated increases in SRP inputs from STW's, particularly from Northampton onwards as the population size in the catchment increases. The change in geology was considered as the Dyrham mudstone changed to the Whitby mudstone formation around the end of water body 3; this may have provided changes in the geochemical relationships of the soils. Using data obtained from the BGS

G-BASE geochemistry survey (Johnson et al., 2005), Cumulative Distribution Functions (CDF) for Mn, Fe and P were plotted for soils formed from the two major soil parent materials (Dyrham and Whitby mudstones) in the Nene catchment to allow comparisons of the concentration ranges. Thus, considering explanation (i) above, only small differences were found in the concentration ranges of P, Fe and Mn from soils formed from the two major soil parent materials (Fig. 5). The largest difference found was that in the Whitby soil where there was an increase in Fe_{Total} and P_{Total} above the 60th percentile compared to the Dyrham soil. Thus these CDF's suggest that the geochemical properties of the soil inputs to the river were broadly similar. The second explanation was based on an increase in arable agriculture within the catchment. Land cover maps of the Nene catchment to the west of Peterborough show that it is predominately arable agriculture. There is a slight increase in the area of improved pasture in the land associated with water bodies 1–2 but arable agriculture is still the major land use (EA, 2014).

The third explanation to be considered was that particle size differences caused changes in element concentration through fining. Firstly, eroding soil has been found to selectively transport fine particles (Sharpley, 1980; Warrington et al., 2009) and this process can also continue within the river. Both these processes concentrate P because fine particles have a greater surface area per unit weight, and oxides are often associated with clay particles. For example Miller et al. (2009) found that P was concentrated up to ~4 times in a selection of artificially eroded soils. By examining the PSD of the Nene sediments in comparison to the catchment soils, it is evident that there has been some loss of fine particles from water bodies 1–3. However, very few of the sediment samples had a PSD finer than the catchment soils, particularly in water bodies 4 6, suggesting that increases in P enrichment were not solely due to an increase in fine particles in the sediment samples. Both PSD data and CDF's were used to examine changes between the soil and sediment. For water bodies 1–3, the shape of the sediment CDF shows that an increase in sediment Fe_{Total} was found in the upper 40% of samples whilst a small increase in P_{Total} occurred compared to the soil parent material (Fig. 6), despite a coarsening of particle size. For the increase in Fe_{Total} compared to the soil, a likely explanation is that the loss of the fine particles from the sediment may have increased the relative concentration of Fe_{Total} because the sand particles were covered in iron oxides. All the samples with the highest Fe_{Total} concentrations in the dataset came from these sand and gravel lag samples and not the homogenised cores (see Section

2.4). The small increase in P may reflect the binding of P to these FeOOH coverings from the soil or some in-situ sorption from the river water.

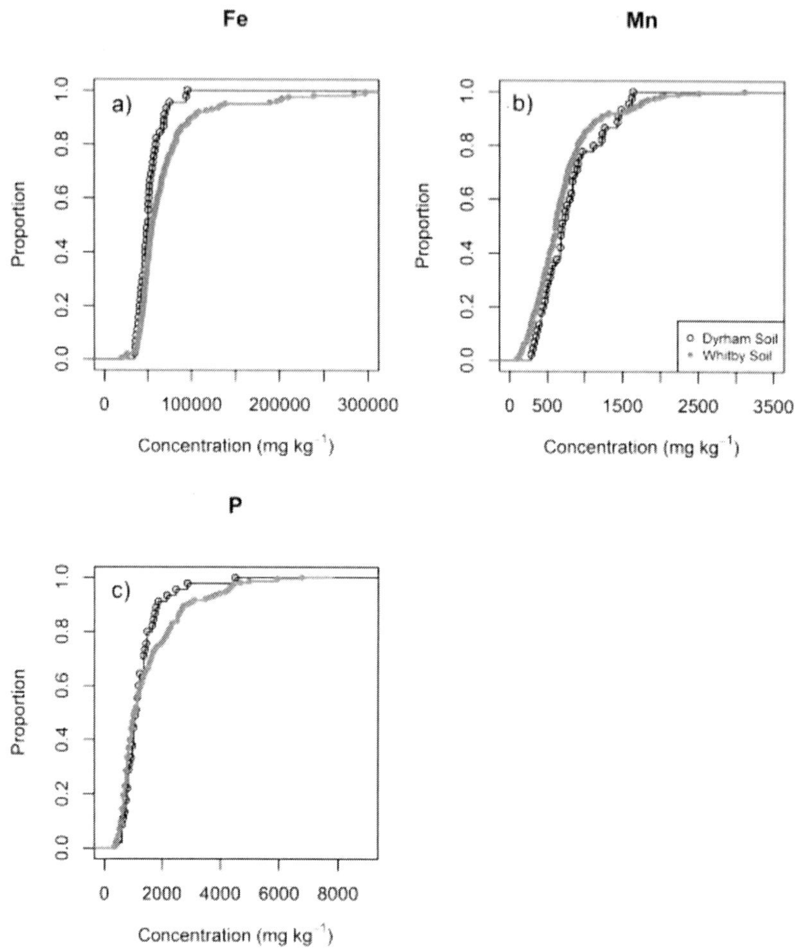

Figure 5. Cumulative Distribution Functions to describe the range of (a) Fe_{Total}, (b) Mn_{Total} and (c) P_{Total} concentrations in soils above the two main soil parent materials, the Dyrham and Whitby mudstones.

For water bodies 4–6, increases in the ranges of Fe_{Total} and Mn_{Total} were confined to the upper 10th percentile in the soil as compared to the sediment (Fig. 7). However, a considerable enrichment in P_{Total} was found in the sediment, as compared to the soil parent material. As the

PSD of the soils is not all that different to many of the sediments in water bodies 4–6, the suggestion is that a large proportion of this enrichment was caused by in-situ sorption of SRP from the water. Whilst agriculture (e.g. fertilisers, direct livestock inputs) is a likely contributor, the probable major source is SRP released via the STW's. The position of STW's that discharge to the river Nene are marked on Fig. 1. In particular it is STW output that is discharged to the Nene from the major urban areas of Northampton, Wellingborough and Kettering that are likely to cause this enrichment. An indication of the size and potential quantity of P discharging from the STW's serving these towns can be gauged by their population sizes of 65,000, 25,000 and 41,000 respectively. By comparison the population of Daventry the biggest town in the Nene catchment in waterbodies 1–3 is 8000 (NVNIA, 2013). Thus, there is a strong suggestion that P enrichment particularly in the sediments of water bodies 4–6 appears to be from in-situ sorption of SRP and the greater urban influence within a still predominantly rural catchment. Further evidence for STW's being the source of the sediment P enrichment is found in the relationship between SRP and Boron. Table 1 reports SRP and Boron concentrations in the river water samples collected at the $Core_D$ locations. Concentrations of SRP were between 0.18 and 4.63 μm P L^{-1}. Boron is used as a whitener in washing powder and has been used as a tracer to determine whether SRP is derived from agriculture or STW's. Typical background concentrations of Boron in river waters are generally <30 μg L^{-1} (Jarvie et al., 2005, Jarvie et al., 2006 and Neal et al., 2010). In this study, despite the small number of sample points there was a positive linear correlation of (r = 0.80) suggesting that the SRP in the river water may be largely linked to discharges from STW's. The increasing concentrations of SRP and Boron were both found with increasing distance from the headwaters of the Nene, and are likely a result of the increasing numbers or size of STW's feeding water into the river system as shown inFig. 1.

Overall the relationship between P_{Total} and Fe_{Total} and Mn_{Total} appears to differ between water bodies 1–3 and water bodies 4–6 because of the likely increase in STW inputs as population size increases and the catchment becomes less rural. A regression model for the prediction of P_{Total} was parameterised. The best fit model was parameterised using Mn_{Total} concentration and the rural/urban catchment influence (with an interaction between them) as predictors and accounted for 78% (adjusted R^2) of the variation in P_{Total} (Table 2); the two lines are for the rural/urban catchment classes for water bodies 1–3 and 4–6 and the two regression equations are as follows:

$$P_{Total}(wb\ 1,2,3) = 4580 + 1.61 * Mn_{Total} \quad (4)$$

$$P_{Total}(wb\ 4,5,6) = 4580 - 3414 + 1.61 * Mn_{Total} + 6.80 * Mn_{Total} \quad (5)$$

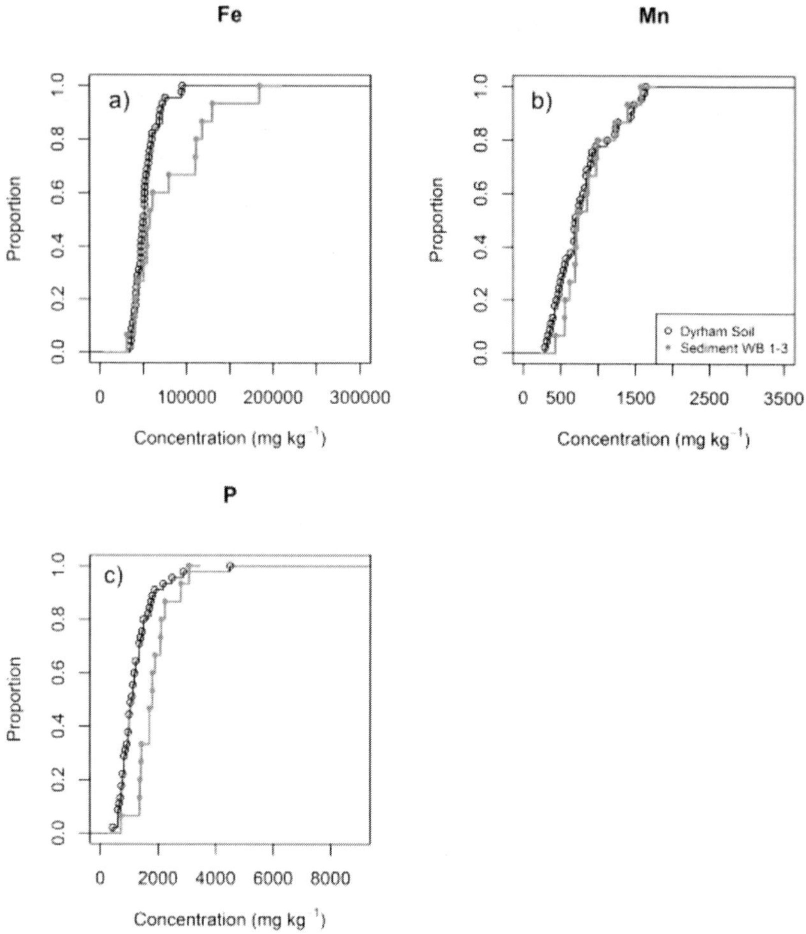

Figure 6. Cumulative Distribution Functions to describe the range of (a) Fe_{Total}, (b) Mn_{Total} and (c) P_{Total}concentrations in soils and sediments (<2 mm) in water bodies 1–3 of the River Nene where the dominant soil parent material is the Dyrham mudstone.

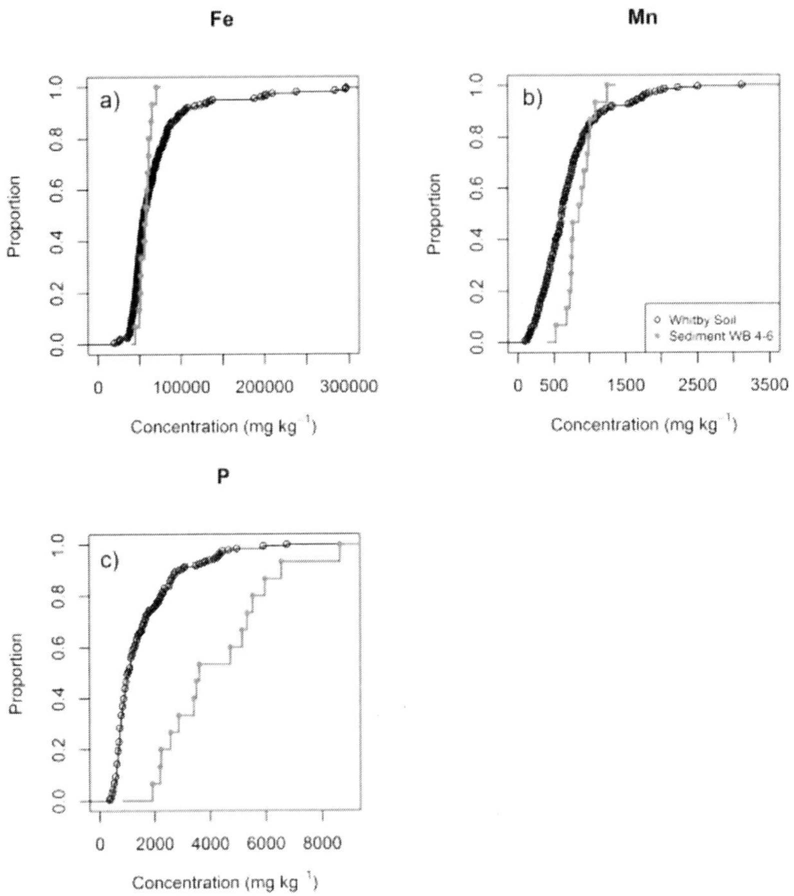

Figure 7. Cumulative Distribution Functions to describe the range of (a) Fe_{Total}, (b) Mn_{Total} and (c) P_{Total}concentrations in soils and sediments in water bodies 4–6 of the River Nene where the dominant soil parent material is the Whitby mudstone.

3.4. P associations with mineral phases and amorphous oxides

SEM and STEM analysis was undertaken on sediments. Minerals phases associated with anaerobic conditions in the sediment were identified in the core samples. These minerals included vivianite ($(Fe^{2+})_3(PO_4)_2 \cdot 8H_2O$), siderite (FeCO3) and Ca–Mn-Carbonate. Fe oxide minerals were associated with clay minerals and mixed MnO*x*–FeOOH coatings were found (See Supplementary Information 2; Fig. 1, Fig. 2 and Fig.

3for images of mineral phases). The concentration and composition of amorphous oxides is likely to be influenced by the dissolution and re-precipitation, and redox reactions of mineral phases during aerobic–anaerobic cycles within the sediment architecture.Table 3 reports concentration ranges and median values of oxalate extractable Al, Fe and Mn. Concentration ranges of $Al_{Oxalate}$, $Fe_{Oxalate}$ and $Mn_{Oxalate}$ were higher in water bodies 4–6 compared to water bodies 1–3. During precipitation of the oxyhydroxides, phosphate ions may become incorporated within their structures. By measuring $P_{Oxalate}$release the importance of amorphous oxides in binding P is demonstrated. Whilst most of the P released from this extraction will be P that has been incorporated within the oxalate extractable oxides, it is possible that a small amount of organic P that binds to the surface of these oxides (Jiang et al., 2015) will be also be released during the extraction. This demonstrates the importance of amorphous oxides to binding both inorganic or organic P species. The mean percentage of P_{Total} that $P_{Oxalate}$ equated to was 31, 59, 74, 89, 94 and 90% for water bodies 1 to 6 respectively indicating that a greater proportion of P was oxalate extractable as distance from the headwaters increases.

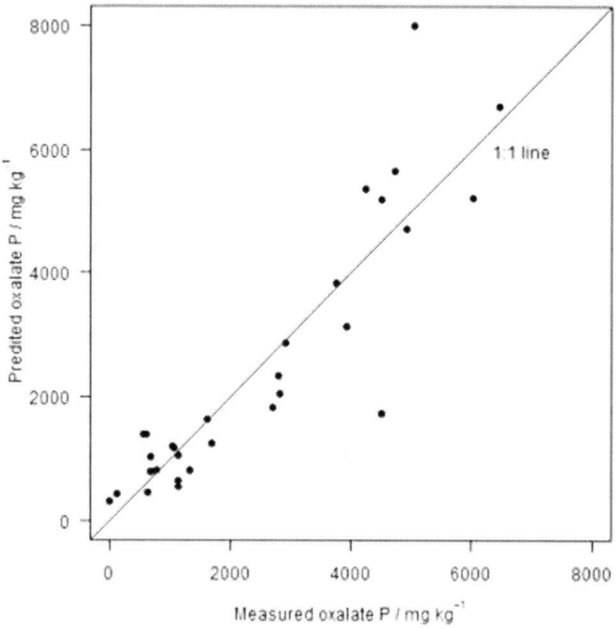

Figure 8. Scatterplot of measured and predicted concentrations of $P_{Oxalate}$ in Nene sediments based on the significant predictors (urban or rural catchment class, the concentration of $Fe_{Oxalate}$ with respect to water body and $Mn_{Oxalate}$ concentration) and their interactions.

Table 1.Parameters obtained from isotherm fitting analysis to calculate Equilibrium Phosphate Concentration (EPC_0) in samples from each of the water bodies of the river Nene based on Equation (1) and EPC_{0Sat}calculated using Equation (2).

Sample	F	k	RSS	EPC_0 (µm P L^{-1})	EPC_{0Sat} (%)	k_d L kg^{-1}	SRP (µm P L^{-1})	Boron (µg L^{-1})
Water body 1	43.9	0.008	0.03	1.31	87.59	33	0.18	41
Water body 2	52.2	0.029	0.44	0.61	180	249	1.65	52
Water body 3	176	0.048	2.33	0.35	444	3743	4.63	53
Water body 4	413	0.027	2.78	1.19	185	4286	3.40	61
Water body 5	403	0.055	9.56	1.57	160	8445	4.35	81
Water body 6	296	0.023	1.50	1.69	126	1555	3.83	79

Table 2. Results of the regression model fitted by ordinary least squares for the prediction of total sediment P concentrations in the River Nene sediments.

	Estimate	Standard error	t-value	*P*-value
Intercept	4580	653	0.702	0.489
Total Mn	1.61	0.717	2.24	0.034
[a]Water bodies 4,5 & 6	−3413	1298	−2.63	0.142
[b]Total Mn: Water bodies 4, 5 & 6	6.80	1.46	4.64	<0.001

[a] Denotes the 'urban' based catchments class for water bodies 4, 5 & 6. The estimate of the regression coefficient (slope) is the difference from the estimated coefficient for the rural class underlying water bodies 1, 2 & 3.
[b] Interaction between Total Mn and WB 1–3 (rural) and WB 4–6 (urban) based catchments.

Table 3. Oxalate extractable concentrations ranges and median concentrations for Al, Fe and Mn in sediments of the River Nene.

	Water bodies 1–3 (mg kg^{-1})	Water bodies 4–6 (mg kg^{-1})
$Al_{Oxalate}$: range	429–2060	987–2373
Median	890	1804
$Fe_{Oxalate}$: range	9324–39,000	15,846–40,543
Median	13,454	32,300
$Mn_{Oxalate}$: range	228–644	275–753
Median	383	515
$P_{Oxalate}$: range	305–1634	1402–7995
Median	828	3842

Linear regression relationships between $P_{Oxalate}$ and $Fe_{Oxalate}$($P_{Oxalate}$ = −947 + 0.139*$Fe_{Oxalate}$; r^2 = 0.50; df = 29; P < 0.001), $Al_{Oxalate}$($P_{Oxalate}$ = −1129 + 2.42*$Al_{Oxalate}$; r^2 = 0.45; df = 29; P < 0.001) and $Mn_{Oxalate}$($P_{Oxalate}$ = −1066 + 7.351 *$Mn_{Oxalate}$; r^2 = 0.25; df = 29; P < 0.01) were found, indicating that all three oxides (mg kg^{-1}) contribute to the sorption and fixation of P. Concentrations of $P_{Oxalate}$ were predicted using regression analysis to assess the factors determining its concentrations in the river sediment. The predictors in the optimum regression model accounted for 78% (adjusted R^2) of the variance (Table 4). In the best fit model (Fig. 8) $Mn_{Oxalate}$ was significant (P < 0.05) as was the term water bodies 4,5,6:$Fe_{Oxalate}$(P < 0.001). This latter term was significant because of the large increase in $Fe_{Oxalate}$concentrations in waterbodies 4–6, compared to waterbodies 1–3 (Table 3) and its role as the dominant P sorbing surface. However, there was strong predictive power demonstrated by $Mn_{Oxalate}$ although its concentration is up to two orders of magnitude lower than that of $Fe_{Oxalate}$. A parameter for differentiating between the more rural and urban catchments was included (Waterbodies 4, 5 & 6) although this was not significant. The two equations predicting $P_{Oxalate}$ (from Table 4) are:

$$P_{Oxalate}(wb\ 1,2,3) = -1404 + 0.046*Fe_{Oxalate} + 3.395*Mn_{Oxalate} \quad (6)$$

$$P_{Oxalate}(wb\ 4,5,6) = -1401 - 1848 + 0.046\ Fe_{Oxalate} + 3.395*Mn_{Oxalate} + 0.1311*Fe_{Oxalate} \quad (7)$$

The degree of P saturation (DPS) can be calculated as a measure of how saturated the P sorbing surfaces are with P. It is calculated using the oxalate concentrations (Moles) because they are the major sorptive surfaces. Thus DPS = $P_{Oxalate}/(Al_{Oxalate} + Fe_{Oxalate} + Mn_{Oxalate})$. The mean concentration of sorbing surfaces was 0.32, 0.37, 0.42, 0.59, 0.69 and 0.58 M for water bodies 1–6 respectively, mirroring the previously described increase in oxalate surfaces as distance increases from the headwaters. Results show that mean DPS for water bodies 1–6 respectively are 5.87, 10.15, 10.19, 20.70, 19.33 and 19.74%.

Table 4. Results of the regression model fitted by ordinary least squares for the prediction of total sediment oxalate extractable P concentrations in River Nene sediments.

	Estimate	Standard error	t-value	*P*-value
Intercept	−1404	531	−1.42	0.167
[a]Water bodies 4,5 & 6	−1848	1354	−1.36	0.1845
$Fe_{Oxalate}$	0.046	0.027	1.68	0.105
$Mn_{Oxalate}$	3.395	1.507	2.25	<0.05[a]
[b]Water bodies 4,5,6:$Fe_{Oxalate}$	0.1311	0.050	2.61	<0.01[b]

[a] Denotes the urban/rural class for waterbodies 4, 5 & 6. The estimate of the regression coefficient (slope) is the difference from the estimated coefficient for the rural (WB1-3)/urban (WB 4–6) influenced catchment class underlying waterbodies 1, 2 & 3.

[b] Interaction between $Fe_{Oxalate}$ and WB 1–3 and 4–6.

3.5. Labile P in sediments

Concentrations of P_{Labile} for the samples taken from each water body are shown in Fig. 3. Mean values for each water body are shown in Supplementary Information 3, Fig. 2. Results show a trend of increasing P_{Labile} concentrations from water body 1 through to water body 6. Concentrations of P_{Labile} varied between ~17 and 100 mg kg^{-1}, these typically being <5% of P_{Total} in all instances with most samples being <2%. P_{Labile}concentrations were examined in the $Core_D$ samples with maximum concentrations being ~60 mg kg^{-1}. There was little evidence of a systematic pattern of P_{Labile} deposition with depth in these six samples from different water bodies, with concentrations of P_{Labile}being in a similar range to those from the homogenised cores.

There were no statistical relationships found between P_{Labile} and LOI, Fe_{Total}, Al_{Total} and Mn_{Total}. Relationships between P_{Labile} and the oxalate extractable elements were examined. The $P_{Oxalate}$ pool (range 0–7995 mg kg^{-1}) was much greater than the P_{Labile}pool (0–100 mg kg^{-1}), demonstrating the degree to which P has been fixed within the amorphous oxide minerals. For the homogenised core samples, no linear relationships were found between P_{Labile} and $Fe_{Oxalate}$ or $Al_{Oxalate}$ although numerous reports have suggested strong relationships in soils and sediments (e.g. Hartikainen et al., 2010). However, Fig. 9 shows a linear regression relationship between P_{Labile} and $Mn_{Oxalate}$ for the homogenised cores that accounted for 63% of the variation in P_{Labile} in sediments (Table 5). In this instance the rural/urban catchment classification was not significant. The mean values of P_{Labile} and $Mn_{Oxalate}$ from the $Core_D$ samples are included in the graph and fall within the homogenised core data.

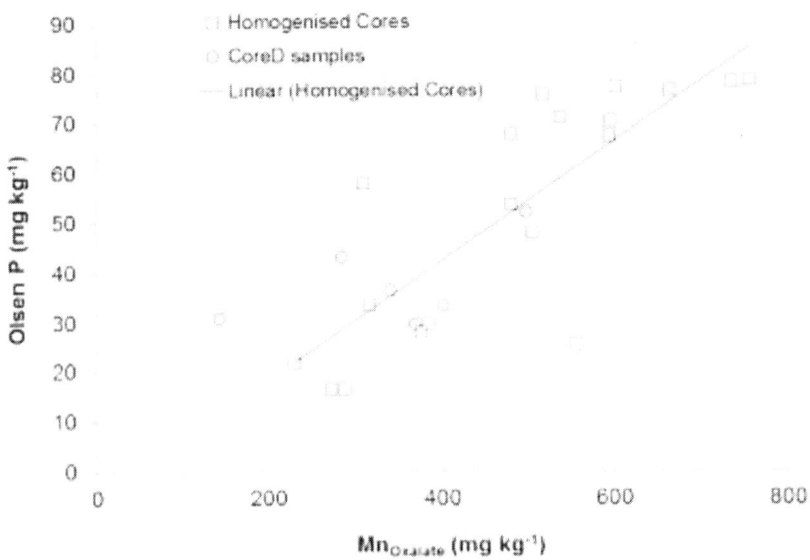

Figure 9. Relationship between P_{Labile} and $Mn_{Oxalate}$ for the homogenised cores and the mean values of P_{Labile} and $Mn_{Oxalate}$ for $Core_D$ samples.

Table 5. Results of the regression model fitted by ordinary least squares for the prediction of P_{Labile} in sediments.

	Estimate	Standard error	t-value	*P*-value
Intercept	−3.973	8.242	−0.48	0.63
$Mn_{Oxalate}$	0.1125	0.017	6.62	<0.001

3.6. Controls on the Equilibrium Phosphorus Concentration

Using the top 5 cm taken from $Core_D$ from each water body, the 'Equilibrium Phosphorus Concentration' was determined. The EPC_0 is primarily a function of the sediment properties that contribute to surface charge such as the particle size distribution, surface area, oxide concentrations, organic matter and water properties such as electrical conductivity (as a proxy for ionic strength) and pH. Concentrations of EPC_0 for the six water bodies of the River Nene were between 0.61 and 1.69 μm P L^{-1} (Table 1) with the greatest values being found in water bodies 5 and 6. Placing these results within the context of other reported results, Hartikainen et al. (2010) found EPC_0 between 0.09 and 10.84 μm L^{-1} for a range of soils and sediments, whilst Jarvie et al. 2005 recorded values of between 0.06 and 5.74 μm L^{-1} (median = 1.36 μm L^{-1}) for sediments in the Hampshire Avon (predominantly chalk) catchment and 0.03 and 6.29 (median = 0.71 μm L^{-1}) for the Wye catchment (mudstone). Stutter and Lumsdon (2008) found EPC_0 values of between 0.03 and 1.75 μm L^{-1} (median = 0.195) for the River Dee catchment where soils are largely derived from granites and acid schists.

Comparison of the EPC_0 with the SRP concentration determined in the water sample taken at each site (Table 1) show that only in water body 1 does the EPC_0 exceed the SRP in the water sample, thus suggesting that the sediment is a source of SRP. However, when the kinetic analysis was undertaken the sample was found to sorb rather than release SRP. This sample was the first to be taken after the floods and the river flow was still very high, so it is possible that equilibrium between SRP and sediment had not yet been reached. Therefore all examined sediments of the Nene were capable of uptake SRP from the water column, and this is similar to the majority of studies from the UK (e.g.House and Denison, 1997, House and Denison, 1998 and Jarvie et al., 2006). The negative values of EPC_{0Sat} after water body 1 (Table 1), suggest that there is considerable under saturation of the EPC_0 with SRP suggesting that the sediment has considerable capacity to absorb further SRP.

The values of EPC_0 were examined in conjunction with the sediment properties that may control the sorption and desorption of SRP. There was no relationship between $P_{Labile}$and EPC_0, suggesting that the P_{Labile} assay used may deplete the whole labile pool and not just the least strongly held part of P_{Labile} that it actively buffering the solid-solution equilibria. There were no correlations between $Al_{Oxalate}$, $Fe_{Oxalate}$ or $Mn_{Oxalate}$, and the values of EPC_0.

3.7. Determining kinetic constants for P uptake

Rate constants were determined for the 0–5 cm segments of Core$_D$ taken from each water body. All sediments tested showed a rapid absorption of SRP (see Tye et al., 2013for graphs), with a pseudo-equilibrium generally being reached within one hour (C_0 inTable 6). This rapid response has been found by others (e.g. Jarvie et al., 2005). The concentration of C_0 (effectively the point of zero net sorption or desorption) was very similar to the value obtained for EPC_0 previously (Table 1). Rate co-efficients (K_r) for SRP uptake varied between 2.36 and 51.71 μmol l^{1-n} g^{-1} h^{-1} (Table 6) for the six water bodies. These values are within the range of previously published uptake rate constants including an average rate constant of 10.6 μmol l^{1-n} g^{-1} h^{-1} (range 0.84–54) for the Hampshire Avon catchment and a mean of 4.96 μmol l^{1-n} g^{-1} h^{-1} (range 0.54–18.9) for the Wye catchment (Jarvie et al., 2005). The values of K_r were examined in respect to surfaces that may sorb SRP. Both $Fe_{Oxalate}$ (r = 0.91) and $Al_{Oxalate}$ (r = 0.91) demonstrate a strong positive correlation with K_r, if the value (51.71 μmol l^{1-n} g^{-1} h^{-1}) for water body 2 is not included, suggesting that these oxides are important in SRP sorption from the water (see Supplementary Information 3; Fig. 3). Investigating why water body 2 does not fit this relationship demonstrates that K_r is determined by a combination of both sediment texture and oxide concentration. This sample had 48% clay (the range in the other samples was between 13 and 42%) but had the second lowest concentrations of $Fe_{Oxalate}$ and $Al_{Oxalate}$. Thus in this instance it was likely that the clay surfaces compensated for the lower $Fe_{Oxalate}$ and $Al_{Oxalate}$ concentrations in the sorption of SRP.

Table 6. Outputs from the kinetic experiments where (K_r) is the kinetic rate constant (μmol^{1-n} l^n g^{-1} h^{-1}), *C0* is the orthophosphate concentration in the overlying water after 24 h (μmol l^{-1}), and *n* is a power term for sediment (0–5 cm) from each of the 6 core$_D$ samples collected from each of the water bodies of the River Nene. Values of K_r, C_0 and n calculated using Equation (3).

	K_r μmol l^{1-n} g^{-1} h^{-1}	C_0	n
Waterbody 1	2.366	1.363	2.05
Waterbody 2	51.714	0.623	2.16
Waterbody 3	10.687	0.662	1.09
Waterbody 4	25.477	1.117	1
Waterbody 5	26.954	1.45	1.26
Waterbody 6	14.113	1.89	2.29

3.8. General discussion

Within the main channel of the river Nene, major sediment inputs include soil and bank erosion, under field drainage and sediments from joining tributaries, whereas diffuse (eroded soil) and point (STW's, field drains) sources are the likely principal sources of P. Subsequent variations in local geomorphology, natural disturbance regimes (e.g. in-flowing tributaries) and human interactions (e.g. land drainage, STW's) combine to produce a continuum of in-stream sediment conditions with respect to the geochemistry and concentrations of P (Naiman et al., 2000 and Fisher et al., 2004). Using a combined knowledge of catchment soil and river sediment geochemistry, an estimate of sediment P enrichment in the Nene, compared to the catchment soils, was made. Results demonstrated that there was little difference in the geochemistry between the major soil parent materials. Recent source apportionment studies have suggested that channel banks and under-field drainage are the greatest contributors to suspended sediment, followed by topsoil and road drainage (Cooper et al. 2015). It was observed during sample collection in the current study that under-field drainage was an important source of sediment into the river along the distance sampled. However, evidence from the sediment PSD analysis and field observations suggests that much of the clay fraction will stay suspended and be washed through to the tidal part the river system. The P load released to the system by this lost sediment is not known. However, it is likely that if it is transported directly from the non-tidal river system it will have remained in oxygenated environments and if sorbed to Fe or Al oxides will remain so, unless utilised by microbes or biota. However, if it is trapped and buried within sediments it may contribute to the P enrichment found in the sediment, possibly undergoing changes in its redox status and being re-precipitated as a mineral phase such as vivianite. Thus considering that there generally is not a large fining of the sediment, no substantial increases in oxide surfaces and that the PSD of sediments is broadly similar to the soils, it is likely that a considerable part of the P enrichment particularly in water bodies 4–6 is derived from in-stream sorption of SRP released from the STW's within the catchment. In the more urban influenced part of the catchment (water bodies 4–6), this enrichment might have caused at least a doubling of P_{Total} compared to typical soil concentrations found within the catchment. This estimate was based on a reasonably recent geochemical survey of the catchment soils and although P concentrations may have declined slightly in the soils (particularly in terms of P_{Labile}) since the geochemical survey was undertaken through improved P fertiliser management, in water bodies 4–6, a substantial shift in the shape of the sediment CDF compared to catchment soils for P was found.

Once sorbed, the key question is whether and how this sediment P is fixed in non-labile forms and how easily it may be released. Where anaerobic conditions have developed within the sediments, mineral phases such as vivianite, or the inclusion of P as a minor constituent of other minerals (e.g. $MnCO_3$), represent forms of P fixation and were identified. However, most P is sequestered, particularly in the lower waterbodies within the oxalate extractable Fe, Al and Mn. For example, by water body 6 up to 90% of $P_{Total}$was associated with the oxalate extractable (mainly amorphous oxides) compared to 31% in water body 1. This was possibly due to the deeper sediments in the lower water bodies and changes in redox status and P speciation. Agudelo et al. (2011) found a similar response and suggested that increases in P associated with amorphous oxides was a result of increased sediment depth changes in redox status and a fining of the particles size as distance from the headwaters increased. Concentration gradients of oxygen with increasing depth in sediment will determine the magnitude of oxidation and oxide speciation. Typically, FeOOH is associated with phosphate sorption in sediments, particularly within the upper oxygenated sections (e.g. 0–10 cm) (Agudelo et al., 2011). Our laboratory measurements of K_r, determined in oxygenated conditions demonstrate the importance of FeOOH and AlOOH. However, within the sediment, there will be a mixture of aerobic–anaerobic zones in the sediments. Oxygen egress from the roots of macrophytes and benthic micro-algae can help form oxides and several studies have demonstrated increased P fixation under macrophyte roots (Jaynes and Carpenter, 1986, Christensen, 1997 and Christensen et al., 1997).

Sediment geochemical properties (e.g. redox change as evidenced by the formation of minerals such as vivianite, clays, oxides) throughout the river length were also shown to determine sediment-water P interactions and the properties (labile P, EPC_0, K_r) that are most likely to influence the 'Ecological Status' of the river ecosystem. The concentration of P_{Labile} was typically <2% of P_{Total}. Concentrations of P_{Labile} in the sediment were found to be as large as 100 mg kg^{-1}, a value considerably larger than recommended for most agricultural soils. There was no correlation found between P_{Labile} and EPC_0, suggesting that the EPC_0 is buffered primarily by the part of P_{Labile} that is most weakly held and rapidly exchangeable. Isotopic exchange experiments using ^{32}P have demonstrated that parts of P_{Labile} can exchange very rapidly or over longer time periods. Thus P_{Labile}sorbed on different mineral or organic phases may respond at different rates to maintain solid ↔ solution equilibria (Furumai et al., 1989). Results demonstrated that values of EPC_0 generally increased but only by a small amount between water bodies 1–3 and water bodies 4–6. Whilst small variations in EPC_0 may occur throughout each water body with time as river conditions and sediment particle size distribution change (Stutter and Lumsdon, 2008), the largest increases in EPC_0 are expected where there is a substantial increase in P_{Labile} as demonstrated in soils (Withers et al., 2009 and Hartikainen, 1991). This is most likely to occur downstream of STW's (Jarvie et al., 2006). In the Nene, the EPC_0 was found to increase slightly where there was a general increase in P_{Labile}. Various interactions may occur that may change whether the sediment becomes a source or sink of P, with those sediments close to the EPC_0 and SRP equilibrium being most at risk. Release of P will occur when EPC_0 > SRP and these conditions can be bought about by (i) the introduction of tertiary P stripping decreasing SRP concentrations downstream of STW's (Jarvie et al., 2006), (ii) the uptake of SRP by biota during summer periods, (iii) changes in texture bought about by changes in river velocity (House and Warwick, 1999) and (vi) the creation of anaerobic conditions through effluent discharge that may aid the release of P, causing eutrophication and decreasing SRP values. Concentrations of SRP in the Nene are of the order of ~6.45 μm P L^{-1} (0.2 mg P L^{-1}), having decreased as a result of the introduction of P-stripping in STW's, from 2 to 3 mg L^{-1} in 1998. Thus, with EPC_0 data in this study of 0.61–1.69 μm L^{-1} P it is likely that SRP will generally exceed EPC_0 under current conditions.

Whilst inorganic P species HPO_4^- and $H_2PO_4^-$ are the species of P used by macrophytes, the potential presence of organic P species (monoester P, diester P, Phytic-like P) in the sediments (not measured) will contribute to P_{Labile} as P is released from organic P species through enzymatic hydrolysis to inorganic P (Wang and Pant, 2010 and Zhu et al., 2013). Thus a consequence of large concentrations of bioavailable P in river sediments is that macrophyte and algal growth are likely to increase. The availability of P and sediment deposition can be closely linked, particularly in relatively slow flowing rivers like the Nene. Rooted aquatic plants have the potential to derive almost all their P requirements from bio-available sediment P reserves (Mainstone and Parr, 2002), although Pelton et al. (1998) suggested that the relative contribution of root uptake to macrophyte P demand varied on the SRP concentration in the overlying water. The increased growth of aquatic plants can lead to increased trapping of sediment.Mainstone and Parr (2002) suggest that high concentrations of labile P can (i) increases re-growth after plant management, (ii) alter the species community structure, favouring species with high growth rates and (iii) reduce root depth, potentially increasing the plants susceptibility to being detached at high river flows and associated sediment remobilization (see section 3.1).

An important issue raised by this work and which is poorly understood but which is essential to sediment-P interactions within river systems is sediment residence time, of which little has been published. After undertaking the sampling on the Nene it was apparent that extended periods of high water flow can remove significant areas of macrophyte communities, along with the sediment that the roots hold in place and the P associated with it. This is a recognised occurrence for the River Nene (Brierley et al., 1989), but may play a vital role in the self-cleaning of the river of legacy P. Thus the sediment remaining in place after these periods of flooding and high flow demonstrate the importance of sediment position, particularly with respect to being protected from rapid river flow. Thus this may lead to possible longer residence times. Therefore more work is required to understand to what extent flooding and high water flow may desilt rivers such as the Nene, and in particular the size and duration of flooding that causes significant desilting to occur. The evidence from this investigation also suggestions that the rate of sediment build up and the associated temporal changes in sediment-P interactions also require further understanding.

In the current work the concentration of MnO*x* was identified as a key predictor in the various species of sediment P modelled, particularly in predicting P_{Total} and P_{Labile}. This is despite the significantly lower concentrations of MnO_x compared to FeO_x limiting the contribution that MnO*x* can make to the binding of P (Bortleson, 1974 and Christensen et al., 1997). It is also generally recognised, that despite its large surface area, its surface chemistry (negative surface charge at near neutral pH) is less suitable for P sorption than that of Fe oxides (Kawashima et al., 1986, Tipping et al., 1984 and Yao and Millero, 1996). In addition, changes in the oxidation state of MnO*x* may limit its ability to sorb P as Mn^{2+} oxyhydroxides have greater sorption capacity than the Mn^{4+} oxyhydroxides which form quickly with exposure to air (Lu and Liao, 1997). Bortleson (1974), who also found strong positive correlations between Mn_{Ox} and P in sediments, suggested that the supply rate and migration of Fe and Mn as influenced by pH and redox conditions caused the iron concentrations, more than the Mn concentrations to vary independently of the final P concentration. Thus the suggestion is that within the $Fe_{Oxalate}$ or Fe_{Total} pools, alongside FeOOH there exists a range of poorly crystalline $Fe^{(II)}$ or mixed $Fe^{(II)}/Fe^{(III)}$ minerals that have different P binding properties. These may include $Fe^{(II)}(OH)_2$ or green rust (a precursor to vivianite) and magnetite (Boughriet et al., 1997). In addition, Crosby et al. (1981) suggested that large differences in P sorption characteristics exist between $Fe^{2+}$and Fe^{3+} derived FeOOH under natural conditions. This difference is a result of the different structures of FeOOH formed from either ferric or ferrous iron as well as their ageing properties.

Thus, the strong predictive power of MnO_x in the models presented in this paper, combined with the knowledge that FeOOH is the more effective P binding surface suggests that, particularly in sediment aerobic–anaerobic transition zones present in the homogenised cores, mechanisms may exist that link the two oxides in the fixation of P. In aquatic aerobic–anaerobic transition environments, MnO*x* has been identified as playing a fundamental role in the process through which Fe^{2+} precipitates to form FeOOH or in the co-precipitation of Mn/Fe oxy-hydroxides (MnO*x*–FeOOH) on which P is later sorbed (Dellwig et al., 2010 and Hongve, 1997). Zhang et al. (2009) and Lu et al. (2014) have both reported that mixed Fe–Mn binary oxides are highly effective at sorbing P. The formation of these MnO*x*–FeOOH phases have been identified in particle analysis from the redox transition zones of ocean waters, where P adsorption forms part of tight element cycling involving Mn–Fe-P (Dellwig et al., 2010). Formation occurs when biogenically produced MnO*x* (Tebo et al., 2004), oxidises Fe^{2+} and Mn(IV) ions are replaced by Fe(III) ions. Postma (1985) suggested that the Fe^{3+} released in this reaction will most likely precipitate as FeOOH on the surface of the MnO*x* particle. This can be followed by immediate adsorption or co-precipitation of P. Van der Zee et al. (2005)suggest that the oxidation of Fe^{2+} by MnO_x may be more common than realised but may not be obvious where there is significant NO_3^-. Mixed MnO*x*–FeOOH-PO_4 oxides in the Nene sediments were identified with both SEM and STEM analysis.

4. CONCLUSIONS

We examined the distribution of P_{Total} and P_{Labile} along a 100 km stretch of the non-tidal river Nene to understand how sediment-P may contribute to river SRP concentrations. Key points identified for sediment management within the Nene Catchment are (i) that the accumulation of sediments generally increases with distance from the headwaters, (ii) the extent of P enrichment appears to be largely influenced by the transition from a rural to a more urbanised catchment where there are increasing population numbers along with larger and more frequent STW's and (iii) that fixation in oxalate extractable phases is a major pathway through which P is fixed within mineral phases. Whilst increases in P_{Labile} were found with distance from the headwaters concentrations were generally <2% of the P_{Total} and did not appear to be a directly related to EPC_0concentration. This work also demonstrated that localised differences in particle size distribution influences P speciation within the broader trends of sediment and P accumulation.

However, the natural clearance of sediment during periods of high water flow, as was found in the sampling undertaken appears to be effective at removing much sediment from the non-tidal river channel, and further assessment of this natural process is needed. Further work should focus on the source of the sediment to the river, particularly in determining the proportions that enter the river via land drains, along with natural and cattle driven bank erosion, these possibly be the greatest sources. In addition, complete budgets of SRP and particulate associated P entering the river from land drains should be considered to assess contributions from these sources. Within this the role of buffer strips needs further investigation as these have been found to be a source of SRP (Stutter et al., 2009). Combining knowledge of the sediment and P budget along with greater understanding of sediment residence time is required to understand more fully the extent of in-stream sorption and release of P.

ACKNOWLEDGEMENTS

The authors would like to thank the Environment Agency for funding the initial work (Grant Number: 30258) and NERC BGS (NEE4608) for funding science budget time for Andrew Tye, Barry Rawlins and Jeremy Rushton. Mick Strutt and Judith Mather provided fieldwork assistance and Alex Bradley produced the map. Simon Chenery and BGS lab staff undertook the total element analysis of the sediments. Murray Lark and Ben Marchant wrote code to fit the EPC_0 and kinetic parameters.

REFERENCES

1. Agudelo, S.C., Melson, N.O., Barnes, P.L., Keane, T.D., Pierzynski, G.M., 2011. Phosphorus adsorption and desorption potential of stream

sediments and field soils in agricultural watersheds. J. Environ. Qual. 40, 144-152.

2. Atkins, 2014. Phosphorus Cycling in Rivers e Environment Agency Report SC120037.

3. Bilotta, G.S., Brazier, R.E., Haygarth, P.M., Macleod, C.J.A., Butler, P., Granger, S., Krueger, T., Freer, J., Quinton, J.N., 2008. Rethinking the contribution of drained and undrained grasslands to sediment-related water quality problems. J. Environ. Qual. 37, 906-914.

4. Bilotta, G.S., Krueger, T., Brazier, R.E., Butler, P., Freer, J., Hawkins, J.M.B., Haygarth, P.M., Macleod, C.J.A., Quinton, J.N., 2010. Assessing catchment-scale erosion and yields of suspended solids from improved temperate grassland. J. Environ. Monit. 12, 731-739.

5. Bortleson, G.C., 1974. Phosphorus, iron and manganese distribution in sediment cores of six Winconsin lakes. Limnol. Oceanogr. 19, 794-801.

6. Brierley, S.J., Harper, D.M., Barham, P.J., 1989. Factors affecting the distribution and abundance of aquatic plants in a navigable lowland river; the River Nene, England. Regul. River 4, 263-274.

7. Boughriet, A., Figueiredo, R.S., Laureynsc, J., Recourt, P., 1997. Identification of newly generated iron phases in recent anoxic sediments: 57Fe Mossbauer and micro Ramon spectroscopic studies. J. Chem. Soc. Faraday Trans. 93 (17), 3209-3215.

8. Brown, A.G., Keough, M.K., Rice, R.J., 1994. Floodplain evolution in the EastMidlands, United Kingdom e the Late glacial and Flandrian alluvial record from the Soar and Nene valleys. Philos. Tran. R. Soc. A 348 (1687), 261-293.

9. Christensen, K.K., 1997. Differences in iron, manganese and phosphorus binding in freshwater sediment vegetated with Littorella uniflora and benthic microalgae. Water Air Soil Poll. 99, 265-273.

10. Christensen, K.K., Andersen, F.Ø., Jensen, H.S., 1997. Comparison of iron, manganese and phosphorus retention in freshwater littoral sediment with growth of Littorella uniflora and benthic microalgae. Biogeochemistry 38, 149-171.

11. Collins, A.L., Stromqvist, J., Davison, P.S., Lord, E.I., 2007. Appraisal of phosphorus € and sediment transfer in three pilot areas identified for the catchment sensitive farming initiative in England: application of the prototype PSYCHIC mode. Soil Use Manag. 23 (S1), 117-132.

12. Cooper, R.J., Rawlins, B.G., Krueger, T., Leze, B., Hiscock, K.M., Pedentchouk, N., 2015. Contrasting controls on the phosphorus concentration of suspended particulate matter under baseflow and storm event conditions in agricultural headwater streams. Sci. Tot. Environ. 533, 49-59.

13. Cotton, J.A., Wharton, G., Bass, J.A.B., Heppell, C.M., Wotton, R.S., 2006. The effects of seasonal changes to in-stream vegetation cover on patterns of flow and accumulation of sediment. Geomorphology 77, 320-334.

14. Cox, B.M., Sumbler, M.G., Ivimey-Cook, H.C., 1999. A Formational Framework for the Lower Jurassic of England and Wales (Onshore Area) British Geological Survey Research Report: RR/99/01.

15. Crosby, S.A., Butler, E.I., Turner, D.R., Whitfield, M., Glasson, D.R., Millward, G.E., 1981. Phosphate adsorption onto iron oxyhydroxides at natural concentrations. Environ. Technol. 2, 371-378.

16. Dellwig, O., Leipe, T., M€ arz, C., Glockzin, M., Pollehne, F., Schnetger, B., Yakushev, E.V., Bottcher, M.E., Brumsack, H.-J., 2010. A new particulate Mn-Fe-P € shuttle at the redoxcline of anoxic basins. Geochim. Cosmochim. Acta 74, 7100-7115.

17. EA, 2014. Water for Life and Livelihoods e a Consultation on the Draft Update to the River Basinb Management Plan: Part 2: River Basin Management Planning Overview and Additional Information. Environment Agency, Bristol, UK.

18. EC, 2000. Directive 2000/60/EC of the European Parliament and of the Council establishing a framework for the community action in the field of water policy. Official J. L 327, 0001-0073 (22/12/2000).

19. Fisher, S.G., Sponsellar, R.A., Heffernan, J.B., 2004. Horizons in stream biogeochemistry: flowpaths to progress. Ecology 85 (9), 2369-2379.

20. Furumai, H., Kondo, T., Ohgaki, S., 1989. Phosphorus exchange kinetics and exchangeable phosphorus forms in sediments. Water Res. 23, 685-691.

21. Hartikainen, H., 1991. Potential mobility of accumulated phosphorus in soil as estimated by the indices of Q/I plots and by extractant. Soil Sci. 152, 204-209.

22. Hartikainen, H., Rasa, K., Withers, P.J.A., 2010. Phosphorus exchange properties of European soils and sediments derived from them. Eur. J. Soil Sci. 61 (6), 1033-1042.

23. Haygarth, P.M., Bilotta, G.S., Bol, R., Brazier, R.E., Butler, P.J., Freer, J., Gimbert, L.J., Granger, S.J., Krueger, T., Macleod, C.J.A., Naden, P., Old, G., Quinton, J.N., Smith, B., Worsfold, P., 2006. Processes affecting transfer of sediment and colloids, with associated phosphorus, from intensively farmed grasslands: an overview of key issues. Hydrol. Process. 20, 4407-4413.

24. Heppell, C.M., Wharton, G., Cotton, J.A.C., Bass, J.A.B., Roberts, S.E., 2009. Sediment storage in the shallow hyporheic of lowland vegetated river reaches. Hydrol. Process. 23, 2239-2251.

25. Hongve, D., 1997. Cycling of iron, manganese and phosphate in a meromictic lake. Limnol. Oceanogr. 42, 635-647.

26. House, W.A., Denison, F.H., Armitage, P.D., 1995. Comparison of the uptake of inorganic phosphorus to a suspended and stream bed-sediment. Water Res 29, 767-779.

27. House, W.A., Warwick, M.S., 1999. Interactions of phosphorus with sediments in the River Swale, Yorkshire, UK. Hydrol. Process. 13, 1103-1115.

28. House, W.A., Denison, F.H., 1997. Nutrient dynamics in a lowland stream impacted by sewage effluent: Great Ouse, England. Sci. Total Environ. 205, 25-49.

29. House, W.A., Denison, F.H., 1998. Phosphorus dynamics in a lowland river. Water Res. 32 (6), 1819-1830.

30. House, W.A., Denison, F.H., 2000. Factors influencing the measurement of equilibrium phosphate concentrations in river sediments. Water Res. 34, 1187-2000.

31. House, W.A., Denison, F.H., 2002. Total phosphorus content of river sediments in relationship to calcium, iron and organic matter concentrations. Sci. Total Environ. 282e283, 341-351.

32. Jarvie, H.P., Withers, P.J.A., Neal, C., 2002. Review of robust measurement of phosphorus in river water: sampling, storage, fractionation and sensitivity. Hydrol. Earth Syst. Sci. 91, 113-132.

33. Jarvie, H.P., Jürgens, M.D., Williams, R.J., Neal, C., Davies, J.J.L., Barrett, C., White, J., 2005. Role of river bed sediments as sources and sinks of phosphorus across two major eutrophic UK river basins: the Hampshire Avon and Herefordshire Wye. J. Hydrol. 304, 51-74.

34. Jarvie, H.P., Neal, C., Withers, P.J.A., 2006. Sewage-effluent phosphorus: a greater risk to river eutrophication than agricultural phosphorus? Sci. Total Environ. 360, 246-253.

35. Jarvie, H.P., Sharpley, A.N., Withers, P.A., Scott, J.T., Haggard, B.E., Neal, C., 2013. Phosphorus mitigation to control river eutrophication: murky waters, inconvenient truths and "postnormal Science". J. Environ. Qual. 42, 295e304.

36. Jaynes, M.L., Carpenter, S.R., 1986. Effects of vascular and nonvascular macrophytes on sediment redox and solute dynamics. Ecology 67 (4), 875-882.

37. Jiang, X., Bol, R., Willbold, S., Vereecken, H., Klumpp, E., 2015. Speciation and distribution of P associated with Fe and Al oxides in aggregate-sized fraction of an arable soil. Biogeosciences 12, 6443-6452.

38. Johnes, P.J., Foy, R., Butterfield, D., Haygarth, P.M., 2007. Land use scenarios for England and Wales: evaluation of management options to support 'good ecological status' in surface waters. Soil Use Manag. 23 (Suppl. 1), 176-194.

39. Johnson, C.C., Breward, N., Ander, E.L., Ault, L., 2005. G-BASE: baseline geochemical mapping of Great Britain and Northern Ireland. Geochem. Explor. Environ. Anal. 5, 347-357.

40. Kawashima, M., Tainaka, Y., Hori, T., Koyama, M., Takamatsu, T., 1986. Phosphate adsorption onto hydrous manganese (King, K.W., Williams, M.R., Fausey, N.R., 2015. Contributions of systematic tile

drainage to watershed-scale phosphorus transport. J. Environ. Qual. 44, 486-494.

41. Lagarias, J.C., Reeds, J.A., Wright, M.H., Wright, P.E., 1998. Convergence Properties of the Nelder-Mead Simplex Method in Low Dimensions. SIAM J. Optim 9 (1), 112-147.

42. Lu, Q., Liao, Z., 1997. Comparative study on characteristics of P fixation by Mn, Fe and Al. Pedosphere 7, 325-330.

43. Lu, J., Liu, H., Zhao, X., Jefferson, W., Cheng, F., Qu, J., 2014. Phosphate removal from water using freshly formed Fe-Mn binary oxide: adsorption behaviours and mechanisms. Colloid Surf. A 455, 11-18.

44. Mainstone, C.P., Parr, W., Day, M., 2000. Phosphorus and River Ecology: Tackling Sewage Inputs. English Nature, Peterborough.

45. Mainstone, C.P., 2010. An Evidence Base for Setting Nutrient Targets to Protect River Habitat. Natural England Research Report NERR034.

46. Mainstone, C.P., Parr, W., 2002. Phosphorus in rivers e ecology and management. Sci. Total Environ. 282e283, 25-47.

47. McDowell, R.W., Sharpley, A.N., Folmar, G., 2003. Modification of phosphorus export from an eastern USA catchment by fluvial sediment and phosphorus inputs. Agric. Ecosyst. Environ. 99, 187-199.

48. McKeague, J.A., Day, J.H., 1966. Dithionite and oxalate extractable Fe and Al as aids in differentiating various classes of soils. Can. J. Soil Sci 46, 13e22.

49. Meadows, I., 2007. Hydrology. In: Allen, P., Boismier, W.A., Brown, A.G., Chapman, A., Meadows, I. (Eds.), Synthetic Survey of the Environmental Archaeological and Hydrological Record for the River Nene from its Source to Peterborough. Northamptonshire Archaeology. Report PNUM 3453.

50. Miller, N., Quinton, J.N., Barberis, E., Presta, M., 2009. Variability in the mobilization of sediment and phosphorus across 13 European soils. J. Environ. Qual. 38, 742-750.

51. Murphy, J., Riley, J.P., 1962. A modified single solution method for the determination of phosphate in natural waters. Anal. Chim. Acta 27, 31-36.

52. Naiman, R.J., Elliot, S.R., Helfield, J.M., O'Keefe, T., 2000. Biophysical interactions and the structure and dynamics of riverine ecosystems: the importance of biotic feedbacks. Hydrobiologia 410, 79-86.

53. Neal, C., Jarvie, H.P., Withers, P.J.A., Whitton, B.A., Neal, M., 2010. The strategic significance of wastewater sources to pollutant phosphorus levels in English rivers and to environmental management for rural, agricultural and urban catchments. Sci. Total Environ. 408, 1485-1500.

54. NVNIA, 2013. A Summary of Access to Natural Greenspace in the Nene Valley Nature Improvement Area.

55. Pelton, D.K., Levine, S.N., Braner, M., 1998. Measurements of phosphorus uptake by macrophytes and epiphytes from the LaPlatte River (VT) using 32P in stream microcosms. Freshw. Biol. 39, 285-299.

56. Postma, D., 1985. Concentration of Mn and separation from Fe in sediments e 1. Kinetics and stoichiometry of the reaction between birnessite and dissolved Fe(ii) at 10C. Geochim. Cosmoschim. Acta 49, 1023-1033.

57. Quinton, J.N., Govers, G., van Oost, K., Bardgett, R.D., 2010. The impact of agricultural soil erosion on biogeochemical cycling. Nat. Geosci. 3, 311-314.

58. Reddy, K.R., Diaz, O.A., Scinto, L.J., Agani, M., 1995. Phosphorus dynamics in selected wetlands and streams of the lake Okeechobee Basin. Ecol. Eng. 5, 183-207.

59. Reid, D.K., Ball, B., Zhang, T.Q., 2012. Accounting for the risks of phosphorus losses through tile drains in a phosphorus Index. J. Environ. Qual. 41 (6), 1720-1729.

60. Sharpley, A.N., 1980. The enrichment of soil phosphorus in runoff sediments. J. Environ. Qual 9, 521-525.

61. Sharpley, A., Jarvie, H.P., Buda, A., May, L., Spears, B., Kleinman, P., 2013. Phosphorus legacy: overcoming the effects of past

management practices to mitigate future water quality impairment. J. Environ. Qual. 42, 1308-1326.

62. Stutter, M.I., Lumsdon, D.G., 2008. Interactions of land use and dynamic river conditions on sorption equilibria between benthic sediments and river soluble reactive phosphorus concentrations. Water Res. 42, 4249-4260.

63. Stutter, M.I., Langan, S.J., Lumsdon, D.G., 2009. Vegetated buffer strips can lead to increased release of phosphorus to waters: a biogeochemical assessment of mechanisms. Environ. Sci. Technol. 43, 1858-1863.

64. Sulcek, Z., Povandra, P., 1992. Methods of Decomposition in Inorganic Analysis. CRC Press Inc., Florida.

65. Tebo, B.M., Bargar, J.R., Clement, B.G., Dick, G.J., Murray, K.J., Parker, D., Verity, R., Webb, S.M., 2004. Biogenic manganese oxides: properties and mechanisms of formation. Annu. Rev. Earth Planet. Sci. 32, 287-328.

66. Tipping, E., Thompson, D.W., Davison, W., 1984. Oxidation products of Mn(II) in lake waters. Chem. Geol. 44, 359-383.

67. Tye, A.M., Hurst, M.D., Barkwith, A.K.A.P., 2013. Nene Phosphate in Sediment Investigation e Environment Project REF:30258. Open Access Report OR/13/ 031.

68. UKTAG, 2012. A Revised Approach to Setting Water Framework Directive Phosphorus Standards.

69. Van der Zee, C., Slomp, C.P., Rancourt, D.G., de Lange, G.J., Van Rapphorst, W., 2005. A Mossbauer spectroscopic study of the iron redox transition in eastern Mediterranean sediments. Geochim. Cosmoschim. Acta 69, 441-453.

70. Wang, J., Pant, H.K., 2010. Enzymatic hydrolysis of organic phosphorus in river bed sediments. Ecol. Eng. 36, 983-968.

71. Warrington, D.N., Mamedov, A.I., Bhardwaj, A.K., Levy, G.J., 2009. Primary particle size distribution of eroded material affected by degree of aggregate slaking and seal development. Eur. J. Soil Sci. 60, 84-93.

72. Williams, J.J.R., Fawthrop, N.P., 1988. A mathematical hydraulic model of the river Nene e a canalized and heavily controlled river. Regul. River 2, 517-533.

73. Withers, P.J.A., Hartikainen, H., Barberis, E., Flynn, N.J., Warren, G.P., 2009. The effect of soil phosphorus on particulate phosphorus in land runoff. Eur. J. Soil Sci. 60, 994-1004.

74. Yao, W., Millero, F.J., 1996. Adsorption of phosphate on manganese dioxide in seawater. Environ. Sci Technol 30, 536-541.

75. Zhang, G.S., Liu, H.J., Liu, R.P., Qu, J.H., 2009. Removal of phosphate from water by a Fe-Mn binary oxide adsorbent. J. Colloid Interface Sci. 335, 168-174.

76. Zhu, Y., Wu, F., He, Z., Guo, J., Qu, X., Xie, F., Giesy, J.P., Liao, H., Guo, F., 2013. Characterisation of organic P in lake sediments by sequential fractionation and enzymatic hydrolysis. Environ. Sci. Technol. 47, 7679-7687.

5

SELENIUM ENRICHMENT IN CARBONIFEROUS SHALES, BRITAIN AND IRELAND: PROBLEM OR OPPORTUNITY FOR SHALE GAS EXTRACTION?

John Parnell [a,*], **Connor Brolly** [a], **Sam Spinks** [a,b], **Stephen Bowden** [a]

[a] School of Geosciences, University of Aberdeen, Aberdeen AB24 3UE, United Kingdom

[b] CSIRO Mineral Resources Flagship, Australian Resources Research Centre, Perth, Australia

ABSTRACT

The Carboniferous Bowland Shale in England, and its correlatives in Ireland, contain anomalously high concentrations of trace elements, including selenium (Se), molybdenum (Mo) and arsenic (As). High levels of these elements reflect high sulphur contents as these elements occur as trace constituents of pyrite. Anomalous Se in particular may have a volcanic provenance, from contemporary volcanic activity and/or drainage from Ordovician volcanogenic sulphide deposits. Following concern over the release of Se and As into groundwater during shale gas extraction in the US, the potential fate of Se and As during any future shale gas extraction from the Bowland Shale merits attention. It is at least an environmental issue that must be managed, but at best it could be an opportunity for extraction of Se in an environmentally sensitive manner.

KEYWORDS

Selenium; Molybdenum; Arsenic; Bowland Shale; Carboniferous; Shale gas

1. INTRODUCTION

The Carboniferous Bowland Shale, and its correlatives in Ireland, is receiving much attention for their potential resources of shale gas (Smith et al., 2010, Jones, 2012,Andrews, 2013 and Taylor et al., 2013). Trace element data help to evaluate the environmental impacts of hydraulic fracturing during shale gas extraction (Chermak and Schreiber, 2014). However, although organic geochemical data is available for the Bowland Shale (Armstrong et al., 1997 and Andrews, 2013), and an assessment of radionuclide flux from shale gas exploitation has been made (Almond et al., 2014), there is no modern published database on trace element geochemistry. This is surprising, given concern over the possible contamination of ground waters by fracking (Myers, 2012, Vidic et al., 2013, Vengosh et al., 2014 and Shonkoff et al., 2014). In particular there is concern over the environmental implications of metals released from shales during gas extraction (Haluszczak et al., 2013 and Chermak and Schreiber, 2014). The need to be at least aware of the trace element chemistry of these particular Carboniferous black shales is evident from long-known selenium toxicity in farm animals from some Irish soils (Fleming and Walsh, 1957 and Rogers et al., 1990), which can be traced to Se enrichments in the underlying black shales (Keily and Fleming, 1969,McGrath and Fleming, 2007 and Fellowes et al., 2013). Livestock health problems have also arisen due to high molybdenum contents in the Irish black shales (Alloway, 2012). Selenium availability is important to plants and animals, but in excess can become toxic, which in the case of human consumption is above 400 μg/day (MacFarquhar et al., 2010), with possible consequences including neurotoxicity, cancer and diabetes (Vinceti et al., 2014 and Sun et al., 2014). Care is therefore often taken to avoid the delivery of elevated concentrations of selenium to the environment, for example through the processing of fossil fuels, including coal storage (Lussier et al., 2003), coal burning (Zeng et al., 2001), and oil refining (Lawson and Macy, 1995). Recently, concern has been expressed over the release of Se to ground waters through shale gas exploration in the USA (Fontenot et al., 2013, reiterated in numerous open access publications, e.g.Hildenbrand et al., 2013, Meyer, 2013 and Schug et al., 2013).

We report here trace element data for a set of 42 samples of the Bowland Shale and its correlatives, from 17 localities across an outcrop width of 600 km (7 England, 1 Wales, 1 Isle of Man, 8 Ireland; Fig. 1; Supplementary Table S1), focusing on selenium (Se), arsenic (As) and molybdenum (Mo). Selenium and As are of strong environmental concern, and Mo is a key indicator of redox conditions as it is enriched in reducing sediments (Algeo and Lyons, 2006). Selenium and Mo are closely related in both the Carboniferous black shales and their soils (Webb et al., 1966 and McGrath and Fleming, 2007). Data from the Barnett Shale, Texas, of comparable age to the Bowland Shale, indicate the release of Se into groundwaters within 2 km of shale gas drilling, at a mean (n = 10) concentration an order of magnitude greater than historical levels, and including individual values greater than the U.S. maximum level permitted in drinking water of 50 μg/l (Fontenot et al., 2013). All 10 values were at or exceeded the more demanding European Union maximum permitted level of 10 μg/l, and there are arguments that the European Union level be lowered further to just 1 μg/l (Vinceti et al., 2013). The same data set showed that the mean As level in 90 water wells in active extraction areas exceeded the maximum permitted level for drinking water (Fontenot et al., 2013). Following environmental concern about exploration in the Barnett Shale, we compare the Bowland Shale and Barnett Shale data.

2. METHODS

The targeted shales span the Dinantian-Namurian boundary. Carboniferous black shales represent one of the best prospects for shale gas resources in the British Isles (Smith et al., 2010). The Upper Bowland Shale (lowest Namurian, especially the Pendleian and Arnsbergian stages E1-E2) is most widespread and correlates with the Edale Shale, Derbyshire, the Ardagh Shale (Ardagh Formation), eastern Ireland, and the Clare Shales (Clare Shale Formation), Cos. Kerry and Clare, western Ireland. The Lower Bowland Shale (uppermost Dinantian, especially the Brigantian stage P1–P2) correlates with the black shales of the Fingal Group, Cos. Dublin and Meath, eastern Ireland, black shales in Co. Leitrim, Northwest Ireland, and in the Isle of Man. In addition, the uppermost stratigraphic sections of the Bowland Shale were deposited in the Chokierian to Marsdenian stages (H-R). Sections of this age include the Holywell Shale, North Wales, and much of the Clare Shales. Detailed stratigraphies are summarized by Waters et al. (2011).

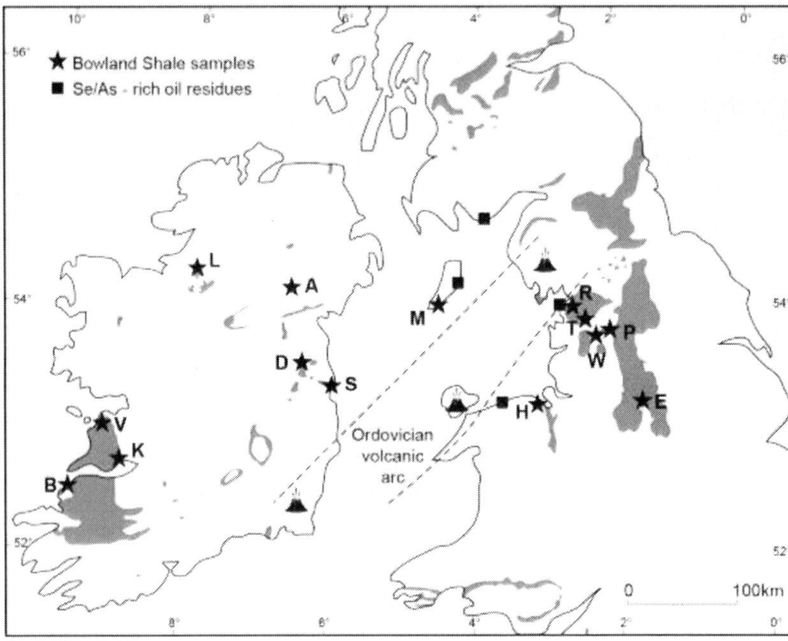

Figure 1. Map of central British Isles, showing outcrop of Namurian rocks, region of Ordovician volcanic basement, sampled localities in Bowland Shale Formation, and localities for Se/As-rich oil residues. A, Altmush; B = Ballybunion; D, Dunshaughlin; E = Edale; H = Holywell; K = Killadysert; L, Co. Leitrim (Glenfarne, Thur Mountain); M = Poyllvaaish, Isle of Man; P = Pendle Hill and Earby; R = Walmsley Bridge; S = Loughshinny; T = Trough of Bowland; V = Lisdoonvarna; W, Whalley and Wiswell.

Trace element contents were measured in shale samples using inductively coupled plasma-mass spectrometry (ICP-MS). Samples of ~30 g rock were milled and homogenised, and 0.25 g digested with perchloric, nitric, hydrofluoric and hydrochloric acids to near dryness. The residue was topped up with dilute hydrochloric acid, and analysed using a Varian 725 instrument. Samples with high concentrations were diluted with hydrochloric acid to make a solution of 12.5 mL, homogenized, then analysed by ICP-MS. Results were corrected for spectral inter-element interferences. The limits of resolution are 0.05 and 10,000 ppm.

3. RESULTS

The data for Mo, Se and As in the Bowland Shale are shown in Table 1, andFig. 2 and Fig. 3. The mean values for Mo, Se and As in the Bowland Shale samples of P1-E2 age are 44.1 ppm, 21.5 ppm and 20.5 ppm respectively. These values are much higher than global mean shale values of 2.6 ppm, 0.6 ppm and 13.0 ppm respectively (Turekian and Wedepohl, 1961). The younger Bowland Shale samples of H-R age have lower Se contents, but comparable Mo and As contents, to the samples of P1-E2 age.

The Mo values for the Bowland Shale range up to 155 ppm. The highest values are recorded from Ballybunion, Co. Kerry, but most of the samples are in the range 20–70 ppm. Mo is characteristically correlated with the organic carbon content in shales (Algeo and Lyons, 2006). This relationship holds for both the Bowland Shale (Fig. 2) and Barnett Shale (Rowe et al., 2008). However, Mo values for the Bowland Shale are much higher than those of the Barnett Shale (Fig. 2). The Mo/TOC ratio varies with deep water renewal time, increasing with turnover rate as shown by data sets for the Black Sea and the Cariaco Basin, Venezuela (Algeo and Lyons, 2006 and Rowe et al., 2008), and is also consistently higher in the Bowland Shale than the Barnett Shale (Fig. 2).

There are no published data for S in the Barnett Shale, but the Se contents of two Barnett Shale samples measured in this study are 8 ppm and 9 ppm. The Mo contents of these samples are typical of a much larger database of Barnett Shale samples (Rowe et al., 2008), implying that the Se contents are also typical. The Se contents of the Bowland Shale clearly range to much higher values than the Barnett Shale samples, with maximum values of 34 ppm 36 ppm and 42 ppm in the Trough of Bowland, Co. Meath and Co. Kerry, Ireland (Fig. 2). A cross-plot of Se and sulphur for the Bowland Shale samples indicates that there is a positive correlation between Se and S (Fig. 3), and that Se substitutes within sulphide minerals, as commonly found in other marine shales (Large et al., 2014). This relationship also indicates that the higher Se contents in the Bowland Shale reflect higher sulphide contents. The two lowest Se values (2.9 ppm, 5.6 ppm) determined in the main Bowland Shale data set are from samples with negligible S contents. The Se values are also much higher than values determined for other black shales (Fig. 2), including the Cambrian Alum Shale, Devonian New Albany Shale and Jurassic Posidonia Shale, which range up to 3.3 ppm, 1.1 ppm and 3.9 ppm respectively (Mitchell et al., 2012). The As content of the Bowland Shale samples range up to 52 ppm (two highest values from Lancashire), and almost all values exceed Barnett Shale values of 8 and 9 ppm. The As values are strongly related to the S content (Fig. 3).

Table 1. Summary geochemical data for Bowland Shale samples.

Full data set	Strat range	n	Se (ppm)	As (ppm)	Mo (ppm)
P1-E2 samples	P1-E2	37			
Mean			21.1	22.8	42.0
Standard Deviation			9.9	10.3	32.5
Q1			15.5	16.6	23.2
Q2 (Median)			18.2	20.2	31.0
Q3			28.8	28.5	61.3
Minimum			2.9	9.0	3.4
Maximum			42.0	51.8	154.5
H-R samples	H-R	5			
Mean			4.6	16.9	25.9
Standard Deviation			3.3	14.5	10.7
Q1			3.3	1.2	22.5
Q2 (median)			3.9	26.4	29.3
Q3			4.1	26.5	30.6
Minimum			1.3	0.8	9.2
Maximum			10.2	29.4	37.7
Regions (localities)			Se range	As range	Mo range
NE Lancs (P,W)	P1-E1	9	3–33	11–52	9–39
NW Lancs (R,T)	P2-E1	6	13–34	16–46	3–68
Derbyshire (E)	E1-E2	3	15–21	21–38	23–50
Isle of Man (M)	P1-P2	3	18–27	11–21	8–41
Cos. Meath/Dublin (A, D,S)	P2-E1	10	8–36	9–41	8–65
Co. Kerry (B)	P1-E1	4	16–42	17–29	68–155
Co. Leitrim (L)	P1-E1	2	12–17	10–20	17–35
North Wales (H)	H-R	2	4–10	26–29	22–31
Co. Clare (K,V)	H-R	3	1–4	1–27	29–38

In summary, both the high sulphur content and rapid turnover of water contributed to the high metal content of the Bowland Shale.

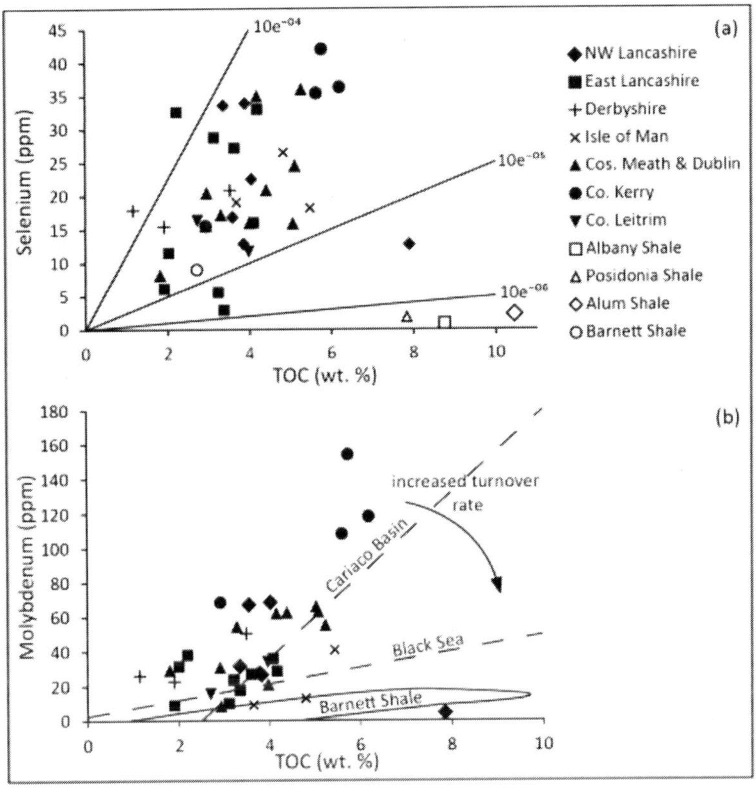

Figure 2. Cross-plots for (A) Se against TOC. Se/TOC ratios for Bowland Shale are higher than for mean values for other black shales (data from Mitchell et al., 2012, except Barnett Shale mean value based on 2 measurements in this study). (B) Mo against TOC. Mo/TOC plot shows Mo and Mo/TOC values for Bowland Shale are consistently higher than in Barnett Shale. Mo/TOC correlation lines for Black Sea and Cariaco Basin show data for Bowland Shale is also high relative to some modern anoxic environments. (Mo data for Barnett Shale and modern environments from Rowe et al., 2008).

4. DISCUSSION

4.1. Anomalous selenium availability

Much of the Se in black shales is associated with organic matter, but Se/TOC ratios vary according to Se availability (Mitchell et al., 2012).

The Se/TOC ratios for the Bowland Shale markedly exceed values for other black shales (Fig. 2), implying anomalous availability. The high Se content of the British and Irish Carboniferous shales may reflect an anomalous source in the watershed. The region is bisected by an Ordovician volcanic arc, from the English Lake District through Anglesey and SE Ireland (Fig. 1), including the volcanic massive sulphide deposits at Parys Mountain, Anglesey, and Avoca, Ireland. In each of these areas, the Ordovician is unconformably overlain by the Carboniferous, and would have been supplying runoff waters into the Carboniferous sea. Stream sediments draining the Ordovician volcanic centres in North Wales today are Se-rich (Fordyce, 2013), so the Ordovician represents a potential source of Se-rich water to the Carboniferous. Contemporaneous tuffs in the late Dinantian-early Namurian of the British Isles indicate that there was also a direct input of volcanic ash into the local Carboniferous sea. Tuff deposition occurred in the Bowland Shale and its time equivalents in central England (Spears et al., 1999), southern Scotland (Monaghan and Pringle, 2004) and Ireland (Legg et al., 1998). The Se content of tuffaceous volcanic rocks can be relatively high (Fordyce, 2013), and the aerosols associated with volcanic activity can be 1 to 3 orders of magnitude richer in Se than associated ash, and add anomalous selenium to the surface environment (Witham et al., 2005 and Floor and Román-Ross, 2012). Interbedded tuffs have been interpreted to indicate a volcanic source of high Se levels in black shales elsewhere (Kulp and Pratt, 2004), and it is very possible that this was also the case in the Bowland Shale. The high As levels in the Bowland Shale may have a similar origin, as volcanic rocks contribute to high As levels in groundwater (e.g. Welch et al., 2000), and the Ordovician volcanogenic sulphide deposits in the palaeo-watershed of the Carboniferous shales are known to release anomalous arsenic into the surface environment today (Mayes et al., 2010).

Clues to the chemistry of waters derived from these sulphur-rich shales can be provided by natural springs. The Clare Shales host spa waters at Lisdoonvarna, Co. Clare; the Fingal Group hosts spa waters at Lucan, Co. Dublin; and two of the biggest spas in England at Harrogate and Buxton are supplied by waters that have passed through the Bowland Shale (Bottrell et al., 1996 and Gunn et al., 2006). Together this represents a notable set of mineralized springs derived from the shales. All four spas are strongly sulphur-rich, and implicate leaching of the pyrite in the shales. More generally, Carboniferous rocks in northern England host numerous 'sulphur springs' (Murphy et al., 2014). As the pyrite is the main residence of Se and As, this implies liberation of Se and As into the groundwaters. The flow of sulphur-rich water through these springs over at least a historical record of hundreds of years implies a long-term alteration of the pyritic sources and long-term liberation of Se.

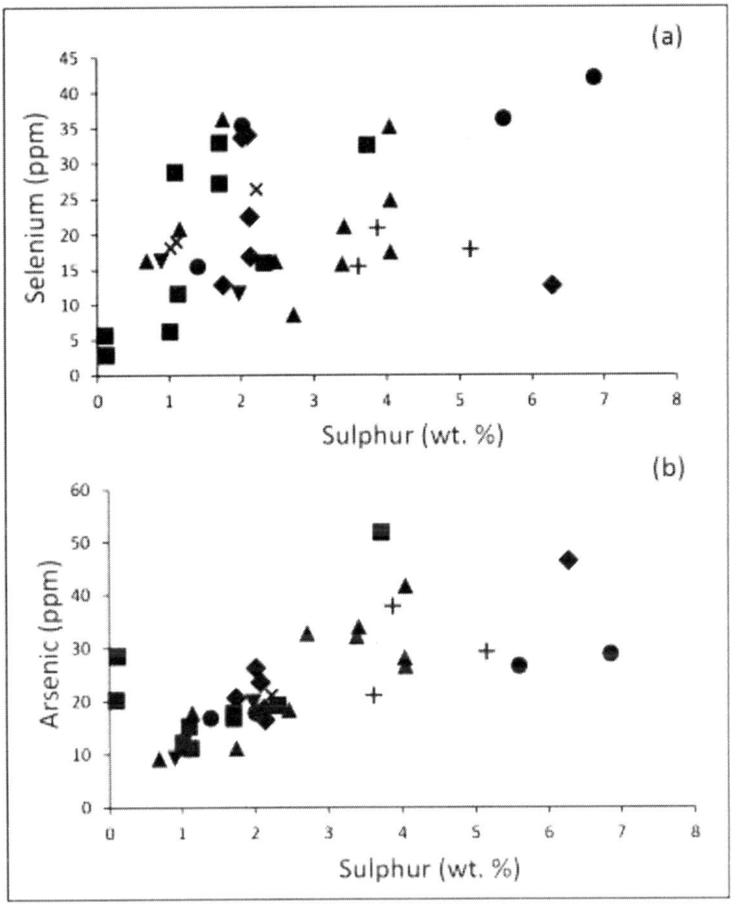

Figure 3. Cross-plots for (A) Se against S, and (B) As against S. Broad trends of increasing Se and As with S are consistent with residence of Se and As within pyrite. Key as in Fig. 2.

Selenium and arsenic may also be associated with hydrocarbons generated from black, organic-rich shales, as evidenced from Se/As-rich by-products in oil refineries (Lawson and Macy, 1995), shale oil processing (Sikonia, 1985) and coal gasification (Liu et al., 2006). The Bowland Shale is a major hydrocarbon source rock in the Irish Sea

(Armstrong et al., 1997). Bituminous oil residues around the margins of the Irish Sea in North Wales, Lancashire and southern Scotland (Fig. 1) all contain traces of selenide mineralization (Parnell, 1988 and unpublished data), and selenium enrichments occur in Carboniferous palaeo-oil reservoirs in the south of the region (Parnell et al., 2015). For example, hydrocarbon residues in Carboniferous sandstones at Heysham, Lancashire (Harrison, 1970) are Se-rich. Oil residues in the Isle of Man additionally contain arsenide mineralization (Parnell, 1988). This implies that the fluids expelled from the Irish Sea Basin are Se- and As-rich, and likely related to hydrocarbon expulsion from the Bowland Shale. The highest Se values in the Carboniferous shales are recorded from Co. Kerry, western Ireland, which is the most thermally mature region in the gas window (Clayton et al., 1989), showing that hydrocarbon generation has not mobilized all of the Se. Notably, base metal sulphide mineralization in the Trough of Bowland and other parts of northern England is also attributed to the release of metals from the Carboniferous shales (Jones et al., 1994). The Clare Shales in Ireland have interacted with groundwater to yield secondary Se mineralization (Francis and Ryback, 1987). Overall, we have a picture of the Bowland Shale as a black shale anomalously rich in trace elements, especially Se and As, and with the potential to release those elements into subsurface fluids.

4.2. Implications

Black shales confer high levels of Se to groundwaters through subsurface leaching (e.g.U.S. Department of Energy, (2011)), leaching of rock waste (e.g. Stillings and Amacher, 2010) and via overlying soils (e.g. Park et al., 2010 and Tuttle et al., 2014). In most cases where groundwaters contain *more* than the maximum level of selenium permitted for drinking water after interaction with black shales, the parent black shale contains much*less* Se than the mean determined here for the Bowland Shale (Mayland et al., 1989,U.S. Department of Energy, 2011 and Fontenot et al., 2013). For example, high levels of dissolved Se occur in water from wells drilled into Cretaceous black shales in Montana with up to 10 ppm Se (Mayland et al., 1989). This maximum value in Montana is exceeded by most of the Bowland Shale analyses. An average of 1 ppm Se for Cretaceous black shales responsible for soil contamination in the southwestern United States (Tuttle et al., 2014) is over an order of magnitude less than in the Bowland Shale. Concern about high Se levels in drinking water have recently spread to western Europe, and some wells in France where the water has been exposed to black shale have had to be abandoned (Bassil et al., 2014). As noted above, the As levels in the Bowland Shale consistently exceed those in the Barnett Shale, whose exploitation has led to groundwater values above the U.S. limit for drinking water. There is a direct implication that the Se and As levels of groundwaters that have interacted with the Bowland Shale require monitoring.

So, should there be concern over these high metal contents in the Bowland Shale, if it is exploited for shale gas? There is awareness of the need to consider such issues in the UK (Stuart, 2012 and Cai and Ofterdinger, 2014), but fortunately a very robust oil and gas regulatory system, covering licence application, planning permission, well construction and environmental permits, makes it very unlikely that the UK would experience the environmental problems encountered in the USA (The Royal Society, 2012).

The concentration of Se, and possibly other trace elements, in extraction waters could also be regarded as an opportunity. Selenium is a rare element for which demand is likely to increase (Moss et al., 2011), but the current supply as a by-product of copper production is in jeopardy as production methods change. Alternative methods of extraction that do not include high temperature (high energy) or toxic chemicals are desirable. Waters produced from Se-rich shales offer new possibilities. An understanding of how to do this can build on methodology being developed to deal with Se-rich waters associated with the storage of coal waste. Some coals, like some black shales, contain anomalous concentrations of Se, leading to environmental problems in several parts of the world, including the Elk River Valley, British Columbia (Lussier et al., 2003). Approaches to sequester Se at low temperature include precipitation onto iron oxides (Wei et al., 2012), precipitation onto activated carbon (Wasewar et al., 2009) and microbial selenate reduction (Tucker et al., 1998). These are all low-energy approaches which could contribute to securing new Se resources, and in the process add incentive to ensuring that extraction waters are processed in an environmentally sensitive manner.

ACKNOWLEDGEMENTS

Research was partly supported by NERC grants (NE/L001764/1 and NE/M010953/1). Mrs. A. Sandison provided skilled technical support.

Barnett Shale was kindly provided by Hu Qinhong. The manuscript was improved by the helpful comments of a reviewer.

REFERENCES

1. Algeo, T.J., Lyons, T.W., 2006. Mo-total organic carbon covariation in modern anoxic marine environments: Implications for analysis of paleoredox and paleohydrographic conditions. Paleoceanography 21

2. Alloway, B.J., 2012. Heavy Metals in Soils. Springer, London. Almond, S., Clancy, S.A., Davies, R.J., Worrall, F., 2014. The flux of radionuclides in flowback fluid from shale gas exploitation. Environ. Sci. Pollut. Res. 21, 12316-12324.

3. Andrews, I.J., 2013. The Carboniferous Bowland Shale Gas Study; Geology and Resource Estimation. British Geological Survey, for Department of Energy and Climate Change (London).

4. Armstrong, J.P., Smith, J., D'Elia, V.A.A., Trueblood, S.P., 1997. The occurrence and correlation of oils and namurian source rocks in the Liverpool Bay-North Wales area. Geol. Soc. Spec. Publ. 124, 195-211.

5. Bassil, J., Naveau, A., Bodin, J., Fontaine, C., Di Tullo, P., Razack, M., Kazpard, V., 2014. The nature of selenium species in the hydrogeological experimental site of Poitiers. Procedia Earth Planet. Sci. 10, 159-163.

6. Bottrell, S.H., Raiswell, R., Leosson, M.A., 1996. The influence of sulphur redox reactions and mixing on the chemistry of shallow groundwaters: the Harrogate mineral waters. J. Geol. Soc. 153, 231-242.

7. London. Cai, Z., Ofterdinger, U., 2014. Numerical assessment of potential impacts of hydraulically fractured Bowland Shale on overlying aquifers. Water Resour. Res. 50, 6236-6259.

8. Chermak, J.A., Schreiber, M.E., 2014. Mineralogy and trace element geochemistry of gas shales in the United States: environmental implications. Int. J. Coal Geol. 126, 32-44.

9. Clayton, G., Haughey, N., Sevastopulo, G.D., Burnett, R.D., 1989. Thermal Maturation Levels in the Devonian and Carboniferous of Ireland. Geological Survey of Ireland, Dublin.

10. Fellowes, J.W., Pattrick, R.A.D., Boothman, C., Al Lawati, W.M.M., van Dongen, B.E., Charnock, J.M., Lloyd, J.R., Pearce, C.I., 2013. Microbial selenium transformations in seleniferous soils. Eur. J. Soil Sci. 64, 629-638.

11. Fleming, G.A., Walsh, T., 1957. Selenium occurrence in certain Irish soils and its toxic effects on animals. Proc. R. Ir. Acad. 58B, 151-166.

12. Floor, G.H., Romı an-Ross, G., 2012. Selenium in volcanic environments: a review. Appl. Geochem. 27, 517-531.

13. Fontenot, B.E., Hunt, L.R., Hildenbrand, Z.L., Carlton, D.D., Oka, H., Walton, J.L., Hopkins, D., Osorio, A., Bjorndal, B., Hu, Q.H., Schug, K.A., 2013. An evaluation of water quality in private drinking water wells near natural gas extraction sites in the Barnett Shale Formation. Environ. Sci. Technol. 47, 10032-10040.

14. Fordyce, F., 2013. Selenium deficiency and toxicity in the environment. In: Selinus, O. (Ed.), Essentials of Medical Geology. British Geological Survey, Keyworth, pp. 373-415.

15. Francis, J.G., Ryback, G., 1987. Chalcomenite from Ballybunnion, Co. Kerry, Eire. Mineral. Mag. 51, 751-752.

16. Gunn, J., Bottrell, S.H., Lowe, D.J., Worthington, S.R.H., 2006. Deep groundwater flow and geochemical processes in limestone aquifers: evidence from thermal waters in Derbyshire, England, UK. Hydrogeology J. 14, 868-881.

17. Haluszczak, L.O., Rose, A.W., Kump, L.R., 2013. Geochemical evaluation of flowback brine from Marcellus gas wells in Pennsylvania, USA. Appl. Geochem. 28, 55-61.

18. Harrison, R.K., 1970. Hydrocarbon-bearing nodules from Heysham, Lancashire. Geol. J. 7, 101e110.

19. Hildenbrand, Z.L., Fontenot, B.E., Carlton, D.D., Schug, K.A., 2013. New Perspectives on the Effects of Natural Gas Extraction on Groundwater Quality. Global Water Forum. http://www.globalwaterforum.org/2013/10/08/new-perspectives-onthe-effects-of-natural-gas-extraction-on-

groundwater-quality/. Jones, D.G., Plant, J.A., Colman, T.B., 1994. The genesis of the Pennine mineralization of Northern England and its relationship to mineralization in central Ireland. In: Fontbote, L., Boni, M. (Eds.), Sediment-hosted Zinc-Lead Ores. Springer, Heidelberg, pp. 198-218.

20. Jones, G.L., 2012. Hydraulic fracturing for shale gas in Ireland - the potential. Proceedings IAH 2012, Session I. Int. Assoc. Hydrologists 13e20.

21. Keily, P.V., Fleming, G.A., 1969. Geochemical Survey of Ireland: Meath-Dublin area. Proc. R. Ir. Acad. 68B, 1-28.

22. Kulp, T.R., Pratt, L.M., 2004. Speciation and weathering of selenium in upper cretaceous chalk and shale from South Dakota and Wyoming, USA. Geochimica Cosmochimica Acta 68, 3687-3701.

23. Large, R.R., Halpin, J.A., Danyushevsky, L.V., Maslennikov, V.V., Bull, S.W., Long, J.A., Gregory, D.D., Lounejeva, E., Lyons, T.W., Sack, P.J., McGoldrick, P.J., Calver, C.R., 2014. Trace element content of sedimentary pyrite as a new proxy for deeptime ocean-atmosphere evolution. Earth Planet. Sci. Lett. 389, 209-220.

24. Lawson, S., Macy, J.M., 1995. Bioremediation of selenite in oil refinery wastewater. Appl. Microbiol. Biotechnol. 43, 762e765.

25. Legg, I.C., Arthur, M.J., Riley, N.J., Reay, D.M., 1998. Geology of the Country around Derrygonnelly and Marble arch. Mem. Geol. Surv. North. Irel. 82 pp. Liu, S., Wang, Y., Yu, L., Oakey, J., 2006. Volatilization of mercury, arsenic and selenium during underground coal gasification. Fuel 85, 1550-1558.

26. Lussier, C., Veiga, V., Baldwin, S., 2003. The geochemistry of selenium associated with coal waste in the Elk River Valley, Canada. Environ. Geol. 44, 905-913.

27. MacFarquhar, J.K., Broussard, D.L., Melstrom, P., Hutchinson, R., Wolkin, A., Martin, C., Burk, R.F., Dunn, J.R., Green, A.L., Hammond, R., Schaffner, W., Jones, T.F., 2010. Acute selenium toxicity associated with a dietary supplement. Archives Intern. Med. 170, 256-261.

28. Mayes, W.M., Potter, H.A.B., Jarvis, A.P., 2010. Inventory of aquatic contaminant flux arising from histrorical metal mining in England and Wales. Sci. Total Environ. 408, 3576-3583.

29. Mayland, H.F., James, L.F., Panter, K.E., Sonderegger, J.L., 1989. Selenium in seleniferous environments. Soil Science Society of America Special Publication no. 23. In: Selenium in Agriculture and the Environment, pp. 15-50.

30. McGrath, D., Fleming, G.A., 2007. Trace Elements and Heavy Metals in Irish Soils. TEAGASC, Dublin.

31. Meyer, T., 2013. New Study Finds High Levels of Arsenic in Groundwater Near Fracking Sites. ProPublica. http://www.propublica.org/article/new-study-findshigh-levels-of-arsenic-in-groundwater-near-fracking-sites. Mitchell, K., Mason, P.R.D., Van Cappellen, P., Johnson, T.M., Gill, B.C., Owens, J.D., Diaz, J., Ingall, E.D., Reichart, G.J., Lyons, T.W., 2012. Selenium as paleooceanographic proxy: a first assessment. Geochimica Cosmochimica Acta 89, 302-317.

32. Monaghan, A.A., Pringle, M.S., 2004. 40Ar/39Ar Geochronology of CarboniferousPermian Volcanism in the Midland Valley, Scotland, 223. Geological Society, London, pp. 219-241.

33. Special Publication. Moss, R.L., Tzimas, E., Kara, H., Willis, P., Kooroshy, J., 2011. Critical metals in Strategic energy technologies. Publications Office of the European Union, Luxembourg.

34. Murphy, P.J., Bottrell, S.H., Parker, K., 2014. Sulphur springs of the Craven Basin, NW England: indicators of natural methane leakage? Q. J. Eng. Geol. Hydrogeology 47, 81-88.

35. Myers, T., 2012. Potential contaminant pathways from hydraulically fractured shale to aquifers. Ground Water 50, 872e882.

36. Park, M., Chon, H.T., Marton, L., 2010. Mobility and accumulation of selenium and its relationship with other heavy metals in the system rocks/soils-crops in areas covered by black shale in Korea. J. Geochem. Explor. 107, 161-168.

37. Parnell, J., 1988. Mineralogy of uraniferous hydrocarbons in Carboniferous-hosted mineral deposits, Great Britain. Uranium 5, 197-218.

38. Parnell, J., Bellis, D., Feldmann, J., Bata, T., 2015. Selenium and tellurium enrichment in palaeo-oil reservoirs. J. Geochem. Explor. 148, 169-173.

39. Rogers, P.A.M., Arora, S.P., Fleming, G.A., Crinion, R.A.P., McLaughlin, J.G., 1990. Selenium toxicity in farm animals: treatment and prevention. Ir. Veterinary J. 43, 151-153.

40. Rowe, H.D., Loucks, R.G., Ruppel, S.C., Rimmer, S.M., 2008. Mississippian Barnett Formation, Fort Worth Basin, Texas: Bulk geochemical inferences and Mo-TOC constraints on the severity of hydrographic restriction. Chem. Geol. 257, 16-25.

41. Schug, K.A., Hildenbrand, Z.L., Fontenot, B.E., 2013 September 11. Elevated heavy metals near natural gas extraction sites in the Barnett Shale. Oil Gas Monit. http://www.oilgasmonitor. com/ elevated-heavy-metals-near-natural-gasextraction-sites-barnett-shale/5899/.

42. Shonkoff, S.B.C., Hays, J., Finkel, M.L., 2014. Environmental public health dimensions of shale and tight gas development. Environ. Health Perspect. 122, 787-795.

43. Sikonia, J.G., 1985. Arsenic management in shale oil upgrading. Environ. Geochem. Health 7, 64e68. Smith, N., Turner, P., Williams, G., 2010. UK data and analysis for shale gas prospectivity. In: Vining, B.A., Pickering, S.C. (Eds.), Petroleum Geology: from Mature Basins to New Frontiers e Proceedings of the 7th Petroleum Geology Conference, pp. 1087-1098.

44. Spears, D.A., Kanaris-Sotiriou, R., Riley, N., Krause, P., 1999. Namurian bentonites in the Pennine Basin, UK e origin and magmatic affinities. Sedimentology 46, 385-401.

45. Stillings, L.L., Amacher, M.C., 2010. Kinetics of selenium release in mine waste from the Meade peak phosphatic shale, Phosphoria Formation, Wooley Valley, Idaho, USA. Chem. Geol. 269, 113e123.

46. Stuart, M.E., 2012. Potential Groundwater Impact from Exploitation of Shale Gas in the UK. British Geological Survey Open Report, OR/12/001. British Geological Survey, Keyworth.

47. Sun, H.J., Rathinasabapathi, B., Wu, B., Luo, J., Pu, L.P., Ma, L.Q., 2014. Arsenic and selenium toxicity and their interactive effects in humans. Environ. Int. 69, 148-158.

48. Taylor, K., Khattab, S., Nolan, K., Redfern, J., Williams, B., Warshauer, S., Hill, J., Armstrong, J., 2013. Sedimentological Characteristics and Shale Gas Potential of Carboniferous Mudstones in Ireland: the Clare

and Northwest Carboniferous Basins. AAPG Search and Discovery Article #90163. AAPG Annual Convention and Exhibition, Pittsburgh. May 19-22, 2013.

49. The Royal Society, 2012. Shale Gas Extraction in the U.K.: a Review of Hydraulic Fracturing. The Royal Society and The Royal Academy of Engineering, London. Tucker, M.D., Barton, L.L., Thomson, B.M., 1998. Reduction of Cr, Mo, Se and U by desulfovibrio desulfuricans immobilized in polyacrylamide gels. J. Industrial Microbiol. Technol. 20, 13-19.

50. Turekian, K.K., Wedepohl, K.H., 1961. Distribution of the elements in some major units of the Earth's crust. Geol. Soc. Am. Bull. 72, 175e192.

51. Tuttle, M.L.W., Fahy, J.W., Elliott, J.G., Grauch, R.I., Stillings, L.L., 2014. Contaminants from Cretaceous black shale: I. Natural weathering processes controlling contaminant cycling in Mancos Shale, southwestern United States, with emphasis on salinity and selenium. Appl. Geochem. 46, 57-71.

52. U.S. Department of Energy, 2011. Natural Contamination from the Mancos Shale. Experimental Sciences Laboratory Report ESL-rpt-2011-01.

53. Vengosh, A., Jackson, R.B., Warner, N., Darrah, T.H., Kondash, A., 2014. A critical review of the risks to water resources from unconventional shale gas development and hydraulic fracturing in the United States. Environ. Sci. Technol. 48, 8334-8348.

54. Vidic, R.D., Brantley, S.L., Vandenbossche, J.M., Yoxtheimer, D., Abad, J.D., 2013. Impact of shale gas development on regional water quality. Science 340, 826-835.

55. Vinceti, M., Crespi, C.M., Bonvicini, F., Malagoli, C., Ferrante, M., Marmiroli, S., Stranges, S., 2013. The need for a reassessment of the safe upper limit of selenium in drinking water. Sci. Total Environ. 443, 633-642.

56. Vinceti, M., Mandrioli, J., Borella, P., Michalke, B., Tsatsakis, A., Finkelstein, Y., 2014. Selenium neurotoxicity in humans: Bridging laboratory and epidemiological studies. Toxicol. Lett. 230, 295-303.

57. Wasewar, K.L., Prasad, B., Gulipalli, S., 2009. Removal of selenium by adsorption onto granular activated carbon (GAC) and powdered activated carbon (PAC). Clean 37, 872-883.

58. Waters, C.N., Somerville, I.D., Jones, N.S., Cleal, C.J., Collinson, J.D., Waters, R.A., Besly, B.M., Dean, M.T., Stephenson, M.H., Davies, J.R., Freshney, E.C., Jackson, D.I., Mitchell, W.I., Powell, J.H., Barclay, W.J., Browne, M.A.E., Leveridge, B.E., Long, S.L., McLean, D., 2011. A Revised Correlation of Carboniferous Rocks in the British Isles. Geological Society Special Report No. 26.

59. Webb, J.S., Thornton, I., Fletcher, K., 1966. Seleniferous soils in parts of England and Wales. Nature 211, 327. Wei, X., Bhojappa, S., Lin, L.S., Viadero, R.C., 2012. Performance of nano-magnetite for removal of selenium from aqueous solutions. Environ. Eng. Sci. 29, 526-532.

60. Welch, A.H., Westjohn, D.B., Helsel, D.R., Wanty, R.B., 2000. Arsenic in ground water of the United States: occurrence and Geochemistry. Ground Water 38, 589-604.

61. Witham, C.S., Oppenheimer, C., Horwell, C.J., 2005. Volcanic ash-leachates: a review and recommendations for sampling methods. J. Volcanol. Geotherm. Res. 141, 299e-26.

62. Zeng, T., Sarfim, A.F., Senior, C.L., 2001. Vaporization of arsenic, selenium and antimony during coal combustion. Combust. Flame 126, 1714-1724.

6

QUALITY ASSESSMENT OF GROUNDWATER IN YENAGOA, NIGER DELTA, NIGERIA

Oki A. O., Akana T. S.

Department of Geology, Niger Delta University, Nigeria

ABSTRACT

This study x-rayed the geochemical characteristics of groundwater quality of the growing City of Yenagoa, the Bayelsa State capital. Water samples were collected from fifteen (15) boreholes. Analysis were carried out on physico-chemical parameters (pH, salinity, electrical conductivity, turbidity, total dissolved solids, alkalinity, hardness, nitrate, chloride, sulphate, bicarbonate, calcium, magnesium, sodium, potassium and iron). Results showed that most results fell within acceptable World Health Organisation (WHO) limits for potable water. p^H in all BH samples were slightly acidic. In addition magnesium was above the limit in all BH samples after comparison with WHO (2006) standard. Data was subjected to Piper's trilinear plot, result from the plot showed the Calcium and Chloride type water was present in the area. This means that the water from the area is good for domestic and industrial use. Although, treatment is required in some areas where the results did not conform to standard before use to prevent health risks.

KEYWORDS

Geochemistry, Aquifer quality, Piper plot, Yenagoa

1. INTRODUCTION

The challenge of ensuring usable water in sufficient quantities to meet the needs of human and ecosystems emerged as one of the primary issues of the 21st century. Poor water quality could cause serious health problems, reduce agricultural productivity and economic prosperity of communities. Problems of this nature have been increasing in scope, frequency, and severity because the demand for water continues to grow while supply of renewable water remains relatively fixed. While it is agreed that water is one of the most important resources with great implications for African development, the freshwater situation in Africa is unfortunately not encouraging. The situation is getting worse as a consequence of population growth, rapid urbanization, increasing agriculture and industrial activities, and lack of adequate capacity to manage freshwater resources. Bayelsa state is a wetland in Niger Delta with possible serious problems of availability of portable water. All the surface water bodies in the state are pollution-laden and hence there is the need to resort to groundwater for consumable water.

Groundwater includes all water found beneath the surface of the Earth. It is derived primarily by percolation of atmospheric or surface water and contained in pore spaces of permeable reservoir rocks (Ayoade, 2003), this makes it a replenishable resource unlike other resources of the Earth. In any geological environment, groundwater chemistry is controlled by the chemistry of the infiltrating water, the chemistry of the porous media including the interstitial cement or matrix of the aquifer, the rate of groundwater flow and the permeability of the aquifer (Offodile, 2002). More than 85% of public water for consumption is sourced from groundwater (Ufoegbune et al, 2009), and this is used for domestic, industrial and agricultural purposes.

2. STUDY AREA DESCRIPTION

This study investigates the quality of groundwater in selected densely populated areas of Yenagoa, the capital of Bayelsa state. Yenagoa lies within latitude 04°4N and 05°02N and longitude 006°15E and 006°24E (fig 1) and is situated in southern part of the Niger Delta of Nigeria. Water is contained in very thick and extensive sand and gravel aquifer within the Benin Formation in Yenagoa. The Benin Formation is the aquiferous layer and all boreholes in the area are drilled into it (Etu-Efeotor and Akpokodje, 1990; Offodile, 2002). Minor intercalations of shale layers give rise to multi-aquifer system, out of which two types have been identified (Etu-Efeotor, 1981). The first (Holocene age) is more prolific and extends to about 60-90m (unconfined) while the second (Oligocene) is less prolific and underlies the first. Multi-aquifer systems have also been identified from lithologic logs of boreholes from other parts of the Niger Delta by Edet (1993); Udom (2006).

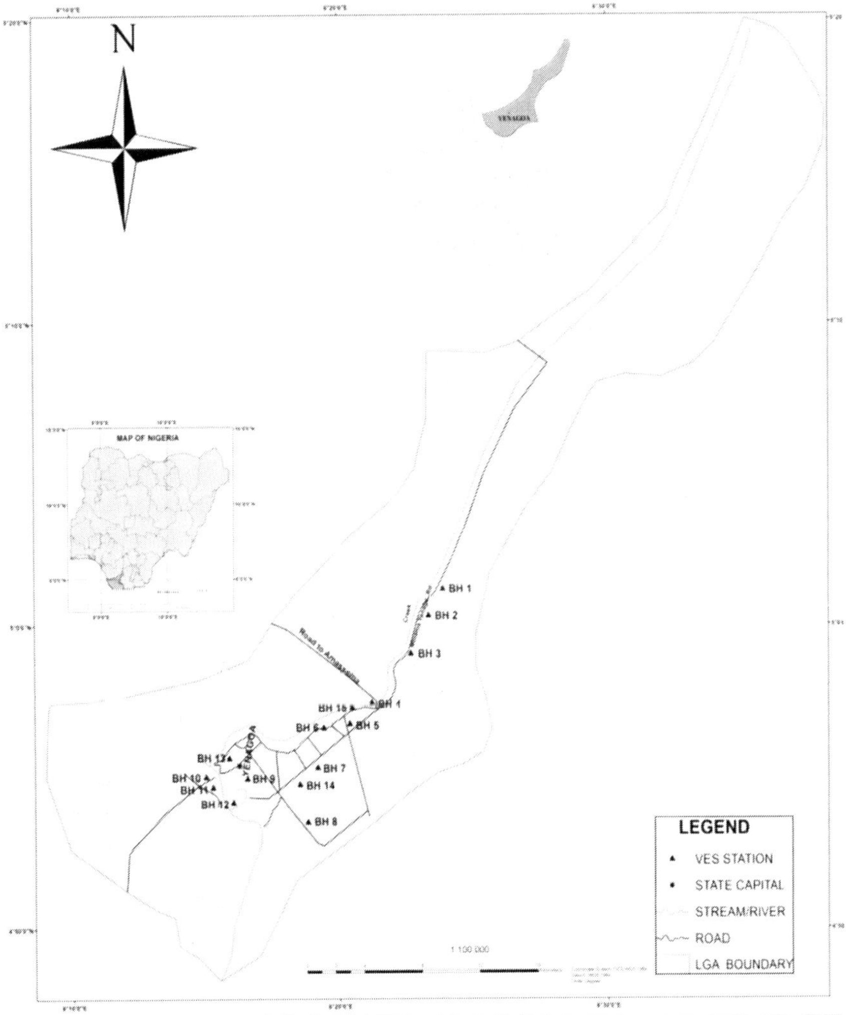

Figure 1. Showing the sampling points on the map of Yenagoa

3. METHODOLOGY

Sterilized water bottles were used to collect representative water samples from boreholes to prevent contamination in this format: BH 1 (Igbogene), BH 2 (Yenegwe), BH 3 (Akemfa), BH 4 (Agudama), BH 5 (Etegwe), BH 6 (Okutukutu), BH 7 (Opolo), BH 8 (Biogbolo), BH 9 (Kpansia), BH 10 (Okaka), BH 11 (Amarata), BH 12 (Swali), BH 13 (Ogbogoro), BH 14 (Ogu) and BH15 (Akaba). Water samples from boreholes were subjected to physico-chemical analysis and results were compared to the World Health Organization Standard for portable water to determine its suitability for domestic and industrial uses. Field analysis was done for pH, conductivity, turbidity and salinity, after which samples were taken to the laboratory within 12 hrs for further physico-chemical analysis. Fifteen samples were collected in the most densely populated towns in Yenegoa local government area at a sample spacing of at least 3 kilometers.

4. RESULT AND DISCUSSIONS

The results of the physico-chemical parameters (table 1) are graphically represented in bar charts from one borehole to another across the study area:

P^H Value

P^H value ranged from 5.6 – 6.91(fig 2) with an average value of 6.30. These values indicated that ground water in the area in slightly acidic, this could be attributed to the abundance of organic matters in the overlying soil and the presence of shale intercalations in the aquiferous coastal plain sands and also due to acid rain caused by prevalent gas flaring in the Niger Delta. However all the samples met the WHO standard for drinking water which is between 6.5 and 8.5.

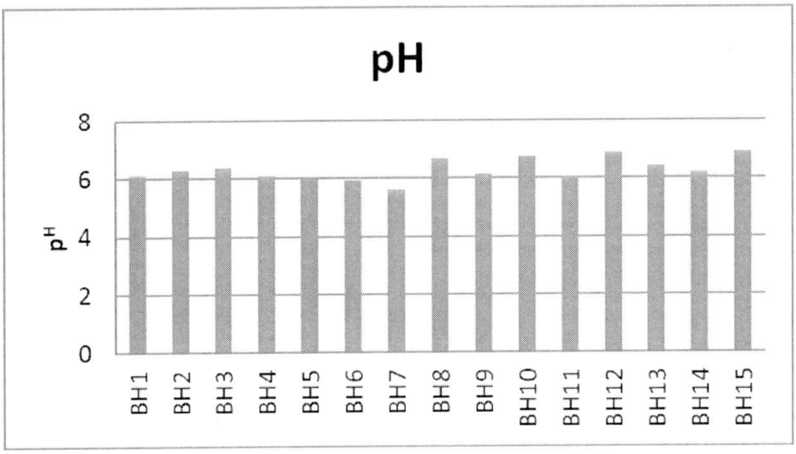

Figure 2. Bar chart showing variation across boreholes

Salinity

The Minimum value was 0.02gil and maximum value was 0.24gil (fig 3) with a mean value of 0.10gil. There is no stipulated limit for WHO standard for drinking water.

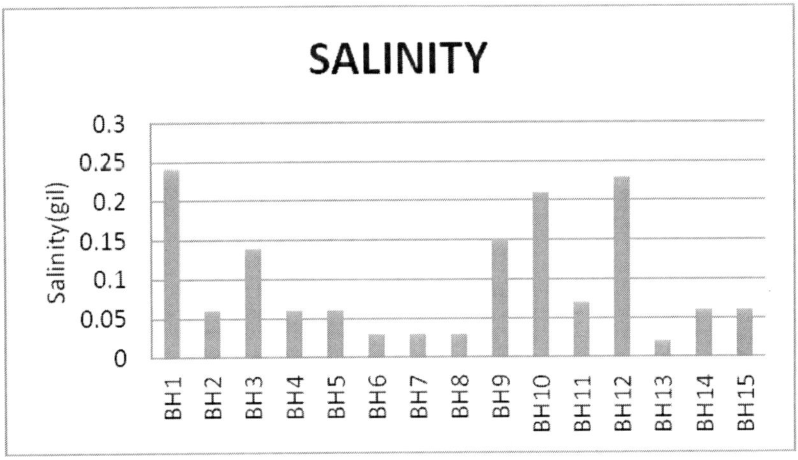

Figure 3. Bar chart showing variation of salinity across boreholes

Table 1. Result from analysis on samples in comparison to World Health Organisation (WHO) standard for portable water

BH	PH	SAL	cond	turb	TDS	TSS	NO_3	Cl	SO_4	HCO_3	TA	TH	Ca	Mg	Na	K	Fe	Mn
1	6.12	0.24	496	2.4	248	0.5	0.36	39	1.4	3	26	25	22.4	6.35	10.86	4.2	0.16	0.02
2	6.3	0.06	164	0.78	82	0.2	0.165	15	0.8	0.4	15	45	8.5	2.48	4	1.46	0.14	0.02
3	6.38	0.14	334	0.4	167	0.02	0.335	21	1.67	1.6	17	18	13.7	3	6.5	1.55	0.24	0.01
4	6.1	0.06	173.1	1.8	86.6	0.05	0.175	14	0.86	1.7	19	37	8.85	2.5	4.85	0.76	0.12	0.02
5	5.99	0.06	163.6	1.76	81.8	0.08	0.165	14	0.82	1.6	3	35	8.6	2.76	4.54	1.2	0.08	0.01
6	5.93	0.03	91.3	0.6	46	0.22	0.094	16	0.48	0.9	5	32	9	2.85	5.2	1.4	0.13	0.02
7	5.6	0.03	84	0.55	42	0.03	0.085	14	0.45	0.8	16	30	7.4	2.38	4.74	1.2	0.18	0.02
8	6.69	0.03	94	0.5	47.5	0.15	0.096	22	0.5	0.9	15	46	12.48	3.62	5.8	1.8	0.17	0.01
9	6.14	0.15	348	2.54	174	0.25	0.348	34	1.75	3.5	16	101	20	5.65	9.95	4	0.2	0.02
10	6.74	0.21	422	0.3	211	0.031	0.42	47	2.1	4.2	15	45	27.86	7.5	13.58	4.65	0.13	0.02
11	6.05	0.07	193.7	1	96.8	0.015	0.204	37	0.96	1.9	18	91	21.48	6.2	9.84	2.62	0.18	0.02
12	6.87	0.23	486	3.65	243	0.4	0.49	23	2.45	4.8	9	33	16.74	4.4	7.6	2.5	0.28	0.01
13	6.43	0.02	77	0.72	38.5	0.072	0.078	16	0.39	0.7	18	27	9.2	2.58	5.4	1.8	0.06	0.02
14	6.2	0.06	160	0.3	80	0.028	0.162	24	0.8	1.6	12	56	14.56	3.8	6	2.26	0.12	0
15	6.91	0.06	172	1.76	86	0.18	0.17	8	0.86	1.4	3	15	6.75	1.76	3.85	0.6	0.18	0.01
MEAN	6.30	0.10	230.58	1.27	115.35	0.15	0.22	22.93	1.08	1.93	13.80	42.4	13.83	3.85	6.85	2.13	0.16	0.02
WHO	6.5-8.5	NS	1000	NS	1000	NS	50	250	400	NS	500	150	75	50	NS	NS	0.3	0.2

SAL-salinity (GIL), COND-conductivity (µs/cm), TURB-turbidity (NTU), TDS-total dissolved solids (mg/l), TSS-total suspended solids (mg/l), TA-total alkalinity (mg/l), TH-total hardness (mg/l), TBC-total bacterial count. Unit of all other ions was in mg/l

Conductivity

Conductivity value ranged from 77µs/cm to 496 µs/cm (fig 4), with an average value of 230.6 µs/cm these values were below 1000µs/cm which was the WHO standard for portable water.

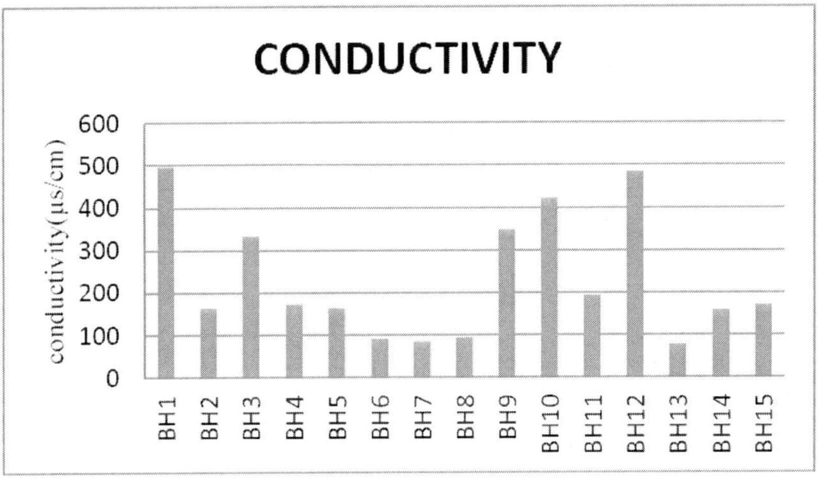

Figure 4. Bar chart showing variation of conductivity across boreholes

Turbidity

The turbidity value ranged from 0.3 NTU to 3.65NTU (fig 5) with an average as 1.27 NTU. There is no guideline for WHO standard for drinking water (2006), causes of high turbidity in water may include the presence of disease causing organism.

Total Dissolved Solids (TDS)

The minimum value of total dissolved solids was 38.5mg/l and maximum value was 248mg/l (fig 6) with the mean value is 115.3mg/l. This values were far below the stipulated value of 1000mg/l by WHO (2006) for drinking water, as such, the water in Yenegoa is good for consumption and irrigation purposes.

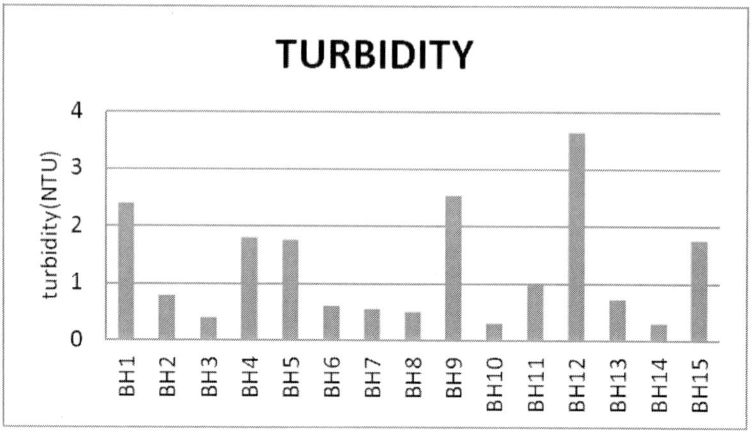

Figure 5. Bar chart showing variation of total suspended solids across boreholes

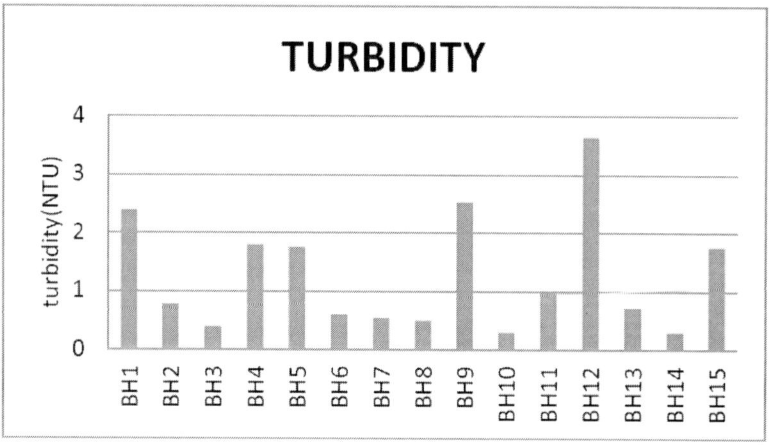

Figure 6. Bar chart showing variation of total dissolved solids across boreholes

Total Suspended Solids (TSS)

The value of total suspended solids ranged from 0.015mg/l to 0.5mg/l (fig 7) with a mean value of 0.15mg/l. No limit is stated by WHO for this parameter.

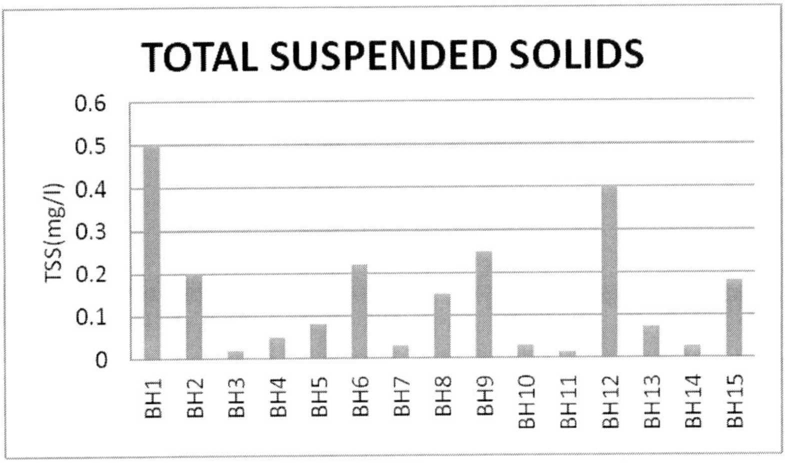

Figure 7. Bar chart showing variation of total suspended solids across boreholes

Nitrate (NO_3)

The minimum value of nitrate is 0. 078mg/l and maximum value was 0.49mg/l (fig 8). This was far below the stipulated value of 50mg/l by WHO (2006) standard for drinking water, hence, the water is recommended for irrigation purposes.

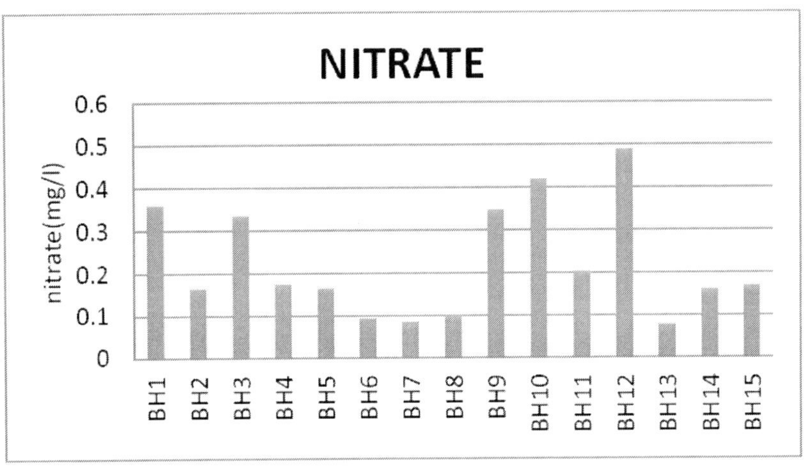

Figure 8. Bar chart showing variation of nitrate across boreholes

Chloride Ion (CL) Content

Chloride in samples ranged from 8mg/l to 47mg/l (fig 9) with mean value of 22.9mg/l. All samples except BH 10 showed values less than 40mg/l stipulated by Trembley et al (1973) which indicate salt-water intrusion. WHO (2006) stipulated 250mg/l as the limit of chloride in drinking water. Primarily, the reason of the limit of chloride for most irrigation processes is placed at 200mg/l.

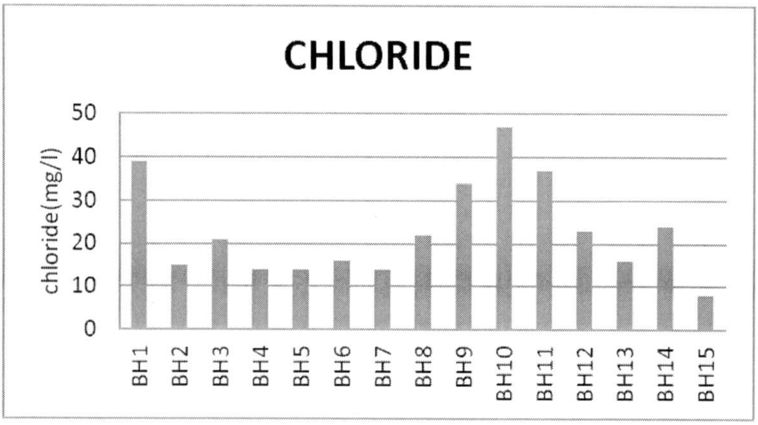

Figure 9. Bar chart showing variation of chloride across boreholes

Sulphate (SO_4) Content

The minimum value of sulphate was 0.39mg/l and the maximum value was 2.45mg/l (fig 10) with a mean value of 1.09mg/l. Sulphate concentration was low in the water compared to WHO stimulated limit of 400mg/l, showing that the water is safe for domestic and industrial purposes.

Bicarbonate (HCO_3) Content

Bicarbonate value ranged from 0.4mg/l to 4.8mg/l (fig 11) and with an average of 1.93gm/l. The source is primarily carbon dioxide in the atmosphere partly from gas and activity of biota in the soil. WHO (2006) presented no limit for this parameter in water and in processing wood chemicals. Upon heating, bicarbonate is changed into steam, carbonate and carbon (iv) oxide. High bicarbonate and alkalinity concentration are undesirable in many industries.

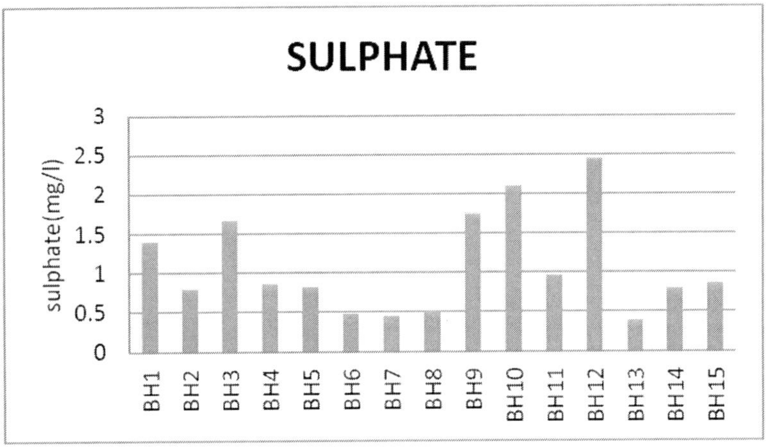

Figure 10. Bar chart showing variation of sulphate across boreholes

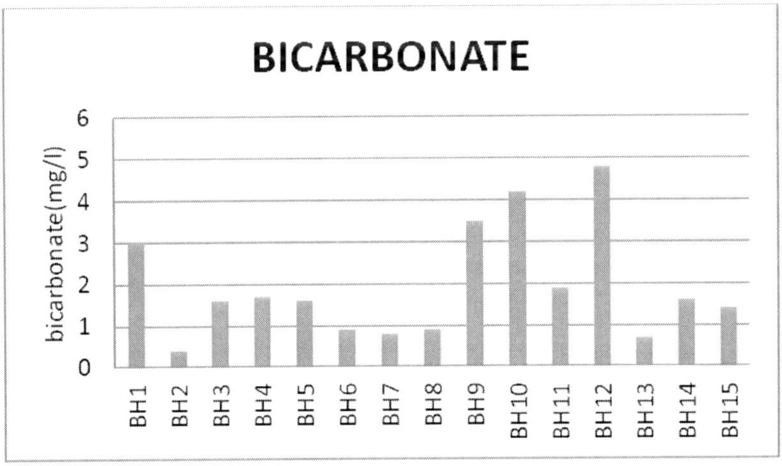

Figure 11. Bar chart showing variation of bicarbonate across boreholes

Total Alkalinity (TA)

The total alkalinity value ranged from 3mg/l to 26mg/l (fig 12). Values were below 500mg/l which is the WHO (2006) stipulated limit for drinking water.

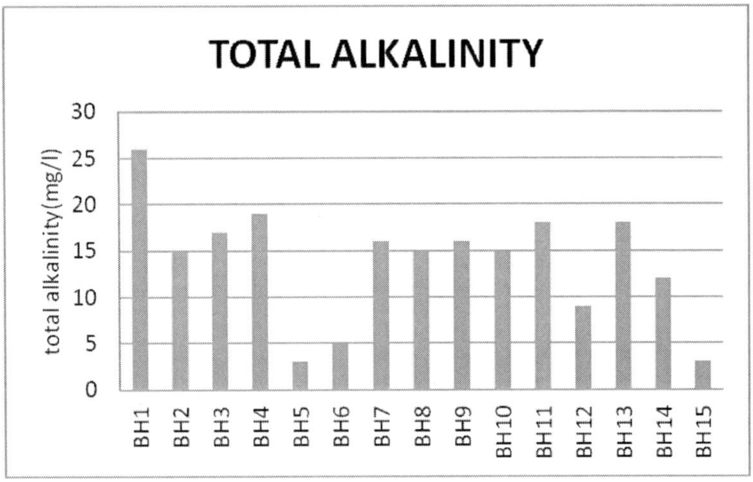

Figure 12. Bar chart showing variation of total alkalinity across boreholes

Total Hardness (TH)

The value of total hardness range from 15mg/l to 101mg/l (fig 13) with a mean value 42.40gm/l. Values were below the stipulated 150mg/l limit by WHO, as such, water sample is safe for drinking.

Calcium Content (Ca)

The value of calcium ranged from 6.75mg/l to 27.86mg/l (fig 14) with a mean value of 13.8mg/l. Values were below the WHO (2006) limit of 75mg/l for safe drinking water.

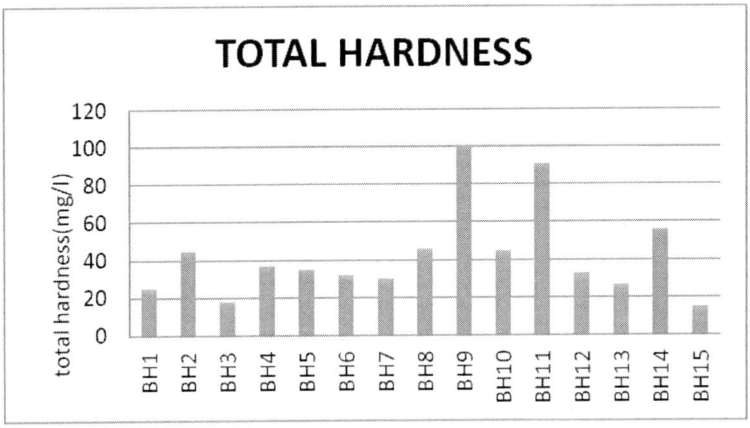

Figure 13. Bar chart showing variation of total hardness across boreholes

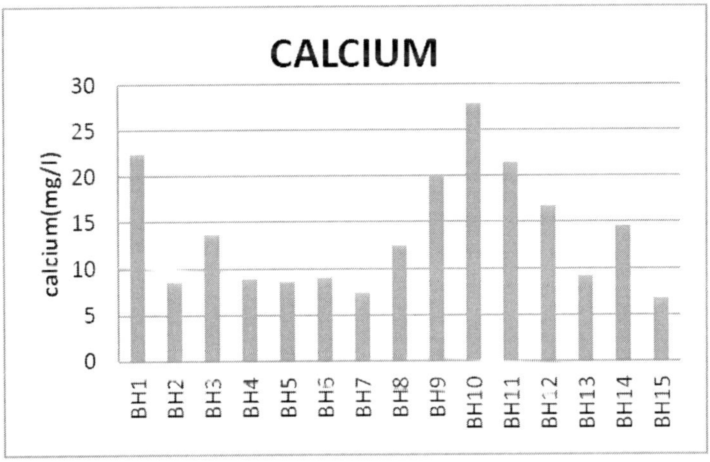

Figure 14. Bar chart showing variation of calcium across boreholeBar chart showing variation of calcium across boreholes

Magnesium (Mg)

The Magnesium value ranged from 1.76mg/l to 7.5mg/l (fig 15) with an average value of 3.85mg/l, all samples were above the stipulated value of 50 mg/l by WHO for water drinking.

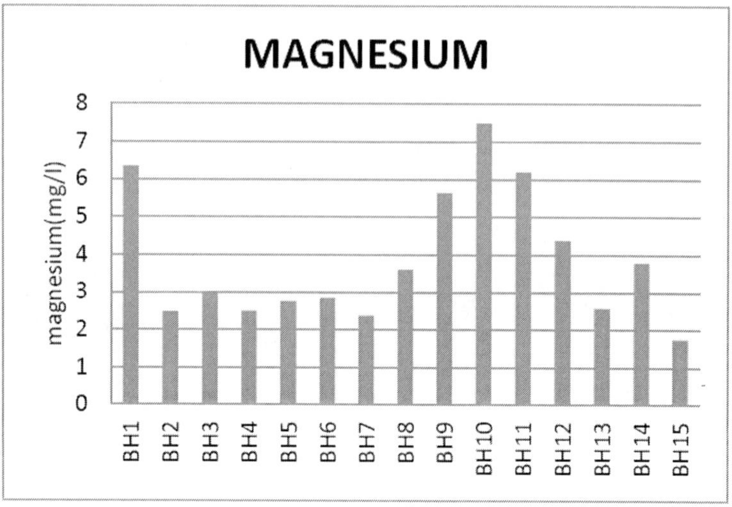

Figure 15. Bar chart showing variation of magnesium across boreholesBar chart showing variation of magnesium across boreholes

Sodium (Na)

The value of Sodium ranged from 3.85mg/l to 13.58mg/l (fig 16) with an average value of 6.85mg/l. There is also no limit by WHO for this parameter. But excess sodium in water may be harmful.

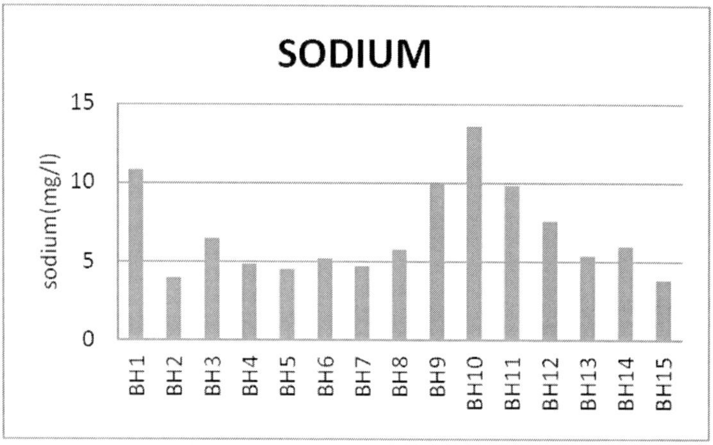

Figure 16. Bar chart showing variation of sodium across boreholes

Potassium (K)

The range of value for Potassium in the water sample analysis is 0.6mg/l to 4.65mg/l (fig 17). There is no guideline for potassium in drinking water by WHO.

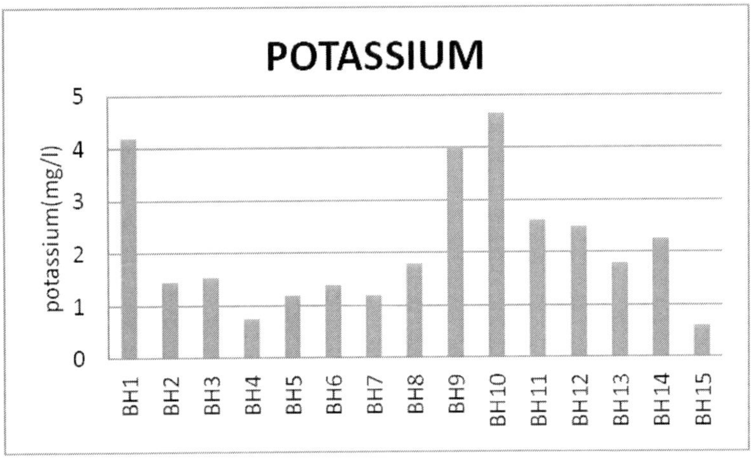

Figure 17. Bar chart showing variation of potassium across boreholes

Iron Content (Fe)

The content of iron in the water samples ranged from 0.06mg/l to 0.28mg/l (fig 18) with an average value 0.02mg/l. These values are below the standard value for iron in drinking water by WHO (2006), as such, the water is suitable for drinking and most industrial purposes with an exception of consumption by boilers, they feed water which requires than less then 0.01mg/l.

Manganese (Mn)

Values ranged from 0.0mg/l to 0.02mg/l (fig 19). These values are far below the stated value of 0.20mg/l for manganese (WHO, 2006) standard for drinking water, as such, water is safe for drinking.

A pie chart (fig 20) showed that the maximum cation found in the water samples was calcium (52%) and potassium is the lowest (8%) in the study area. Also, considering the concentration of anions, chloride was highest (88%), and the lowest was sulphate (4%). The concentrations of these major cations and anions falls within the allowable limits slated by WHO for portable water.

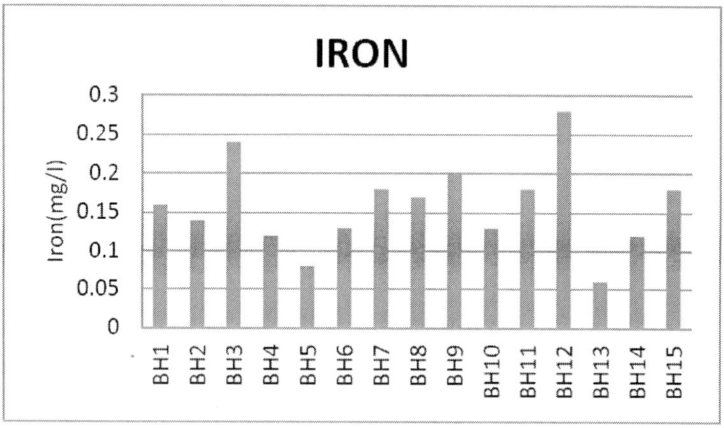

Figure 18. Bar chart showing variation of iron across boreholes

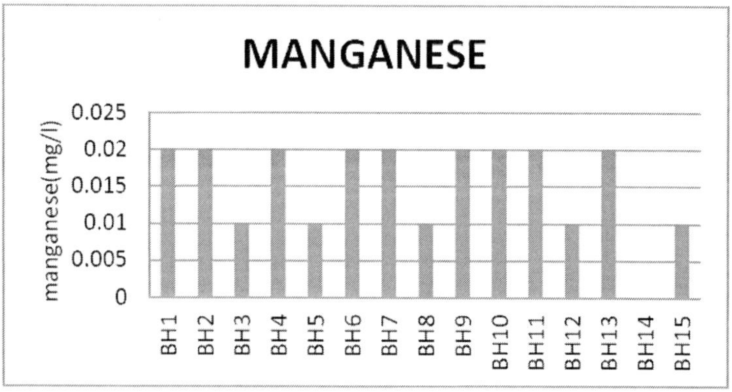

Figure 19. Bar chart showing variation of manganese across boreholes

5. PIPER TRILINEAR DIAGRAM

Trilinear plotting systems developed by Piper (1944) were used in the study of the water chemistry and quality. Piper diagrams are a combination of cation and anion triangles that lie on a common baseline. A diamond shape between them is used to replot the analysis as circles whose areas are proportional to their Total Dissolved Solids (TDS). The position of an analysis that is plotted on a Piper diagram can be used to make a tentative conclusion as to the origin of the water represented by the analysis. The diamond part of a piper diagram (fig 21) may be used to characterize different water types. Generally, groundwater can be divided into four basic types according to their placement near the four corners of the diamond. Water that plots at the top of the diamond is high in both Ca^{2+} + Mg^{2+} and Cl^{-} + SO_4^{2-}, which results in an area of permanent hardness. Water that plots near the left corner is rich in Ca^{2+} + Mg^{2+} and HCO_3^{-} and is the region of water of temporary hardness. Water plotted at the lower corner of the diamond is primarily composed of alkali carbonates (Na^{+}+ K^{+} and HCO_3+CO_3^{2-}). Water lying near the right-hand side of the diamond may be considered saline (Na^{+}+ $K^{+}$and Cl^{-} + SO_4^{2+}). The water types in the study area were thus designated according to the area in which they occur on the diagram segments as explained in the table (fig 22 and table 2).

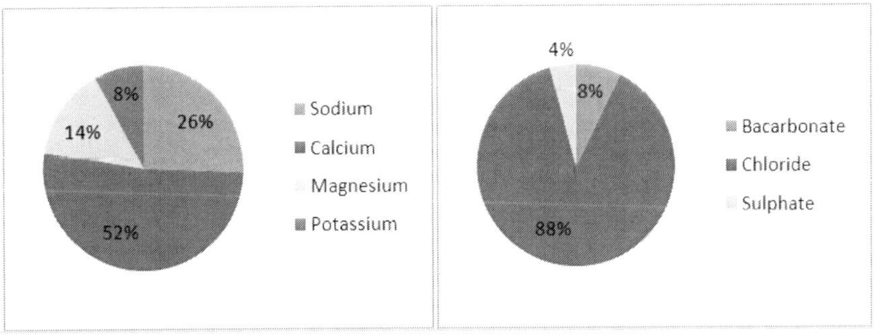

Figure 20. A pie chart showing the mean concentration of major cations and anions in the area

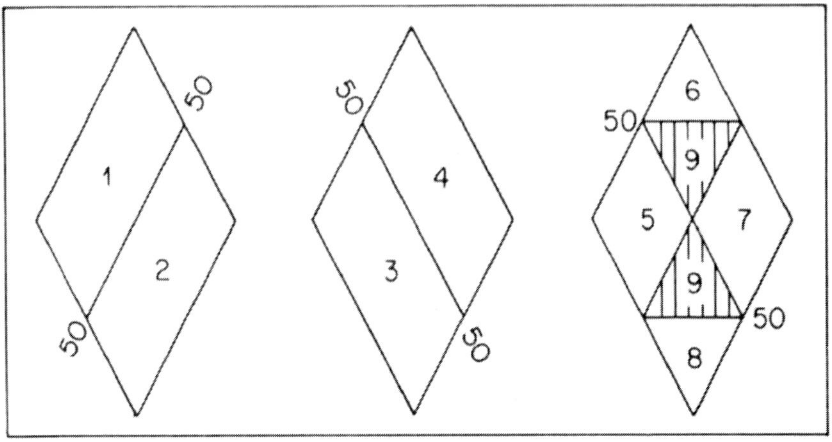

Figure 21. Subdivisions of the diamond-shaped field of the Piper diagram

Table 2. Characteristics of borehole water in Yenagoa on the basis of Piper trilinear diagram

Subdivision of the diamond	Characteristics of the corresponding subdivisions of diamond shaped field	No. of Borehole within the subdivisions of the diamond field
1	Alkaline earth (Ca+Mg)Exceed alkalis (Na+K)	ALL(1-15)
2	Alkalis exceed Alkaline Earths	NIL
3	Weak Acids (CO_3+HCO_3) exceed strong Acids (SO_4+Cl)	NIL
4	Strong acids exceeds weak acids	ALL(1-15)
5	Magnesium bicarbonate type	NIL
6	Calcium-chloride type	ALL(1-15)
7	Sodium-chloride type	NIL
8	Sodium-bicarbonate type	NIL
9	Mixed type (no cation-anion exceed 50%)	NIL

6. CONCLUSIONS

Variations observed in the chemistry of water samples drawn from various boreholes may be attributed to various processes both natural and artificial, e.g., the chemistry of the aquifer material, the rate of groundwater flow, permeability of the aquifer, land use for irrigation and waste disposal e.t.c. Analysis showed that all parameters from all the borehole samples were within the WHO limits for potable water except p^H and magnesium. p^H in all BH samples showed water was slightly acidic, also magnesium was above the stipulated limit in all BH samples when compared with WHO (2006). The water type by Piper's trilinear plot was characterized by calcium-chloride type hydro-chemical facie. Water hardness is caused primarily by the presence of cations such as calcium and magnesium and anions such as carbonate, bicarbonate and chlorides, this implied water in the area is relatively hard.

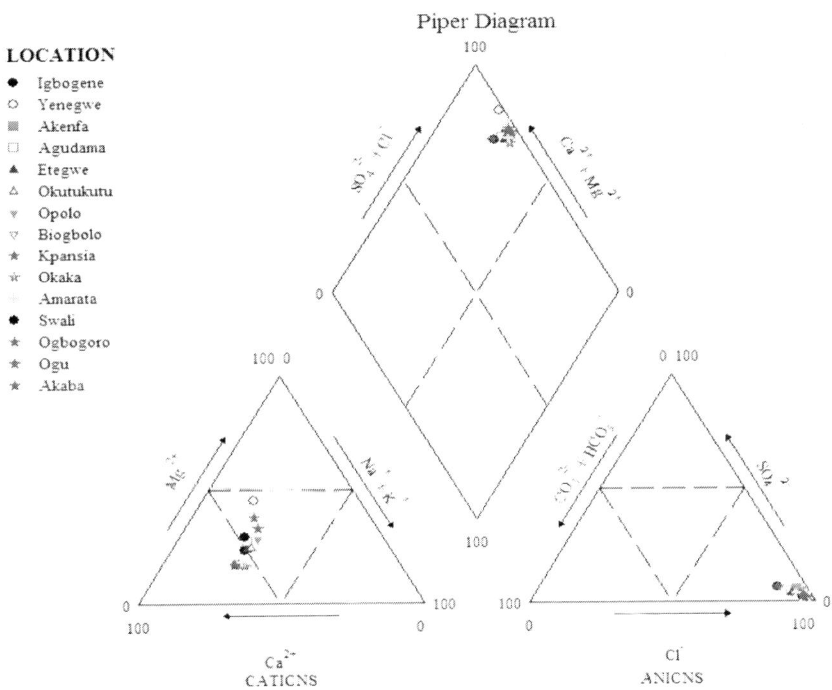

Figure 22

Water analysis data should be collected continuously in the study area, the data would aid the building of hydro-geochemical models, which would in turn be used to make informed predictions about changes in groundwater quality. In areas where the chemistry of water does not fall within the permissible limit for potable water set by the World Health Organization (WHO) treatment was recommended.

REFERENCES

1. Ayoade, J. O. Tropical Hydrology and Water Resources. Macmillan Ltd. 2003; p. 276.

2. Edet, A.E. Groundwater Quality Assessment in parts of Eastern Niger Delta. Nigeria. Environ. Geol. 1993; 22(1): 41-46.

3. Etu-Efeotor, J.O. Preliminary Hydrogeochemical investigations of Sub-Surface waters in parts of the Niger Delta Nig. J. Min. Geol. 1981;18(1): 103-105.

4. Etu-Efeotor, J. O and Akpokogje. E. G. Aquifer systems of the Niger Delta. Journal of Mining and Geology.1990; 26 (2): 279 – 285.

5. Offodile, M. E. Groundwater study and development in Nigeria. Mecon Engineering Services Ltd, Jos, Nigeria. 2002; Pp 239 – 345.

6. Piper, A. M. A graphical procedure in the geochemical interpretation of water analysis. American Geophysical Union Transactions. 1944; 25: 914 – 923.

7. Tremblay, J. J., D Cruz, A and Auger, H. Salt water Intrusion in the Summerside Area, P. E. I. Groundwater. 1973;11: 42 – 49.

8. Udom, G.J. Quality status of groundwater in Yenagoa and its environs, Bayelsa State, Nigeria. J. Sci. Ind. Stud. 2006;4(1): 45-51.

9. Ufoegbune, G. C., Lamidi, K. I., Awomeso, J. A, Eruola, A. O., Eruola, A. O., Idowu, O. A and Adeofun, C. O. Hydro-geological characteristics and groundwater quality assessment in some selected communities of Abeokuta, Southwest Nigeria. Journal of Environmental Chemistry and Ecotoxicology. 2009; 1:010-022.

10. World Health Organization. Guidelines for Drinking Water Quality: Incorporating 1st and 2nd Addlenda. Vol.1 Recommendations 3rd Edition Geneva. 2006.

7

CHEMISTRY OF SUPRAGLACIAL PONDS IN THE DEBRIS-COVERED AREA OF LIRUNG GLACIER IN CENTRAL NEPAL HIMALAYAS

Maya P. Bhatt 1,3 • Nozomu Takeuchi 2 • Miguel F. Acevedo 3

1 Institute for the Biogeochemistry and Marine Chemistry, KlimaCampus - University of Hamburg, Bundesstrasse 55, 20146 Hamburg, Germany

2 Department of Earth Sciences, Chiba University, Chiba 263-5822, Japan

3 Department of Electrical Engineering, Institute of Applied Sciences, University of North Texas, Denton, TX 76203-5017, USA

ABSTRACT

Glacial meltwater samples were collected from 15 supraglacial ponds at elevations ranging between 3989 and 4292 m above sea level in the debris-covered area of Lirung glacier in central Nepal Himalayas, from November 2010 to October 2011 on a bimonthly basis. We examined elevation trends, and their control by seasonality, dominant geochemical processes, and seasonal dynamics in these supraglacial ponds. Concentration of some parameters showed a decreasing trend with elevation but with high variability and modulation by seasons, being more prevalent in the post-monsoon. Sulfate is the only parameter showing significant elevation trends regardless of the season, although the trend is also affected by the monsoon. The concentrations of major chemical solutes were highest during post-monsoon season and lower in the pre-monsoon and monsoon seasons, indicating influence of seasonality on dissolution, dilution, and hydrological processes. Contribution to chemical loads from marine aerosols to these pond water appeared to be minor except for sodium and magnesium. Chemical weathering of carbonate and silicate minerals fueled by dissolution of pyrite appeared as the dominant geochemical processes.

KEYWORDS

Supraglacial ponds, Weathering, Central Himalaya, Sulfide minerals, Debris covered, Lirung glacier

1 INTRODUCTION

1.1 Background

Water from ice melt from the Himalayas constitutes the major water resources in Nepal; most of the major rivers of Nepal, including Karnali, Koshi, and Narayani, originate from glaciers in the Himalayas. For example, the headwater of Narayani is located at Langtang glacier in central Nepal. Himalayan glaciers are melting rapidly due to continued rising of Earth's temperature causing hydrological changes in the river headwater. Particularly in the debris-covered area of valley glaciers, a number of supraglacial ponds or lakes are formed by ice melt and sometimes collapse due to rock and moraine falls to the ponds or lakes. Many glacial lakes are reported in the debris area of the glaciers around the world as well as in Himalayas (Lliboutry et al. 1977; Kirkbride 1993; Takeuchi and Kohshima 2000; Sakai et al. 1998; Bhatt et al. 2007; Sakai and Fujita 2010; Sakai 2012; Benn et al. 2012; Takeuchi et al. 2012).

Supraglacial ponds on Himalayan glaciers are increasing in size; for example, the Tsho Rolpa glacier lake in the Gaurishankar Himalaya in Nepal has increased its size more than sevenfold in the last four decades (WWF-Nepal 2005). Rise in temperature increases the probability of glacial lake outburst flood (GLOF) events in the region, in turn, causing catastrophic consequences in downstream communities (Yamada and Sharma 1993). These events could result in loss of lives, natural resources, and infrastructure and property worth millions of dollars (Yamada and Sharma 1993; Richardson and Reynolds 2000; ICIMOD-

UNEP 2002; Kattelmann 2003). There are 2323 glacier lakes in Nepal of which 20 are potentially dangerous, and 12 GLOF events have already been observed in Nepal since 1960 (ICIMOD 2011).

Thick debris cover resists the ablation process (Sakai et al. 2000), and the occurrence of high turbidity (suspended sediments) in supraglacial ponds within debris-covered area plays a great role in glacier melting (Takeuchi et al. 2012). Debris-covered glaciers exhibit more retreat than the other glaciers in the central Himalaya region; this is probably due to the presence of thin debris and highly turbid ponds as they absorb more heat and accelerate the ablation processes within the debris area (Sakai et al. 2000; Takeuchi et al.2012).

Based on observations during 2000–2008, Scherler et al. (2011) reported that the majority of glaciers within Hindu Kush Himalayan region are retreating. In the worlds' glaciers, retreating rate varies with the landscape position and the actual rates and mechanisms are still subject to debate. Low-elevation glaciers in central Himalaya are retreating with a rapid rate probably due to deposition of black carbon based on the visual observation. However, the contribution to black carbon seems insignificant in melting most glaciers in high Himalayan glaciers (Ming et al. 2013).

The first denudation studies from Nepal Himalaya reported high rates of chemical erosion (Bhatt et al. 2000). At the outlet point of Lirung debris-covered glacier in central Nepal Himalaya, chemical erosion rate is higher than those reported for the Antarctic, Arctic, and European glaciers (Sharp et al. 1995; Wadham et al. 1998; Bhatt et al. 2000; Green et al. 2005). Debris may have played a great role in the release of solute from this catchment. Uplift and erosion, tectonic recycling of sediment, drives high denudation rate in many active mountain belts (Stallard 2000; Stallard and Edmond 1983; Gaillardet et al. 1999; West et al. 2005; Bhatt and McDowell 2007; Gabet 2007) together with climatic conditions of that basin (McDowell and Asbury 1994; White and Blum 1995; Hartmann 2009; Bookhagen 2010; Gabet et al. 2010; Hirschmiller et al. 2014). Air temperature and precipitation control glacier melting and hydrolysis conditions of glaciered basins, thus modulating dissolution rates of the minerals of such basins according to the seasons (Bhatt et al. 2000,2007). Water chemistry and dominance of weathering processes in a basin primarily depend on the underlying bedrocks and soil types of the basin.

Chemistry of glacial meltwater, supraglacial ponds, glacier lakes, and ice core studies have been documented from the Himalayan region of Nepal

(Kamiyama 1984; Watanabe et al. 1984; Jenkins et al.1995; Reynolds et al. 1995; West et al. 2002; France-Lanord et al. 2003; Quade et al. 2003; Evans et al.2004; Tipper et al. 2006; Barnard et al. 2006; Bhatt et al. 2000, 2007, 2008, 2009; Wolff-Boenisch et al.2009; Lupker et al. 2012), yet it is important to attain a better understanding of the spatial and temporal variability of water chemistry and its relationship with weathering processes in this region.

Bhatt et al. (2007) conclude that the dominant processes determining the chemical composition of glacier pond water within debris area of Lirung glacier are sulfide oxidation coupled with carbonate dissolution and chemical weathering of aluminosilicate.

The oxidation of pyrite can be written as (Drever 1988)

$$FeS_2 + 3.75O_2 + 3.5H_2O \rightarrow Fe(OH)_3Precip + 2SO_4^{2-} + 4H^+$$

Carbonate and dolomite dissolution takes place as follows:

$$CaCO_3 + CO_2(aq) + H_2O \rightarrow Ca^{2+} + 2HCO_3^-$$

$$CaMg(CO_3)_2 + CO_2(aq) + 2H_2O \rightarrow Ca^{2+} + Mg^{2+} + 4HCO_3^-$$

The major aluminosilicate dissolution reactions taking place within debris area of Lirung glacier constructed based on the pioneering work of Garrels and Mackenzie (1967) for Na-feldspar, Ca-feldspar, K-feldspar, and biotite are as follows:

$$2NaAlSi_3O_8 + 2CO_2 + 3H_2O \rightarrow Al_2Si_2O_5(OH)_4 + 2Na^+ + 2HCO_3^- + 4SiO_2$$

$$CaAl_2Si_2O_8 + 2CO_2 + 4H_2O \rightarrow Al(OH)_3 + 2Ca^{2+} + 2HCO_3^- + 2SiO_2$$

$$KAlSi_3O_8 + 2CO_2 + 3H_2O \rightarrow Al_2Si_2O_5(OH)_4 + 2K^+ + 2HCO_3^- + 4SiO_2$$

$$KMg_3AlSi_3O_{10}(OH)_2 + 7CO_2 + 3.5H_2O \rightarrow Al_2Si_2O_5(OH)_4 + 3Mg^{2+} + K^+ + 7HCO_3^- + 2SiO_2$$

In Bhatt et al. (2007), we studied water from supraglacial ponds in the debris-covered area of Lirung glacier during the pre-monsoon to post-monsoon period of 1996 and compared to the water at the glacier outlet. This debris-covered area is a very dynamic environment, and the pond sizes and positions change based on weather, landslides, and avalanches. Therefore, we choose to research the same area 15 years later (2010–2011) and measure water quality parameters bimonthly during a full year in order to better understand the seasonality of geochemical processes. In addition, we included measurements of inorganic carbon at the end of the study because of lack of reports of variation in inorganic carbon species from this region.

There were many changes observed regarding size and position of lakes within the basin. In this paper, we report our findings of this second sample collection and the data represent now a unique picture of a system changed after the major earthquake of April 2015; landslides and avalanches wiped out the Langtang village and severely destroyed the base camp at Kyangjin, and hundreds of people lost their lives in the region. The size and position of lakes drastically changed after the earthquake within the debris area.

1.2 Objectives and Hypotheses

The main objectives of this new study are to confirm our previous results regarding dominant geochemical processes and to further examine the variations among ponds and the effects of seasonality on the geochemical processes in this debris-covered glacier system. Our major research question is how does seasonality affects the hydrological regime of this system and elevation patterns of water constituents among the ponds. To study this question, we propose four hypotheses:

- *Research Hypothesis 1* There are elevation gradients for some water quality parameters in the ponds showing increasing concentration with decreasing elevation. The corresponding null hypothesis H0 to test is that there is no trend in elevation for all of the solutes.

- *Research Hypothesis 2* These elevation gradients would be more clearly marked during the pre-monsoon and post-monsoon seasons, and become more variable during the monsoon when the flows are higher. The corresponding null hypothesis H0 to test is that elevation trends are the same regardless of season.

- *Research Hypothesis 3* The main source of sulfate ions is from the oxidation of pyrite which fuels the dissolution process of minerals such as carbonates (providing calcium) and aluminosilicates (providing magnesium). The corresponding null hypothesis H0 to test is that there is no relationship between sulfate and base cations calcium and magnesium.

- *Research Hypothesis 4* There is a characteristic temporal pattern showing minimum values at the pre-monsoon, followed by increasing concentrations in the monsoon due to runoff, and a maximum at the post-monsoon when water freezes. Subsequent melting later in the post-monsoon leads to decrease in concentration to the pre-monsoon. The corresponding null hypothesis H0 to test is that there is no difference in values between post-monsoon and pre-monsoon for all water parameter.

2 STUDY AREA

2.1 Site Description

The Langtang mountain range (28°13′01.9″N, 85°33′42.0″E) is located in the central Nepal Himalayas (Fig. 1). The Lirung glacier is a debris-covered glacier in the Langtang valley and about 60 km north of the capital city Kathmandu. The highest point in this range is a peak of Mt. Langtang Lirung, 7234 m a.s.l. The study area was the debris-covered area on Lirung glacier ranging between 3989 and 4292 m a.s.l. in elevation (Fig. 1). The total basin area of the Lirung glacier is 13.8 km^2 of which 33 % has high relief and is covered with bare rocks, 16 % correspond to debris-covered glaciers (nonvegetated and covered with rocks, boulders, cobbles, and talus with thin soils), and the rest (51 %) is glacier covered (Fukusima et al. 1987). Human influence is more intense in lower-elevation sites, but limited to seasonal settlement and livestock grazing.

2.2 Geologic Setting

The geology of the Himalayan basin consists of four units from north to south separated by major thrust systems: the Tethyan Sedimentary Series (TSS), the High Himalaya Crystalline (HHC), the Lesser Himalaya (LH), and the Siwaliks (Chabaux et al. 2001; France-Lanord et al. 2003). The Langtang valley lies in a complex transition zone between the HHC metasediments in the south and TSS in the north. The Langtang valley is primarily underlain by high-grade metamorphic rocks with traces of igneous rocks including migmatities, gneisses, schists, phyllites, and granites (Inger and Harris 1992). The glacier bedrock consists of biotite, quartz, plagioclase with minor amount of muscovite, alkali feldspar, ilmenite, and sillimanite based on the X-ray fluorescence analysis of rock samples from the debris area of Lirung glacier (Bhatt et al. 2008). The Cryothents, cryumbrepts, and lithic types of soils are dominant in the high-elevation area of Langtang valley (SD 1984). The presence of sulfide-bearing minerals such as pyrite, galena, sphalerite, and chalcopyrite enhances the dissolution rates of minerals within the basin (Bhatt et al. 2007). The gray phyllites and grilstones with conglomerates with white massive quartzites, basic intrusion, gray calcareous slates and carbonates, and thick beds of gray siliceous dolomites are found in the high mountain region of central Nepal Himalaya (DMG 1980, 1994).

3 MATERIALS AND METHODS

3.1 Sample Collection

Glacier meltwater sampling was carried out at the 15 supraglacial ponds in the debris-covered area from fall 2010 to fall 2011 (Fig. 1; Table 1). Ponds P1 through P13 were measured on a bimonthly basis. Two additional ponds, P14 and P15, located at higher elevation were measured in October 2011. Based on visual observation of turbidity, three types of ponds were found within the debris area: blue type—very clear water, yellow type—slightly turbid water, and green type—highly turbid water with ice walls. We assessed color of each pond qualitatively based on combination of these three basic water colors (Table 1).

Each water sample was filtered through a pre-combusted glass microfiber filter (Whatmann GF/F) with a pore size 0.45 μm in the field for the dissolved organic carbon (DOC) samples and polycarbonate microfiber filter with a pore size of 0.45 μm for the major ions and dissolved silica analysis. Supraglacial pond water samples were collected in 100 mL acid-washed polyethylene bottles and kept refrigerated in Kathmandu and then frozen to transport to the Institute of Biogeochemistry and Marine Chemistry of the University of Hamburg for analysis. Dissolved silica samples were refrigerated only. The DOC samples were taken in 30 mL glass vials in which 30 μL of phosphoric acid was added just after the filtration in the field. The alkalinity samples were collected in 200 mL polyethylene bottles in which 200 μL of mercuric chloride was added just after sampling to prevent microbial activity. Pre-weighted filter papers were used to measure the suspended sediment (SS) in each sample.

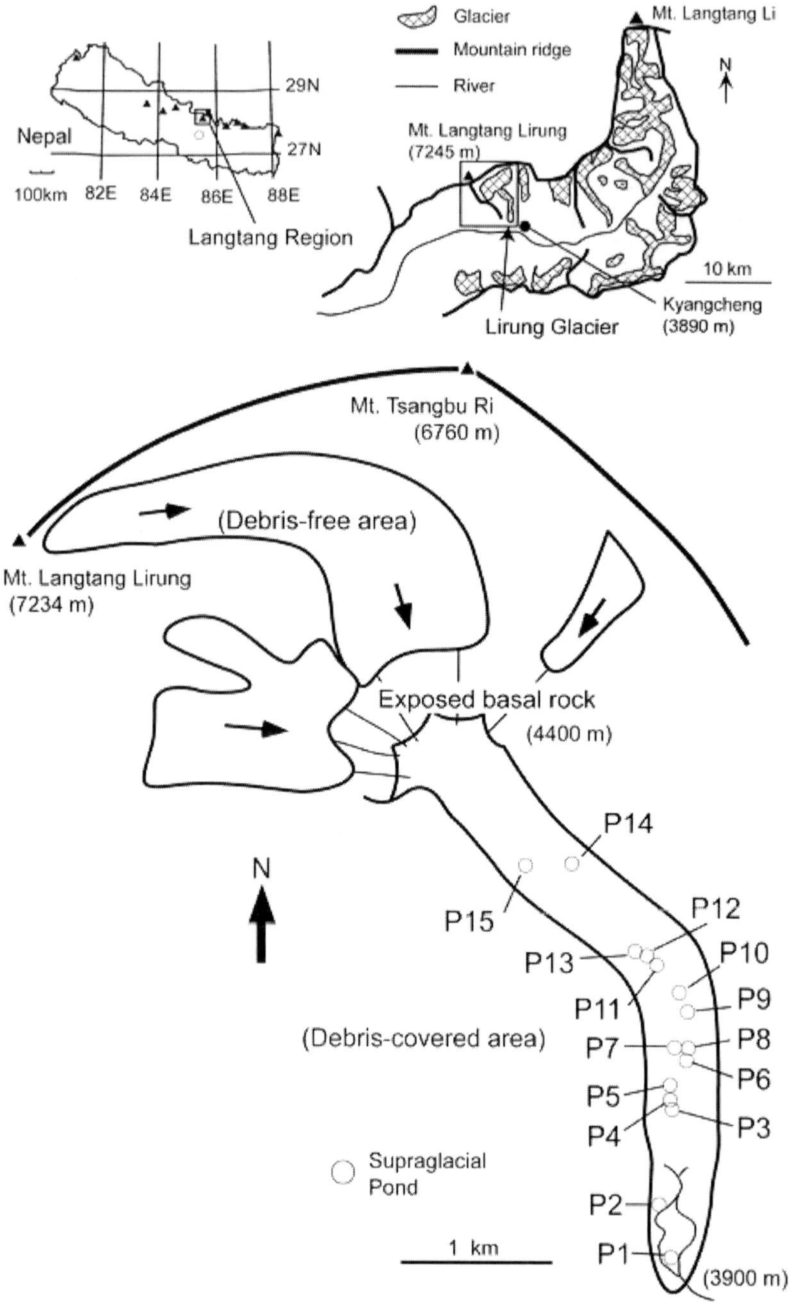

Figure 1. Location of Lirung glacier in central Nepal Himalaya with sampling sites of supraglacial ponds in debris-covered area of the glacier

Table 1. Description of supraglacial ponds within Lirung debris area in Langtang valley in central Nepal Himalaya

Sample ID	Color	Site	Information	Latitude	Longitude	Elevation (m)	pH	Water Temp (°C)	Air Temp (°C)	EC (μS cm^{-1})
P-1	Yellow	LHS	Outlet flow	28°13′1[illegible].1″	85°33′40.5″	3986	7.48	8.5	10.5	138
P-2	Blue	LHS	Isolated	28°13′16.9″	85°33′41.8″	3987	7.54	9.5	12.1	126
P-3	Blue–green	LHS	Frozen	28°13′29.9″	85°33′43.5″	4006	8.07	6.2	12.7	106
P-4	Light blue	LHS	Isolated	28°13′31.2″	85°33′43.6″	4012	7.61	2.5	12.5	130
P-5	Light blue	LHS	Flow water	28°13′32.0″	85°33′43.9″	4012	7.74	2.8	13.8	139
P-6	Light blue	RHS	Isolated/rocks	28°13′40.3″	85°33′42.8″	4029	7.78	2.8	14.1	121
P-7	Yellow	Central	Rock flow	28°13′42.8″	85°33′44.0″	4032	7.6	3.3	16.2	139
P-8	Blue–green	RHS	Stones/ice	28°13′42.1″	85°33′47.4″	4068	nd	2.3	16.3	nd
P-9	Green	Central	Rock/no debris	28°13′52.1″	85°33′47.3″	4087	8.24	6	11.6	45
P-10	Green	RHS	Ice cliffs	28°13′56.5″	85°33′46.0″	4102	8.13	2.7	14.8	30
P-11	Yellow	RHS	Turbid	28°14′02.9″	85°33′42.6″	4133	7.69	6.3	18.2	22
P-12	Green–yellow	RHS	Turbid/ice	28°14′03.3″	85°33′42.1″	4136	7.92	6.3	18.2	20
P-13	Yellow	LHS	Rocks turbid	28°14′03.6″	85°33′40.8″	4139	7.73	5.4	13.8	21
P-14	Green yellow	RHS	Ice cliffs	28°14′14.2″	85°33′20.4″	4292	7.56	7.5	14	nd
P-15	Green yellow	LHS	Ice cliffs	28°14′23.8″	85°33′16.2″	4292	7.78	7	14	43

LHS and RHS indicate left- and right-hand sides of the debris area, respectively. No data are indicated by nd. Physical–chemical parameters were measured only in October 2011 and are only indicative of fall conditions

In the field, water temperature, electrical conductivity (EC) and pH were measured by thermometer, EC meter (Hanna), and pH meter (Hanna), respectively, for all ponds in October 2011.

3.2 Analytical Methods

SS was measured by gravimetric methods applied to the 47-mm GF/F and polycarbonate microfiber filter after drying in a vacuum oven at 40 °C for 48 h. Alkalinity was determined by acid titration method (APHA1995). Detection limits for SS and alkalinity were 1 and 0.1 mg L^{-1}, respectively. Dissolved silica (SiO_2) was analyzed by DR 3800 spectrophotometer (HACH Company) with the standard molybdenum blue 8185 method.

Major cations (Na^+, K^+, Mg^{2+}, Ca^{2+}, and NH_4^+) and anions (F^-, Cl^-, NO_2^-, NO_3^-, and SO_4^{2-}) were determined by cation (Compact IC pro-cation) and anion (Compact IC pro-anion) chromatography, respectively. Anions were measured with a Metrohm self-regenerating suppressor MSM (50 mmol $L^{-1} H_2SO_4$) and analytical column Metrosep A SUPP 5-250. Cations were measured with a Metrohm 881 (C4) cation separation column 150 mm long and conductivity detection. Analytical errors were <2 % for Cl^-, NO_3^-, SO_4^{2-}, Na^+, K^+, Mg^{2+}, Ca^{2+}, and SiO_2, <4 % for NH_4^+, and <5 % for PO_4–P. Total dissolved nitrogen (TDN) and DOC were measured with high-temperature Pt-catalyzed combustion using a Shimadzu TOC-V_{CSH} with a total nitrogen measuring unit. Dissolved organic nitrogen (DON) was calculated by subtracting inorganic nitrogen from TDN. Marine aerosol contribution to different elements was corrected by using molar ratios of different elements as reported in earlier studies (Keene et al. 1986; McDowell et al. 1990; Millot et al. 2002; Bhatt and McDowell 2007). Marine aerosol contribution to pond waters was corrected for Na^+, K^+, Mg^{2+}, Ca^{2+}, and SO_4^{2-} by using concentration of chloride as reference species and the sea-salt-corrected chemical species represented with an asterisk.

3.3 Statistical Analysis

Relationship of concentration of chemical parameters with pond category such as color is examined by boxplots and ANOVA. To test the null for hypothesis 1, elevation patterns are examined by linear regression. These regressions are performed by season in order to test the null of hypothesis 2. Relationship of calcium and magnesium with sulfate was examined by linear regression to test the null of hypothesis 3. The average of major parameters over all ponds was calculated per month to emphasize temporal variations. Points at nonsampled months (December, February, April, June, and August) are linearly interpolated for each pond and then averaged over all ponds. This interpolation is for visualization purposes only. We did not include ponds 14 and 15 in the seasonal analysis because these ponds were measured only in October 2011. To test the null corresponding to hypothesis 4, we used paired *t* test of maxima and minima values in the year across ponds for each water parameter. Weathering patterns are visualized by ternary diagrams for anions and cations. All statistical analysis and graphics were performed by the R program version 3.2.1 (R Development Core Team 2015). Ternary diagrams were prepared using the ggtern package of R.

4 RESULTS

4.1 Overview and Composition of Major Ions

In October 2011, water temperature of observed ponds was 4.73 °C in average and ranged from 1.6 to 7.85 °C; such variation was due to the difference in sampling time and ice presence among the ponds. Air temperature ranged from 7.0 to 18.2 °C, and variation is linked with the difference in sampling time. SS for those months averaged 43 mg L^{-1} and ranged from 16 to 141 mg L^{-1}. SS was much higher at pond P1 which is close to the Lirung outlet point (which collects the main water flow of the debris-covered area) and much lower in ponds with ice cliffs and the ponds in higher elevation sites (Table 2).

Composition of major cations was $Ca^{2+} \gg Na^{+} > K^{+} > NH_4^{+} > Mg^{2+}$; however, concentration varies widely due to specific site conditions (Table 2). The major anion compositions were alkalinity $> SO_4^{2-} \gg Cl^{-} > NO_3^{-} \gg F^{-}$. The charge balance error (CBE) was less than 8 % for samples P1 to P6 in low-elevation sites, but it was less than 13.7 % for all samples within the debris area. The cause for this discrepancy is not clear, but we assume it may be due to unmeasured organic acids.

Table 2 Annual average chemical compositions of supraglacial ponds within Lirung debris area in Langtang valley in central Nepal Himalaya during fall 2010 to winter 2011

Pond ID	SS (mg L^{-1})	DOC (mg L^{-1})	TDN (mg L^{-1})	Alk (mg L^{-1})	F (mg L^{-1})	Cl (mg L^{-1})	NO_3 (mg L^{-1})
P-1	141 ± 249	4.1 ± 6.6	0.4 ± 0.1	13.4 ± 8.8	0.2 ± 0.3	1.7 ± 3.8	0.4 ± 0.4
P-2	31 ± 35	1.2 ± 0.3	0.3 ± 0.1	11.7 ± 7.1	0.10 ± 0.04	0.4 ± 0.3	0.3 ± 0.4
P-3	55 ± 66	1.3 ± 0.3	0.4 ± 0.3	11.7 ± 9.4	0.08 ± 0.04	1.7 ± 3.9	0.6 ± 0.8
P-4	24 ± 31	1.3 ± 0.4	0.4 ± 0.2	10.3 ± 5.7	0.09 ± 0.04	0.3 ± 0.2	0.3 ± 0.5
P-5	31 ± 25	1.6 ± 0.9	0.5 ± 0.2	10.0 ± 9.2	0.07 ± 0.03	3.5 ± 7.5	0.3 ± 0.3
P-6	19 ± 11	1.5 ± 0.7	0.4 ± 0.1	7.7 ± 6.4	0.09 ± 0.05	1 ± 2	0.35 ± 0.44
P-7	23 ± 18	1.6 ± 1.0	0.3 ± 0.1	8.0 ± 4.6	0.11 ± 0.06	0.3 ± 0.1	0.3 ± 0.5
P-8	17 ± 8	2.3 ± 1.9	0.4 ± 0.3	9.0 ± 6.0	0.09 ± 0.05	0.5 ± 0.3	0.4 ± 0.5
P-9	25 ± 17	1.5 ± 1.2	0.2 ± 0.1	5.7 ± 2.7	0.07 ± 0.03	0.3 ± 0.3	0.24 ± 0.18
P-10	95 ± 176	2.2 ± 1.5	0.7 ± 0.6	10.7 ± 5.9	0.08 ± 0.06	0.3 ± 0.3	1.2 ± 1.4
P-11	32 ± 28	1.6 ± 1.1	0.3 ± 0.2	5.0 ± 2.8	0.07 ± 0.04	0.5 ± 0.6	1.0 ± 1.8
P-12	60 ± 106	1.5 ± 0.9	0.4 ± 0.2	7.0 ± 2.4	0.09 ± 0.05	0.6 ± 0.7	0.5 ± 0.7
P-13	50 ± 64	2.9 ± 3.6	0.3 ± 0.3	6.7 ± 2.7	0.08 ± 0.03	0.3 ± 0.3	0.8 ± 1.3
P-14	16	2.34	0.280	2	0.03	0.3	nd
P-15	27	1.44	0.241	2	0.02	0.2	nd

Pond ID	SO_4 (mg L^{-1})	NH_4 (mg L^{-1})	Na (mg L^{-1})	K (mg L^{-1})	Mg (mg L^{-1})	Ca (mg L^{-1})	Si (mg L^{-1})
P-1	6.6 ± 3.7	0.8 ± 1.2	2.7 ± 3.6	2.1 ± 1.6	0.5 ± 0.2	8.1 ± 3.8	1.1 ± 0.4
P-2	7.2 ± 6.2	0.7 ± 1.0	1.8 ± 1.9	1.7 ± 0.7	0.7 ± 0.3	8.7 ± 6	0.9 ± 0.4
P-3	3.1 ± 2.6	1.4 ± 1.4	2.5 ± 3.8	1.8 ± 1.8	0.4 ± 0.1	6.1 ± 2.2	0.8 ± 0.6
P-4	5.4 ± 4.8	0.9 ± 0.8	1.6 ± 1.5	1.3 ± 0.7	0.5 ± 0.2	8.8 ± 5	1.1 ± 0.4
P-5	2.9 ± 3.3	0.4 ± 0.2	3.7 ± 6.2	2.9 ± 3.4	0.4 ± 0.3	7.7 ± 8.5	0.6 ± 0.5
P-6	5.3 ± 6.1	1.1 ± 1.0	2.2 ± 2.6	1.5 ± 1.2	0.6 ± 0.3	6.6 ± 4.8	0.7 ± 0.5
P-7	5.5 ± 6.0	0.16 ± 0.18	1.6 ± 1.6	1.7 ± 1.2	0.7 ± 0.5	10.7 ± 8.6	1.1 ± 0.6
P-8	2.5 ± 2.9	0.13 ± 0.11	1.6 ± 1.5	1.2 ± 0.6	0.4 ± 0.2	6.7 ± 4.3	0.9 ± 0.4

P-9	1.5 ± 0.8	0.2 ± 0.3	1.1 ± 1.4	1.1 ± 0.4	0.2 ± 0.1	4.2 ± 3.1	0.7 ± 0.1
P-10	1.9 ± 2.1	0.12	1.3 ± 1.4	1.2 ± 1.0	0.2 ± 0.2	4.9 ± 3	0.7 ± 0.5
P-11	2.4 ± 3.5	0.06	1.6 ± 1.9	1.4 ± 1.2	0.7 ± 0.5	5.3 ± 5	0.8 ± 0.6
P-12	4.3 ± 3.5	0.08 ± 0.05	1.6 ± 1.3	2.1 ± 1.7	0.6 ± 0.3	6.8 ± 5.4	0.8 ± 0.4
P-13	2.8 ± 2.3	nd	1.5 ± 1.3	1.2 ± 0.7	0.4 ± 0.3	4.9 ± 2.3	1.1 ± 1.0
P-14	0.3	nd	0.13	0.66	nd	0.75	0.31
P-15	0.4	nd	0.25	0.46	nd	1.13	0.28

Ponds P1 through P13 were measured on a bimonthly basis and averaged (the standard deviation is based on $n = 7$ months for each pond). Ponds P14–P15 located at higher elevation were measured only in October 2011 ($n = 1$ month for each pond). No data are indicated by nd. Number of observations per parameter is $7 \times 13 + 2 = 93$

Average concentrations of $^*Na$, $^*K$, $^*Mg$, $^*Ca$, and $^*SO_4$ after sea-salt correction and elemental ratios are presented in Table 3. The contribution to concentrations of $^*Na$, $^*K$, $^*Mg$, $^*Ca$, and $^*SO_4$ from rock weathering appeared as 75.8, 99.0, 90.2, 99.7, and 98.5 % to their total concentrations, respectively. This result suggests that lithology has dominant control on the chemistry of pond waters, while contribution to Na and Mg from sea-salt aerosols is minor. Contribution to marine aerosols varied with the specific site locations of the ponds and showed highest contribution in pond P5 and lowest contribution in pond P7 showing influence of landscape position. Calcium appeared as a dominant ion among all measured chemical species and released mostly (99.7 %) from rock and soil dissolution. The sum of sodium and potassium after sea-salt correction contributes 32 %, and the sum of calcium and magnesium after sea-salt correction contributes 68 % to the total sum of cations among all the ponds.

Phosphate concentration was not detected in several ponds. Fluoride contributes 0.7 % to total sum of anions. Average nitrate concentration was 0.52 mg L^{-1} and contributed 4 % to the total sum of anions. Sulfate concentration after sea-salt correction appeared as a dominant anion with average concentration of 3.4 mg L^{-1} and ranged 0.25–7.21 mg L^{-1}. Sulfate contributes 25 % to total sum of anions. Average alkalinity was 7.4 mg L^{-1} with the range 2–13.4 mg L^{-1} and contributes 64.6 % to total sum of anions.

4.2 Hypothesis 1: Elevation Trends

In general, color varied with elevation (ANOVA p value = 0.017), progressing from blue to green and then green to yellow, except yellow that occurs from low to middle elevation (Fig. 2a). The pH of the ponds (measured in October 2011) was basic, showing a significant variation with pond color being highest for blue–green and green ponds (Table 1; Fig. 2b).

Boxplots variation in pond elevation and some water parameters measured in October 2011 with pond color. *Scatter plots* some water parameters measured in October 2011 versus elevation

Table 3 Annual average chemical compositions of major chemical species and elemental ratios after sea-salt correction in supraglacial ponds within Lirung debris area in Langtang valley in central Nepal Himalaya during fall 2010 to winter 2011

Sample ID	*SO4 (mg L^{-1})	*Na (mg L^{-1})	*K (mg L^{-1})	*Mg (mg L^{-1})	*Ca (mg L^{-1})	*Sbcat (mg L^{-1})	*Ca/*Na (molar)	*Mg/Na (molar)	*K/*Na (molar)	*SO_4/*Na (molar)	Si/(*Na + *K) (meq/meq)
P-1	6.48 ± 3.85	1.80 ± 1.76	2.05 ± 1.53	0.42 ± 0.36	8.02 ± 3.79	12.28 ± 1.42	2.56 ± 1.24	0.22 ± 0.19	0.67 ± 0.51	2.59 ± 1.57	1.15 ± 0.54
P-2	7.21 ± 6.21	1.57 ± 1.75	1.70 ± 0.77	0.49 ± 0.44	8.65 ± 6.04	12.42 ± 2.59	3.16 ± 1.98	0.30 ± 0.24	0.64 ± 0.24	3.30 ± 2.54	1.12 ± 0.70
P-3	2.98 ± 2.68	1.50 ± 1.71	1.79 ± 1.70	0.13 ± 0.28	6.05 ± 2.14	9.47 ± 0.82	2.32 ± 0.72	0.08 ± 0.15	0.70 ± 0.58	1.43 ± 1.12	1.05 ± 0.75
P-4	5.41 ± 4.85	1.40 ± 1.39	1.31 ± 0.65	0.33 ± 0.31	8.5 ± 4.95	11.80 ± 2.13	3.58 ± 2.05	0.22 ± 0.21	0.55 ± 0.28	2.76 ± 2.51	1.53 ± 0.85
P-5	2.77 ± 3.46	1.74 ± 2.25	2.85 ± 3.24	0.04 ± 0.39	7.60 ± 8.32	12.23 ± 3.39	2.50 ± 2.12	0.02 ± 0.16	0.96 ± 0.85	1.14 ± 1.10	0.59 ± 0.42
P-6	5.23 ± 6.17	1.62 ± 1.80	1.52 ± 1.13	0.27 ± 0.40	6.62 ± 4.78	10.02 ± 1.93	2.34 ± 1.52	0.16 ± 0.21	0.55 ± 0.37	2.32 ± 2.44	0.85 ± 0.62
P-7	5.50 ± 5.99	1.47 ± 1.59	1.74 ± 1.18	0.49 ± 0.52	10.73 ± 8.63	14.41 ± 3.79	4.19 ± 3.11	0.32 ± 0.31	0.70 ± 0.44	2.69 ± 2.70	1.46 ± 0.85
P-8	2.48 ± 2.91	1.31 ± 1.35	1.21 ± 0.59	0.14 ± 0.25	6.66 ± 4.30	9.33 ± 1.84	2.91 ± 1.82	0.10 ± 0.18	0.54 ± 0.26	1.36 ± 1.54	1.37 ± 0.69
P-9	1.44 ± 0.82	0.77 ± 1.19	0.90 ± 0.52	0.06 ± 0.13	4.19 ± 3.10	5.93 ± 1.37	3.10 ± 1.49	0.07 ± 0.10	0.68 ± 0.26	1.33 ± 0.49	1.67 ± 0.26
P-10	1.95 ± 2.05	1.13 ± 1.26	1.24 ± 1.02	0.08 ± 0.17	4.91 ± 3.04	7.36 ± 1.21	2.48 ± 1.38	0.07 ± 0.13	0.64 ± 0.48	1.23 ± 1.17	1.28 ± 1.09
P-11	2.39 ± 3.46	1.31 ± 1.61	1.34 ± 1.17	0.16 ± 0.36	5.25 ± 5.03	8.06 ± 2.06	2.30 ± 1.79	0.11 ± 0.21	0.60 ± 0.43	1.31 ± 1.54	1.26 ± 0.86
P-12	4.24 ± 3.53	1.20 ± 1.38	2.05 ± 1.64	0.30 ± 0.39	6.73 ± 5.37	10.29 ± 2.18	3.21 ± 2.22	0.23 ± 0.26	1.00 ± 0.70	2.53 ± 1.83	1.03 ± 1.03
P-13	2.74 ± 2.33	1.38 ± 1.22	1.03 ± 0.75	0.23 ± 0.33	4.21 ± 3.14	6.85 ± 1.24	1.75 ± 1.48	0.16 ± 0.26	0.44 ± 0.36	1.42 ± 1.37	1.76 ± 2.65
P-14	0.25	0.09	0.66	nd	0.75	1.50	4.68	nd	4.19	1.96	2.12
P-15	0.39	0.13	0.46	nd	1.13	1.72	5.15	nd	2.14	2.21	2.36

Ponds P1 through P13 were measured on a bimonthly basis and averaged (the standard deviation is based on $n = 7$ months for each pond). Ponds P14–P15 located at higher elevation were measured only in October 2011 ($n = 1$ month for each pond). Number of observations per parameter is $7 \times 13 + 2 = 93$

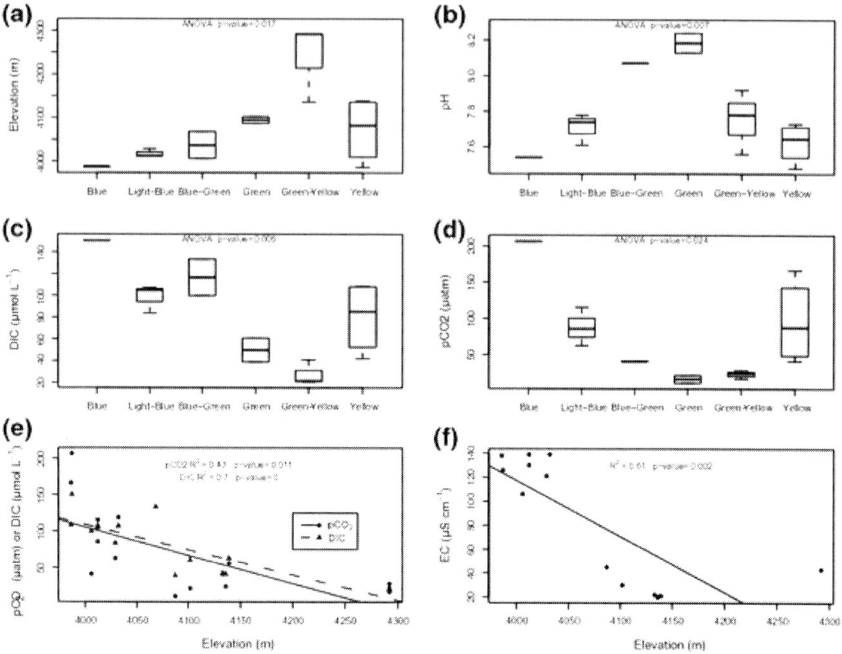

Figure. 2

The concentration of inorganic carbon species measured in October 2011 is compiled in Table 4. Among the inorganic carbon species, bicarbonate has the dominant control. Dissolved inorganic carbon (DIC) has higher concentration in blue-type ponds followed by yellow and green types of ponds (Fig. 2c). Similarly to DIC, pCO_2 has higher concentration in blue-type ponds followed by yellow and green types of ponds (Fig. 2d). A blue-type isolated pond (P2) showed highest amount of DIC in which bicarbonate and carbonate also appeared highest on this pond, while the CO_2 and pCO_2 appeared highest in pond P8 in the middle section. This pond, P8, was covered with ice, and there was big ice cliff fall on the pond just before sampling, so the CO_2 and pCO_2 appeared exceptionally high in this pond. Based on the pCO_2 data, all the ponds except P8 are unsaturated with carbon dioxide content.

DIC showed significant decreasing trend of concentration with increasing elevation with $R^2 = 0.75$, $p$value $= 3.6 \times 10^{-5}$ (Fig. 2e). The contribution of pCO_2 to the total DIC also decreases with increasing elevation. After discarding the value in P8 as no data, a linear regression of pCO_2 versus elevation yielded low coefficient of determination but clear trend ($R^2 = 0.38$, p value = 0.011). These results allow to reject the null H0 corresponding to hypothesis 1.

Table 4. Inorganic carbon species in supraglacial ponds within Lirung debris area in Langtang valley in central Nepal Himalaya during winter 2011

Sample ID	pH	Alkalinity (μmol L^{-1})	DIC (μmol L^{-1})	CO_2 (μmol L^{-1})	HCO_3 (μmol L^{-1})	CO_3 (μmol L^{-1})	pCO_2 (μatm)
P-1	7.68	99.9	108.7	9.4	98.9	0.11	166.4
P-2	7.72	139.9	150.4	11.3	138.6	0.18	206.5
P-3	7.29	99.9	99.6	2.5	96.4	0.37	40.9
P-4	7.28	99.9	107.3	8.2	98.8	0.12	115.6
P-5	7.74	99.9	104.9	6.0	98.5	0.16	85.6
P-6	7.78	79.9	83.7	4.4	79.0	0.14	62.8
P-7	7.60	99.9	107.3	8.1	98.8	0.12	119.0
P-8	nd	99.9	133.5	33.7	99.6	0.03	473.3
P-9	8.24	40.0	38.8	0.7	37.8	0.20	11.0
P-10	8.13	59.9	60.7	1.5	58.9	0.22	21.4
P-11	7.69	40.0	42.0	2.5	39.4	0.06	40.9
P-12	7.92	40.0	40.8	1.5	39.2	0.10	24.1
P-13	7.73	59.9	63.0	3.5	59.4	0.09	55.8
P-14	7.56	20.0	21.4	1.7	19.8	0.02	28.2
P-15	7.78	20.0	20.6	1.0	19.6	0.04	16.7

EC decreases significantly with increased elevation (Table 1; Fig. 2f) also allowing to reject the H0 corresponding to hypothesis 1. EC has higher concentration in blue-type ponds than green type of ponds; however, EC in yellow ponds varies widely and had values as high as in blue ponds and as low in green ponds. This indicates that EC is more a function of elevation than of color (not shown in the figure; ANOVA, nonsignificant p value = 0.228).

Most major ions showed slightly higher concentration in the lower-elevation sites than in the high-elevation sites except nitrate that shows increasing trend (Tables 2, 3). Elevation trends of the annual averages reported in these tables are significant for some parameters: alkalinity, *Ca, *SO_4, *Na, and NO_3 (Fig. 3), allowing rejecting the H0 for hypothesis 1. Thus, there are elevation trends for some water parameters.

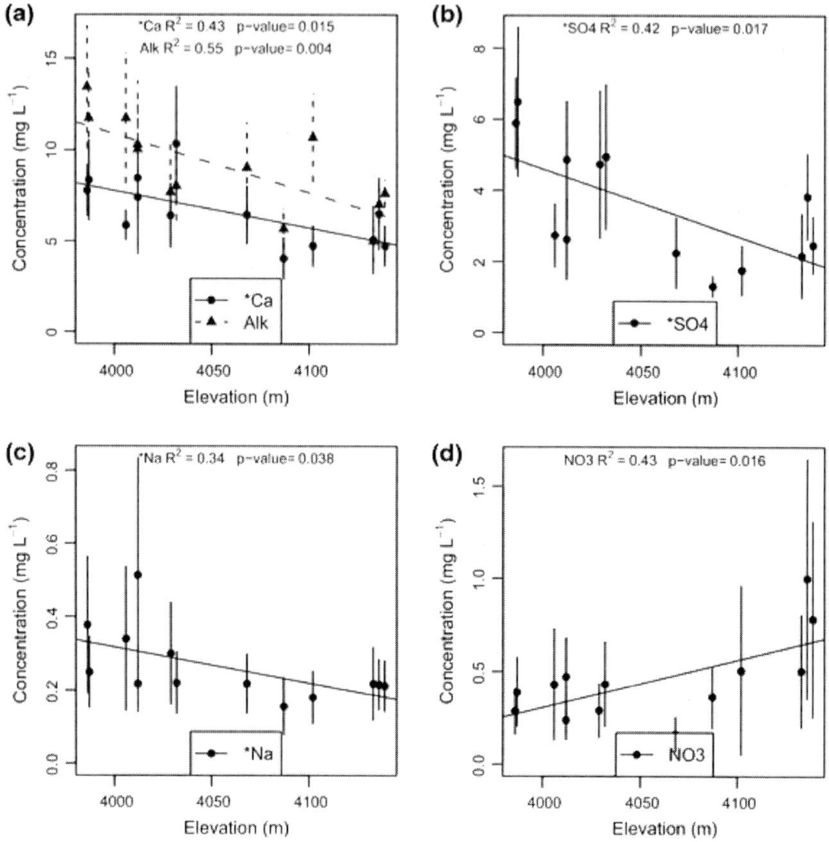

Figure 3. Altitudinal gradient of the annual average of those chemical parameters that resulted in significant variation with elevation. All trends are negative except nitrate that shows an increasing trend

There was no pattern of variation in major ions with respect to color of the pond, except for $*SO_4$ which exhibits the lowest concentration in purely green ponds. Sulfate is relatively significant ($p = 0.064$) variation with color being lowest for blue–green and green ponds (graph not shown).

Table 2 indicates that the lowest Si concentration was observed in the uppermost pond (P15) and the highest in a central pond (P7) in which the main flow mixes with pond water; this water appeared highly turbid due to the presence of cobbles and soils from moraine falls. In general, the silicon concentration showed no decreasing trends of concentration with increasing elevation. DOC and TDN did not show a significant increasing trend with elevation.

4.3 Hypothesis 2: Seasonal Changes in Elevation Trends

Sulfate shows significant differences regardless of season (Fig. 4a, c, d) with quite low p values 0.006 and 0.013 for pre-monsoon and post-monsoon. It is also significant for the monsoon months but shows an increasing trend and less significant p value = 0.053. These results allow to reject the H0 corresponding to hypothesis 2, and thus, some elevation trends vary with the seasons. *Mg varies the same way in the pre-monsoon (Fig. 4b) correlating with sulfate as further explored later. Several parameters, F, alkalinity, *K, and *Ca, showed a significant trend with elevation only for the post-monsoon season (Fig. 5), demonstrating again the seasonal influence of elevation patterns and supporting hypothesis 2.

4.4 Hypothesis 3: Dominant Geochemical Processes

The sea-salt-corrected cations *Ca and *Mg showed a significant relationship with sulfate for all seasons (Fig. 6 a, c, e). The slope for *Ca is about 20 times higher than *Mg. The coefficient of determination is high for the pre-monsoon and post-monsoon months (Fig. 6a, e), suggesting that both parameters are released by similar geochemical process. In fact, there is a strong correlation between *Mg + *Ca and *SO_4 for the pre-monsoon (Fig. 6b). The relationship between *Mg + *Ca and *SO_4 is also relatively significant for post-monsoon months (Fig. 6f) but not for monsoon months (Fig. 6d). These results allow to reject the H0 corresponding to hypothesis 3, and thus, we conclude that there is a relationship between oxidation of pyrite and dissolution of carbonates and aluminosilicates.

The contributions to sodium and potassium after sea-salt correction accounted for 32 % of the total sum of base cations suggesting more contribution to these ions due to intense dissolution of silicate minerals within the basin. Ratio of silicon to sum of sodium and potassium after sea-salt correction Si/(*Na + *K) averages ~1.5 although it is highly variable (Table 3), indicating silicate weathering in this system. There is high variability variation in this ratio Si/(*Na + *K) with no significant trend with elevation (regression with R^2 = ~0.0, p value = 0.89).

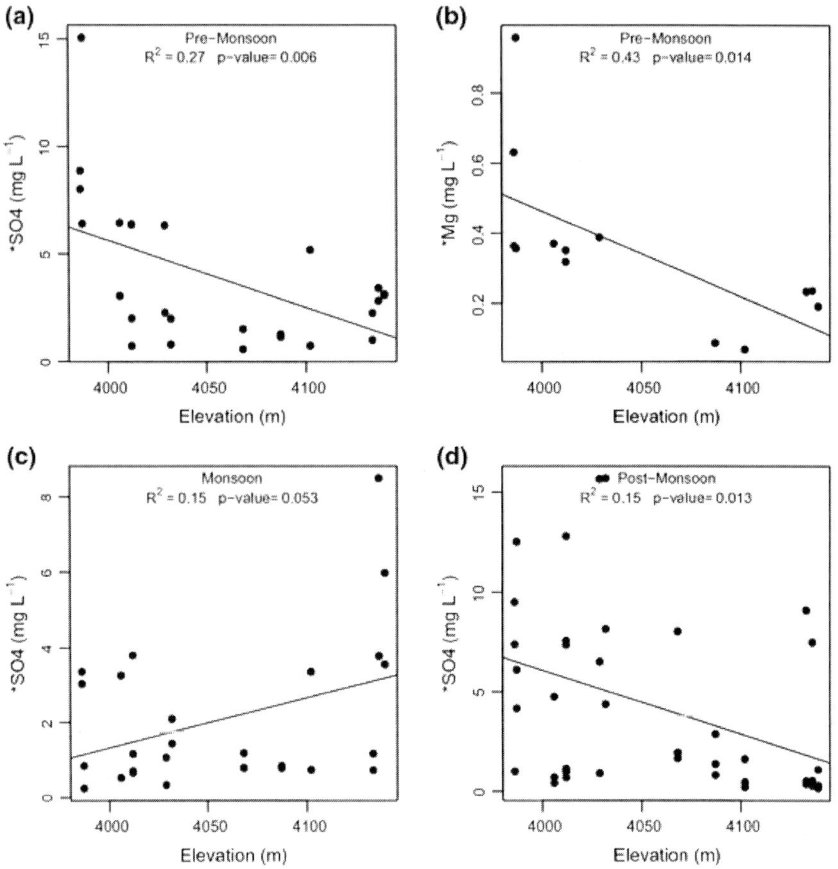

Figure 4. Altitudinal variations in sulfate and magnesium. Sulfate shows a significant trend for all seasons but weaker during monsoon. Mg is significant only for the pre-monsoon

Figure 7a shows a ternary diagram for major cations (*Ca, *Na, *K). We do not include *Mg because all values are very low and in this case dolomite does not contribute significantly. Most of the points fall within the calcium–potassium line suggesting carbonate and plagioclase dissolution as the predominant geochemical processes. Many points stretch toward the sodium apex especially during post-monsoon season in lower-elevation sites, indicating the role of hydrolysis and temperature to enhance the dissolution of aluminosilicates, primarily plagioclase.

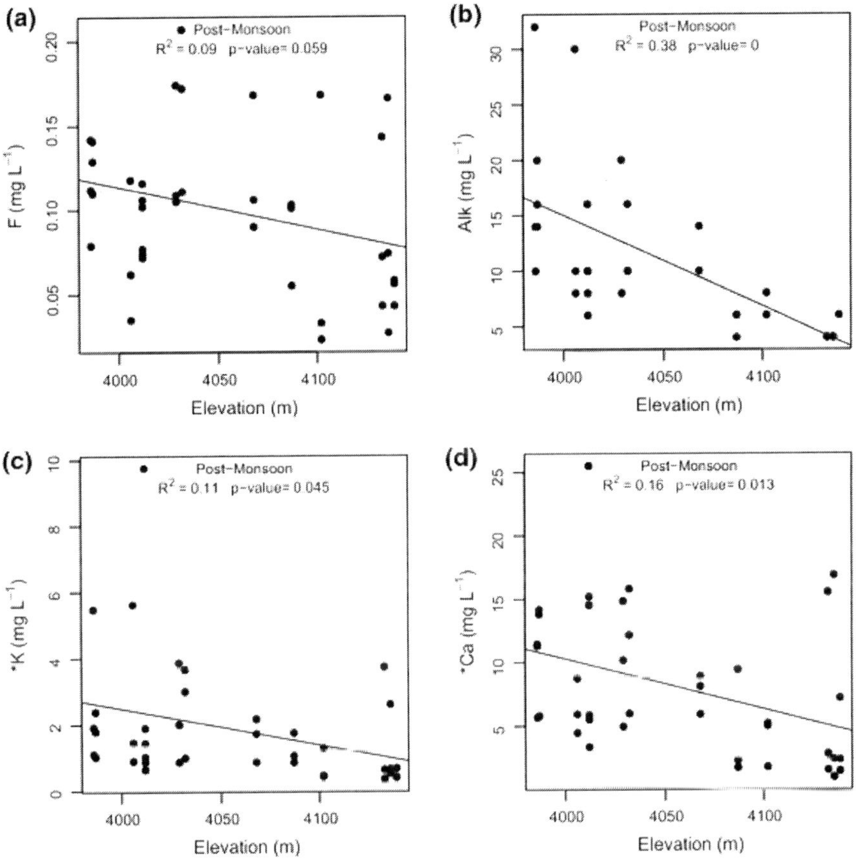

Figure 5. Altitudinal variations in some chemical parameters that resulted significant during post-monsoon

Figure 7b shows a ternary diagram using alkalinity, Cl, and $^*SO_4$ to represent composition of major anions. Most of the points lying along the sulfate alkalinity line suggesting clearly that the oxidation of pyrite coupled with the dissolution of carbonate and aluminosilicate primarily plagioclase appeared as dominant geochemical processes in the area.

Some of the points scattered toward chloride apex especially during monsoon and post-monsoon season, suggesting contribution from evaporative minerals due to favorable hydrolysis condition.

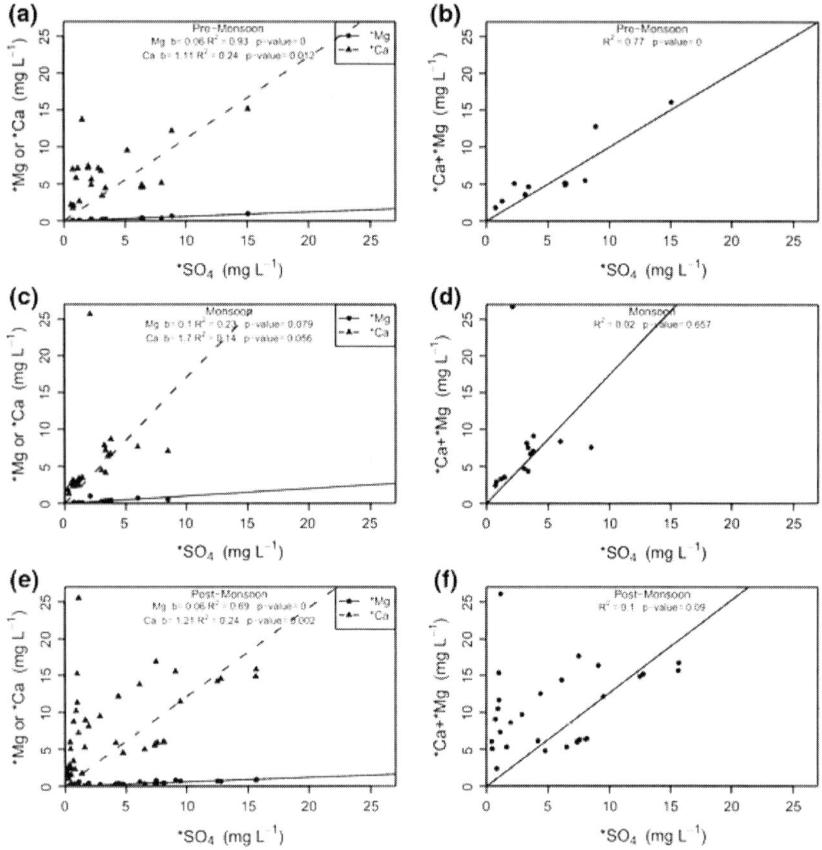

Figure 6. Relationship between sulfate and calcium, magnesium, and calcium plus magnesium illustrating relative contributions and variation by season

The contribution of sea-salt to sulfate is negligible in the central Himalayan region. In average, sulfate contributes nearly 25 % to total anions and alkalinity contributes less than 64.6 % to total anions among the ponds in the debris area.

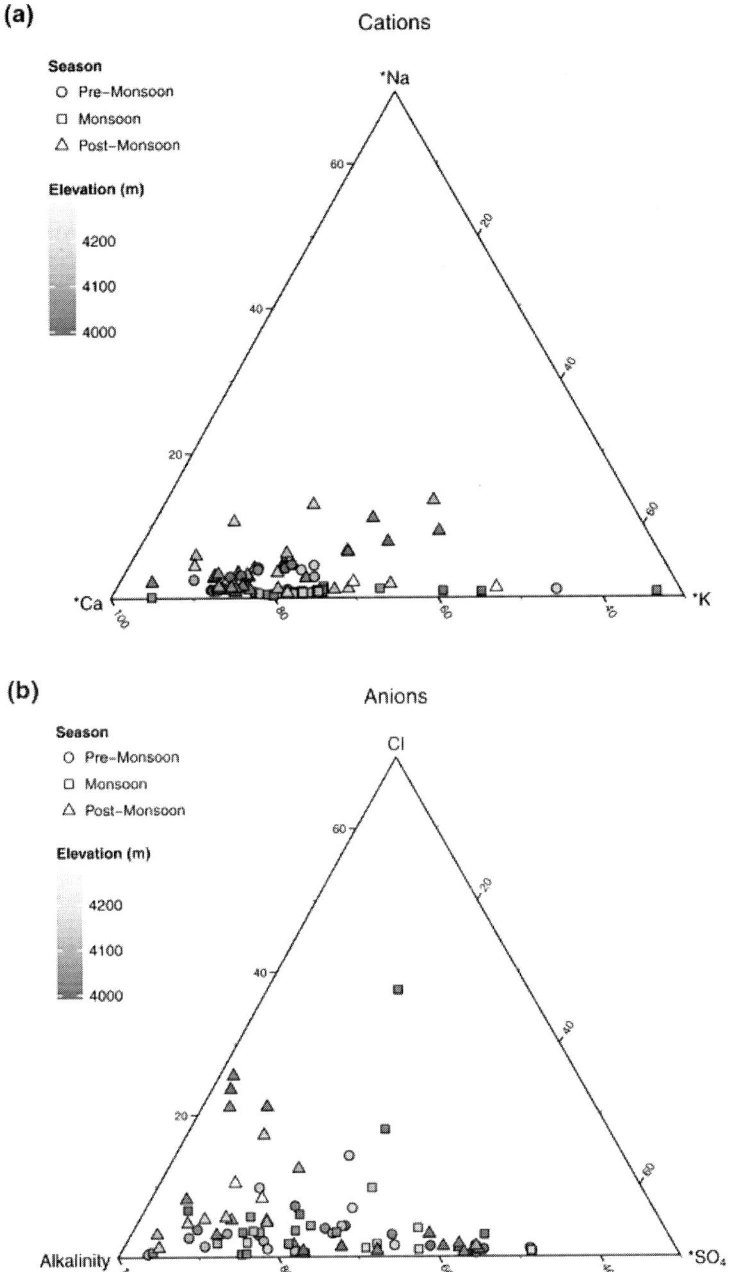

Figure7. Ternary diagrams by season (*symbol type*) and elevation (*fill shade*). **a** Sea-salt-corrected cations *Ca, *Na, and *K. **b** Alkalinity, Cl, and sea-salt-corrected $*SO_4$

4.5 Hypothesis 4: Seasonal Variability of Chemical Compositions

Figure 8 shows high concentrations for December or January when water discharge is low and surface is frozen, thus rejecting ions in the ice and increasing concentration in water. Then, concentration decreases by dilution when water starts to melt and leading to a minimum in the pre-monsoon months of May and June. Subsequently, there is an increase with the monsoon due to the runoff.

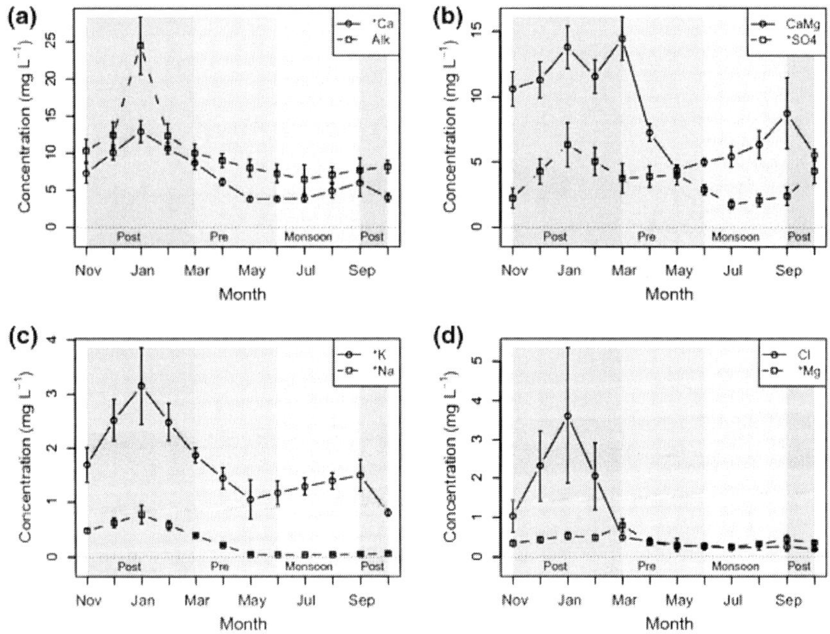

Figure 8. Average major parameters (ions) over all ponds to emphasize temporal variations. Points at nonsampled months (Dec, Feb, April, June, and August) are linearly interpolated for each pond and then entered into the average. Interpolated points are only used for visualization, not analysis. Seasons are indicated by *shading* and *labeled* at the *bottom* using the following convention: pre-monsoon: March–May; monsoon: June–September, post-monsoon October–February

In general, all the major base cations showed highest concentration during the post-monsoon season followed by pre-monsoon and monsoon season. Paired *t* test of maxima and minima values in the year across ponds for each water parameter yielded significant differences for all parameters. The *p* values obtained are SS 0.014, alkalinity 1.2×10^{-5}, F 0.006, Cl 0.026, NO_3 0.0006, *SO_4 1.2×10^{-5}, DOC 0.02, TDN 0.0003, *Na 0.0005, *K 0.0002, *Mg 3.6×10^{-6}, *Ca 6.8×10^{-6}, and Si 2.3×10^{-7}. These results allow to reject the H0 corresponding to hypothesis 4, and therefore, we conclude that there are significant differences between the maximum and minimum values observed in the year.

Calcium, sodium, potassium, and magnesium follow the seasonal pattern (Fig. 8a, c, d), except that higher magnesium is delayed to the end of the post-monsoon (Fig. 8d). Not shown in the figure, it is worth noting that pond P7 showed exceptionally high concentration of base cations after sea-salt correction during the monsoon, which is probably due to moraine fall and high hydrolysis condition. During monsoon, moraine fall events were observed frequently and water becomes highly turbid. Pond P7 is located at the main flow of the debris, and hence, the availability of fresh minerals and high hydrolysis condition at this site accelerates the dissolution process.

In general, anions also showed highest concentration during the post-monsoon followed by pre-monsoon and monsoon, as illustrated by alkalinity, sulfate, and chloride (Fig. 8a, b, d). Silicon also showed higher concentration during post-monsoon decreasing toward the pre-monsoon; but then, it follows an increasing trend which we ascribe to rock fall, moraine fall, and change in hydrological regime (Fig. 9b).

Other parameters also exhibit high concentrations in the post-monsoon, but the highest concentration of DOC, TDN, and nitrate was delayed to the end of the post-monsoon (Fig. 9b, c). DOC shows lowest concentration in monsoon months possibly because of the effect of dilution. Not shown in the figure, it is worth noting that pre-monsoon DOC values showed an increasing trend with altitude except for pond P1 near Lirung outlet. SS exhibits a second peak during the monsoon probably due to high discharge and higher frequency of landslides, thus causing the flows to carry higher loads of suspended matter (Fig. 9a). The high and very variable values of SS for December are interpolated and cannot be interpreted as certain. TDN shows a similar increase in the monsoon months (Fig. 9c). Not illustrated by the figure, pond P10 showed the highest concentration of TDN during pre-monsoon and monsoon seasons, while the isolated pond P2 showed the lowest concentration during post-monsoon season.

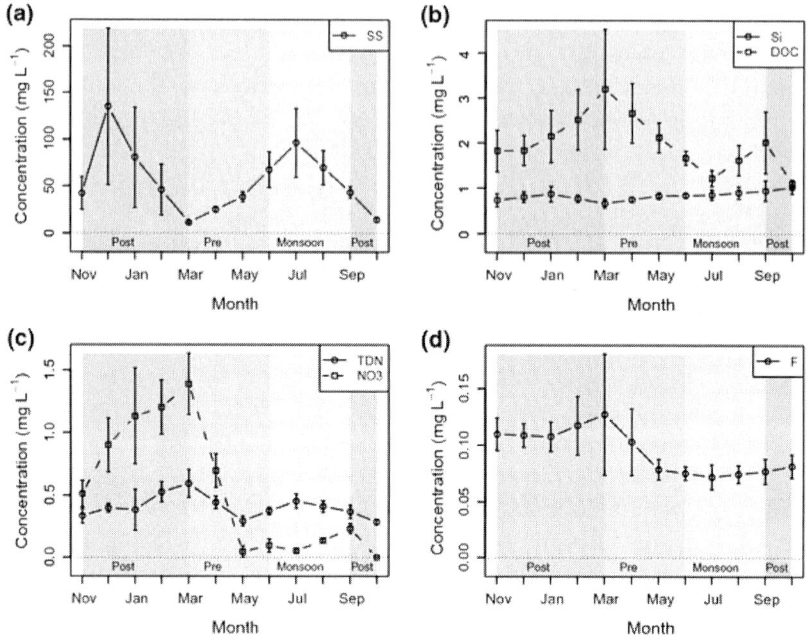

Figure 9. Average SS and chemical parameters over all ponds to emphasize temporal variations. Points at nonsampled months (Dec, Feb, April, June, and August) are linearly interpolated for each pond and then entered into the average. Interpolated points are only used for visualization, not analysis. Seasons are indicated by *shading* and *labeled* at the *bottom* using the following convention: pre-monsoon: March–May; monsoon: June–September, post-monsoon October–February

Nitrate showed strong seasonal variation pattern with highest concentration in post-monsoon followed by pre-monsoon and monsoon values (Fig. 9c). This variation pattern is probably due to the influence of livestock and change in flow regime due to seasons. Fluoride showed the same seasonal variation trend (Fig. 9d).

5 DISCUSSION

5.1 Overview and Composition of Major Ions

It is well-known that sea-salt-contributed aerosols impact major ions observed in watersheds (Stallard and Edmond 1981; McDowell et al.

1990; Bhatt and McDowell 2007); therefore, we evaluated these impacts in our study area. We found that calcium and sodium are the least and the most impacted element by marine aerosols, respectively, and that the contribution of marine aerosols to the total dissolved load appears to be minor advocating the lithological control on dissolved matter as reported by earlier studies (Dalai et al. 2002; Bhatt et al. 2008, 2009). The high contribution to sodium and potassium after sea-salt correction among the ponds within high Himalaya region is due to the higher content of silicate minerals in the basin compared to the lower elevation of the landscape (Bhatt et al. 2009).

The contribution to sodium and potassium after sea-salt correction accounts for 32 % of the total sum of base cations, suggesting that the contribution to these ions is mainly due to intense dissolution of silicate minerals within the basin. The contribution to sodium and potassium after sea-salt correction resulted much higher than those in the low-elevation mid-mountains and Tarai region in southern part of Nepal probably due to the variation in lithology of the basin (Bhatt et al. 2009).

We found sulfate concentration many times higher than in most of the largest river systems of the world and in the fastest weathering regimes of a silicate terrane on the earth surface (McDowell and Asbury 1994; Meybeck and Ragu 1997; Gaillardet et al. 1999; Bhatt and McDowell 2007; Tartai et al. 1998; Bhatt et al.2014a).

Takeuchi et al. (2012) found that various organisms such as insects, copepods, and algae live in ponds in the debris-covered area of this glacier. These authors found insects and algae in the very clear type of pond (blue type in our classification), insects only in the slightly turbid type of pond (yellow type in our classification), and no living organisms in highly turbid type of pond surrounded by ice walls (green type of pond in our classification). Thus, we suggest that the distribution of microorganisms by type of pond is mainly related to inorganic carbon, which varies according to pond color. Blue types of pond formed many years ago and are isolated from the main flow of the debris, whereas highly turbid green types of ponds formed recently and are mostly surrounded by big rocks, boulders, and cobbles. Yellow types of ponds are within the main channel and erosion-prone area; whereas the green types of ponds formed recently within the debris-prone area but are surrounded with ice walls.

5.2 Hypothesis 1: Elevation Trends of Measured Chemical Parameters

Some chemical parameters showed changes with elevation in this steep gradient (300 m elevation change in a distance of 2 km). Higher concentrations of major solutes and dissolved inorganic carbon species in the lower-elevation ponds are probably due to the influence of hydrological flow path, high residence time (comparatively flat surface), higher temperature (absence of big ice walls), and abundant availability of fresh reactive minerals due to physical erosion transported through the debris area. Altitudinal control of solute release is documented in other systems, for example, in the southern Swiss Alps (Drever and Zobrist 1992), in the Santa Rosa Mountains in the USA (Riebe et al. 2004), along the Himalayan river systems in central Nepal (Bhatt et al. 2009), along the Himalayan rivers (France-Lanord et al. 2003), and in the Everest region of Nepal (Baumler and Zech 1994).

TDN and DOC did not show elevation trends. Variations in TDN and ammonium are primarily due to the variation in livestock influence and human seasonal settlement. DOC variation may be due to spatial variability of organic-rich sediments and DOC thermal degradation. The increase in organic carbon in the sediment reservoir is due to Himalaya weathering affecting the carbon cycle in the region (France-Lanord and Derry 1997). The consumption of CO_2 by the silicate weathering in the Ganga and Yamuna watersheds in the Himalayan region is balanced by the release of CO_2 through weathering of organic-rich materials like black shale from the basin further suggests the presence of organic-rich materials in the high Himalaya region (Dalai et al. 2002). Furthermore, it has been documented that erosion from the Himalayan basin delivers high amount of organic carbon to the Bay of Bengal (Ittekkot 1988; Subramanian and Ittekkot 1991; France-Lanord and Derry 1997).

In this high Himalaya basin, the pCO_2 appeared low since CO_2 is quickly absorbed by the minerals during the dissolution processes because the glacier meltwater is diluted and reaction rate is fast. Bicarbonate appeared as a dominant inorganic carbon species, suggesting that the rate of dissolution controls carbonate chemistry.

Silicon did not show significant elevation trends but showed exceptional values in ponds 7 and 13. High concentration of silicon at the central pond is linked directly with high temperature and high-flow conditions, and the abundant availability of fresh reactive mineral surfaces. Some earlier studies have documented the role of temperature and runoff on the dissolution of silicate minerals (Amiotte Suchet and Probst 1993;

McDowell and Asbury 1994; Dorn and Brady 1995; White and Blum 1995; Ludwig et al. 1998; Gaillardet et al. 1999; France-Lanord et al. 2003; Krishnaswami and Singh 2005; Wolff-Boenisch et al. 2009). The lower concentration of silicon was observed in the ponds that were covered with ice cliffs and rocks, whereas the higher concentration was observed in the ponds that are covered with moraine walls and those conducting the main water flow. A variation in mineral surface area, controlled by physical denudation rates, plays a great role on the variability in silicon concentration within the basin. Navarre-Sitchler and Brantley (2007) found that weathering advance rates vary over seven orders of magnitude when compared across multiple scales.

5.3 Hypothesis 2: Seasonal Changes in Elevation Trends

There are more instances of elevation patterns in the post-monsoon season than in the other seasons. Elevation trends are modulated by seasonality possibly because higher flows in the monsoon redistribute concentration among ponds and freezing in the winter months, which occurs in the post-monsoon, accentuates concentration differences. Subsequent melting will diminish concentration differences in most instances. Sulfate shows significant elevation differences regardless of season; however, the patterns are opposite and less significant in the monsoon.

Sulfide oxidation is also dependent on the hydrolysis condition of the ponds; hence, the more production of sulfate in lower-elevation sites releases more solutes and bicarbonate due to enhanced dissolution process with high runoff. The high Himalayan region has high content of sulfur-bearing minerals, but the hydrolysis condition is lower than the low-elevation region (France-Lanord et al. 2003; Bhatt et al. 2009).

5.4 Hypothesis 3: Weathering Extent and Controlling Parameters

Generally, it appeared that sulfide oxidation has a dominant role in the dissolution processes within high Himalaya region based on the high sulfate content and appearance of rusty precipitation among some ponds and appearance of rocks within debris area supporting the presence of pyrite. Dominance of calcium and sodium among base cations and the dominance of sulfate and alkalinity mainly as bicarbonate among anions suggest that dissolution of carbonate and aluminosilicate coupled with sulfide oxidation within the basin appeared as a dominant geochemical process which controls the major solutes. Sulfide oxidation coupled with carbonate dissolution has been observed in subglacial drainage basins within Himalaya and other glacierized basins of the world (Tranter and Raiswell 1991; Bartarya 1993; Tranter et al. 1993; Fairchild et al. 1994; Galy and France-Lanord 1999; Hasnain and Thayyen 1999; Pandey et al. 1999; Chen et al. 2002; Hodson et al. 2002; Millot et al. 2003; Bhatt et al. 2000, 2007, 2008, 2009, 2014a; Wolff-Boenisch et al.2009).

The strong relationship between *Mg and *Ca and $*SO_4$ indicates that these ions are derived from related processes. Our study area is underlain by high-grade metamorphic rock consisting biotite, quartz, and plagioclase. Biotite $KMg_3AlSi_3O_{10}(OH)_2$ appears as dominant silicate mineral, so the source of *Mg is the biotite dissolution. During the process of pyrite oxidation, the dissolution process of aluminosilicates including biotite and carbonate also becomes intense. So we observe the strong relationship between $*SO_4$and *Mg, and $*SO_4$ and *Ca + *Mg. The additional Mg could be contributed from the dissolution of dolomites as thick beds of gray silicious dolomites are found in high mountain region in central Nepal. The source of calcium is primarily from Ca-plagioclase $CaAl_2Si_2O_8$ and carbonate dissolution within the basin.

The calculated average ratios of Si/(*Na + *K) among the ponds within Lirung debris areas appeared lower than the calculated ratios from Langtang-Narayani river system in central Nepal Himalayas (unpublished data) and much lower than the calculated ratios from different positions of the Rio Icacos (which drains a tropical rain forest) that has the fastest documented weathering rates of any silicate terrain in the world (McDowell and Asbury 1994; White and Blum 1995; Bhatt and McDowell 2007). The wide variation in Si/(*Na + *K) ratio in our study suggests spatial variation in weathering extent within the study basin. During the dissolution process, the ratio of Si/(*Na + *K) for Na-feldspar to beidellite gives 1.7, to kaolinite gives 2, and to gibbsite gives 3, and K-feldspar to illite gives 3 (Huh et al. 1998). Dalai et al. (2002) found Si/(*Na + *K) average ratios of 1.2 and also with wide variation along the Yamuna river system suggesting low extent of silicate weathering within drainage basin. The ratio of Si/(*Na + *K) was very small in Siberian Rivers compared to the streams of the Guayana Shield and streams of Puerto Rico (Huh et al. 1998; Edmond et al.1995; Bhatt and McDowell 2007). The high extent of weathering rate within tropical regions such as the Guayana Shield and Puerto Rico is probably due to regimes of high precipitation and high temperature.

5.5 Hypothesis 4: Seasonal Variability of Measured Chemical Parameters

In general, the seasonal pattern for major anions and cations was similar, showing the highest concentration during the post-monsoon, a decrease toward the pre-monsoon, and slight increase in the monsoon. During the post-monsoon season, the temperature drops and glacier melting decreases, and as a consequence, flow decreases and concentration becomes higher than during the pre-monsoon and monsoon seasons. An exception to the general pattern is Mg, which had a delayed peak toward the end of the post-monsoon, which was also observed for DOC, TDN, and nitrate. Another exception to the general pattern is the increasing trend of silicon with time probably due to favorable hydrolysis condition and abundant availability of fresh reactive mineral surfaces due to intense physical erosion. Furthermore, pyrite oxidation also becomes more intense under favorable hydrolysis condition and high physical erosion-prone area which releases abundant sulfate and supply protons that accelerate the dissolution process.

SS and TDN showed a different seasonality trend to major chemical species, since there was an increase in concentration during the monsoon season. We explain this effect by high discharge and higher frequency of landslides, thus causing the flows to carry higher loads of suspended matter. Also, increased flow carries contribution from decomposed vegetation and livestock manure from nearby area. Higher SS concentrations occur in the post-monsoon but are highly variable; a high post-monsoon concentration was observed at pond P1 close to outlet point due to mixing of highly turbid main flow to the pond.

DOC, TDN, and nitrate showed a slightly different seasonal variation trend from major ions (except Mg) since the highest concentration is delayed to the latest part of the post-monsoon season. The high concentration of DOC, TDN, and nitrate in later part of the post-monsoon is primarily due to inputs from livestock (yaks, horses, sheep) activities and decomposed shrub and grass from nearby area.

5.6 Comparison of Average Chemistry with Same and Different Basins of the World

The average chemical compositions of the supraglacial ponds in the debris-covered area of Lirung glacier are compiled with the chemical compositions of the same and different basins of the world (Table 5). The concentrations of DOC in supraglacial ponds appeared nearly the same as river water samples in high-altitude areas of the same region but the export rate of DOC calculated threefold higher in low-elevation sites of the Langtang-Narayani river system than the high-elevation sites in the same area (unpublished data). The DOC of these supraglacial ponds is threefold lower than a pond in the Kathmandu valley, Nepal, and tenfold lower than a heavily urbanized basin (Bagmati River) in the Kathmandu valley, probably due to difference in vegetation, anthropogenic inputs, and temperature (Bhatt et al. 2013, 2014b).

The alkalinity of supraglacial ponds appeared comparable with the Tatra basin of Europe and lower than the Colorado Rocky Mountain area and Cascade lake in Washington of USA, Khumbhu basin within Everest region of Nepal and Lirung outlet point of the same region of Nepal and much lower than the lakes of Spiti and Pandoh basins of India (Axtman and Stallard 1995; Reynolds et al. 1995; Musselman et al. 1996; Henriksen et al. 1997; Tartari et al. 1998; Bhatt et al. 2000, 2007; Das and Dhiman 2003; Borowiak et al.2006; Anshumali and Ramanathan 2007). The alkalinity of glacier meltwater at the Lirung out point is quite comparable with the Southern Cascade lakes and lakes from rocky mountain areas in Colorado, which is slightly higher than the Tatra basin in Europe and much lower than Pandoh Lake and Spiti basin in India. The difference in alkalinity is primarily related with flow regime of the basins—high alkalinity appeared with the high-flow condition area. The extraordinarily high concentration of alkalinity in the lakes of Spiti and Pandoh basins is probably due to the availability of carbonates and high temperature than the other basins in the same region and in the world. The measured nitrate is much higher than the Southern Cascade lakes and rocky mountain areas in Colorado, comparable with other Himalayan basin in Nepal, and much lower than Pandoh Lake probably due to variation in aquatic species. The extraordinarily high concentration of nitrate in Pandoh Lake is probably due to the input from anthropogenic sources. The concentration of sulfate appeared much higher in the Himalayan region than the North American and European basins probably due to variation in sulfide-bearing minerals.

Table 5. Comparison of overall average chemical compositions of supraglacial ponds within Lirung debris area in Langtang valley with other high-altitudes lakes of same and different basins of the world

Measured parameters	[a]Cascade Lake USA	[b]Rocky Mountai USA	[c]Tatra basin Europe	[d]Pandoh Lake India	[e]Spiti basin India	[f]Khumbhu basin Nepal	[g]Khumbhubasin Nepal	[h]LirungOutlet Nepal	[i]Lirung basin Nepal	This study Nepal
SS ($mg\ L^{-1}$)	nd	nd	nd	nd	nd	nd	nd	35.0	50.3	46.38
pH	nd	6.62	nd	7.13	9.09	nd	7.4	8.0	7.97	7.72
DOC ($mg\ L^{-1}$)	nd	nd	nd	nd	nd	nd	nd	nd	nd	1.90
TDN ($mg\ L^{-1}$)	nd	nd	nd	nd	nd	nd	nd	nd	nd	0.39
Alk ($mg\ L^{-1}$)	13.01	16.5	9.6	80.5	103	16	15.0	16.8	10.03	8.99
F ($mg\ L^{-1}$)	nd	nd	nd	nd	nd	nd	nd	nd	nd	0.09
Cl ($mg\ L^{-1}$)	nd	0.10	0.2	2.4	8	0.09	0.07	0.15	0.12	0.88
NO_3 ($mg\ L^{-1}$)	0.06	0.39	nd	45.7	nd	1.20	0.25	1.00	0.54	0.52
SO_4 ($mg\ L^{-1}$)	0.78	1.78	3.8	8.2	203	5.51	4.99	5.06	1.05	3.95
NH_4 ($mg\ L^{-1}$)	nd	0.02	nd	nd	nd	nd	0.00	0.02	0.05	0.50
Na ($mg\ L^{-1}$)	0.14	0.51	0.4	3.8	12	0.47	0.44	0.70	0.15	1.91
K ($mg\ L^{-1}$)	0.71	0.22	0.2	2.1	1	0.54	0.51	0.83	0.54	1.63
Mg ($mg\ L^{-1}$)	0.17	0.49	0.2	3.3	32	0.33	0.34	0.40	0.10	0.49
Ca ($mg\ L^{-1}$)	1.21	2.99	3.0	18.0	50	4.36	4.37	4.98	2.44	6.89
Si ($mg\ L^{-1}$)	0.48	nd	nd	nd	1.45	0.95	0.63	1.10	0.56	0.87

[a] Data from Axtman and Stallard (1995)
[b] Data from Musselman et al. (1996)
[c] Data from Henriksen et al. (1997) and Borowiak et al. (2006)
[d] Data from Anshumali and Ramanathan (2007)
[e] Data from Das and Dhiman (2003)
[f] Data from Reynolds et al. (1995)
[g] Data from Tartari et al. (1998)
[h] Data from Bhatt et al. (2000)
[i] Data from Bhatt et al. (2007)

Base cations showed higher concentration in the Himalayan basins than the North American and European basins possibly due to variations in carbonate rocks and temperature. Silicon concentration appeared comparable in the same basin, Khumbhu basin in the Everest region of Nepal, higher than Southern Cascade lakes, and much lower than the Spiti basin in India. The concentrations of major solutes among supraglacial ponds appeared lower than the concentration of rivers in the high-altitude region of the same area advocating the role of hydrolysis condition and climate primarily temperature controls the dissolution rates of the minerals. The variation in minerals, hydrolysis condition, and climate primarily temperature controls the spatial variability in chemistry of high-altitude ponds and lakes in Himalayas, Europe, and North American regions.

6 CONCLUSIONS

Using four research hypotheses, we have examined elevation trends, and their control by seasonality, dominant geochemical processes, and seasonal dynamics in the supraglacial ponds in the debris-covered area of Lirung glacier in central Nepal Himalayas. We have been able to reject the null H0 for all hypotheses. Concentration of some parameters showed a decreasing trend with elevation but with high variability and modulation by seasons, being more prevalent in the post-monsoon. Sulfate shows significant elevation trends regardless of the season, although the trend is also affected by the monsoon. Chemical concentration of all parameters resulted highest during the post-monsoon, decreasing toward the pre-monsoon, and slight increases in the monsoon season, advocating the major control of seasonality on chemical concentrations. Sulfate resulted strongly correlated with magnesium and with calcium, as well as with the sum of magnesium and calcium. This result indicates that the dominant geochemical processes are chemical weathering of carbonate and silicate minerals fueled by dissolution of pyrite. The contribution to marine aerosols appeared to be relatively small in the total mass of solutes in the debris-covered area. The variations in flow condition, presence of ice, availability of fresh reactive mineral surfaces—physical erosion, variation in temperature, and site-specific conditions are prime causes for the variation in chemical compositions among the pond types.

ACKNOWLEDGMENTS

The authors would like to thank Tom Jäppinen, Lisette Kretzschmann, and Ronny Lauerwald for their help in laboratory and analysis work. The authors are grateful to Kedar Rijal, Anil Shrestha, and Bijaya Pokhrel for their help and cooperation in field work. Thanks to Quinga Tamang, Mingma Tamang, Deepak Bhatt, Gyanendra Pant, Gyan Shrestha, Rajan Subedi, and other members of the field campaign for their help and cooperation during the sampling. The authors would like to thank the editor for suggestions and three anonymous reviewers for their insightful comments which improve the paper greatly. This research work was supported through the Cluster of Excellence 'CliSAP' (EXC177), KlimaCampus-University of Hamburg, funded through the German Science Foundation (DFG).

CONFLICTS OF INTEREST

Disclosure of potential conflicts of interest.

ETHICAL STANDARD

We maintain the integrity of the research and its presentation achieved by following the rules of good scientific practice. We have followed the principles of ethical and professional conduct.

REFERENCES

1. Amiotte Suchet P, Probst JL (1993) Modeling of atmospheric CO_2 consumption by chemical weathering of rocks—application to the Garonne, Congo and Amazon Basins. Chem Geol 107(3–4):205–210

2. Anshumali, Ramanathan AL (2007) Seasonal variation in the major chemistry of Pandoh Lake, Mandi District, Himachal Pradesh, India. Applied Geochemistry 22(8):1736–1747

3. APHA, American Public Health Association (1995) Standard method for the examination of water and waste water. 19th Edition. Edited by Eaton AD, Clesceri LS, Greenberg AE. American Public Health Association, 1015 Fifteenth Street, NW. Washington, DC 20005, USA

4. Axtman EV, Stallard RF (1995) Chemical weathering in the South Cascade Glacier basin, comparison of subglacial and extra-glacial weathering. In: Tonnessen KA, Williams MW, Tranter M (eds) Biogeochemistry of seasonally snow-covered catchments. IAHS Publ. No. 228, pp 431–439

5. Barnard PL, Owen LA, Finkel RC, Asahi K (2006) Landscape response to deglaciation in a high relief, monsoon-influenced alpine environment, Langtang Himal, Nepal. Quaternary Sci Reviews 25(17–18):2162–2176

6. Bartarya SK (1993) Hydrochemistzy and rock weathering in a subtropical lesser Himalayan river basin in Kumaun, India. J Hydrol 146(1–4):149–174

7. Baumler R, Zech W (1994) Soils of the high-mountain region of eastern Nepal—classification, distribution and soil formation process. Catena 22(2):85–103

8. Benn DI, Bolch T, Hands K, Gulley J, Luckman A, Nicholson LI, Quincey D, Thompson S, Toumi R, Wiseman S (2012) Response of debris-covered glaciers in the Mount Everest region to recent warming, and implications for outburst flood hazards. Earth-Sci Rev 114:156–174

9. Bhatt MP, McDowell WH (2007) Controls on major solutes within the drainage network of a rapidly weathering tropical watershed. Water Resour Res 43:W11402.

10. Bhatt MP, Masuzawa T, Yamamoto M, Sakai A, Fujita K (2000) Seasonal changes in dissolved chemical composition and flux of melt water draining from Lirung glacier in the Nepal Himalayas. In: Proceedings of a workshop on Debris-Coved Glaciers held at Seattle, Washington, USA. IAHS Publ no 264, pp 277–288

11. Bhatt MP, Masuzawa T, Yamamoto M, Takeuchi N (2007) Chemical characteristics of pond waters within the debris area of Lirung Glacier in Nepal Himalaya. J Limn 66(2):71–80

12. Bhatt MP, Masuzawa T, Yamamoto M, McDowell WH (2008) Chemical weathering in central Himalaya: dissolved silica dynamics in glacier meltwater. International conference on hydrology and climate change in mountainous area (ICHCC), SOHAM—UNESCO, pp 162–179

13. Bhatt MP, Masuzawa T, Yamamoto M, Gardner KH (2009) Spatial variations in chemical compositions along Langtang-Narayani river system in central Nepal. Environ Geol 57:557–569.

14. Bhatt MP, Bhatt S, Gaye B (2013) Controls on pond water chemistry within Kathmandu, Valley, Nepal. Int J Lakes Rivers 6(2):153–172

15. Bhatt MP, Bhatt S, Gaye B (2014a) Controls on Gosaikunda Lake Chemistry within Langtang National Park in High Himalaya, Nepal. Int J Geosci 5:1100–1115. Bhatt MP, McDowell WH, Gardner KH, Hartmann J (2014b) Chemistry of heavily urbanized Bagmati river system in Kathmandu Valley, Nepal: export of organic matter, nutrients, major ions, silica and metals. Environ Earth Sci.

16. Bookhagen B (2010) Appearance of extreme monsoonal rainfall events and their impacts on erosion on the Himalaya. Geomat Nat Hazards Risk 1(1):37–50

17. Borowiak D, Polkowska Z, Przyjazny A (2006) The hydrochemistry of high altitude lakes in selected mountain ranges of central and southern Europe. Limnol Rev 6:21–30

18. Chabaux F, Riotte J, Clauer N, France-Lanord C (2001) Isotopic tracing of the dissolved U fluxes of Himalayan Rivers: implications for present and past U budget of the Ganges-Brahmaputra system. Geochim Cosmochim Acta 65(19):3201–3217

19. Chen J, Wang F, Xia X, Zhang L (2002) Major element chemistry of the Changjiang (Yangtze River). Chem Geol 187:231–255

20. Dalai TK, Krishnaswami S, Sarin MM (2002) Major ion chemistry in the headwaters of the Yamuna river system: chemical weathering, its temperature dependence and CO_2 consumption in the Himalaya. Geochim Cosmochim Acta 66(19):3397–3416

21. Das BK, Dhiman SC (2003) Water and sediment chemistry of high Himalayan lakes in the Spiti Valley: control on weathering, provenance and tectonic setting of the basin. Environ Geol 44:717–730

22. DMG, Department of Mines and Geology (1980) Geological map of Central Nepal. Department of Mines and Geology, His Majesty's Government of Nepal, Kathmandu, Nepal

23. DMG, Department of Mines and Geology (1994) Geological map of Nepal. Department of Mines and Geology, His Majesty's Government of Nepal, Kathmandu, Nepal

24. Dorn RI, Brady PV (1995) Rock-based measurement of temperature dependent plagioclase weathering. Geochim Cosmochim Acta 59:2847–2852

25. Drever JI (1988) The Geochemistry of Natural Waters, 2nd edn. Prentice Hall, Englewood Cliffs

26. Drever JI, Zobrist J (1992) Chemical weathering of silicate rocks as a function of elevation in the southern Swiss Alps. Geochim Cosmochim Acta 56:3209–3216

27. Edmond JM, Palmer MR, Measures CI, Grant B, Stallard RF (1995) The fluvial geochemistry and denudation rate of the Guayana Shield in Venezuela, Columbia and Brazil. Geochim Cosmochim Acta 59:3301–3325

28. Evans MJ, Derry LA, France-Lanord C (2004) Geothermal fluxes of alkalinity in the Narayani river system of central Nepal. Geochem Geophys Geosyst 5(8):Q08011.

29. Fairchild IJ, Bradby L, Sharp M, Tison JL (1994) Hydrochemistry of carbonate terrains in alpine glacial settings. Earth Surf Proc Land 19(1):33–54

30. France-Lanord C, Derry L (1997) Organic carbon burial forcing of the carbon cycle from Himalayan erosion. Nature 390:65–67

31. France-Lanord C, Evans M, Hurtrez JE, Riotte J (2003) Annual dissolved fluxes from central Nepal rivers: budget of chemical erosion in the Himalayas. Comptes Rendus Geosci 335(16):1131–1140

32. Fukusima Y, Kawashima K, Suzuki M, Ohta T, Motoyama K, Kubota H, Yamada T, Bajracharya OR (1987) Runoff characteristics in three glacier-covered watershed of Langtang Valley, Nepal Himalaya. Bull Glacier Res 5:11–18

33. Gabet EJ (2007) A theoretical model coupling chemical weathering and physical erosion in land-slide dominated landscapes. Earth Planet Sci Lett 264:259–265

34. Gabet EJ, Wolff-Boenisch D, Langner H, Burbank DW, Putkonen J (2010) Geomorphic and climatic controls on chemical weathering in the High Himalayas of Nepal. Geomorphology 122:205–210

35. Gaillardet J, Dupre B, Louvat P, Allegre CJ (1999) Global silicate weathering and CO_2 consumption rates deduced from the chemistry of large rivers. Chem Geol 159(17):3–30

36. Galy A, France-Lanord C (1999) Weathering processes in the Ganga-Brahmaputra basin and the riverine alkalinity budget. Chem Geol 159(1–4):31–60

37. Garrels RM, Mackenzie FT (1967) Origin of chemical compositions of some springs and lakes. Equilibrium concepts in natural water system, Advanced Chemistry Series 67. American Chemical Society, Washington DC, pp 222–242

38. Green WT, Stage BR, Preston A, Wagers S, Shacat J, Newell S (2005) Geochemical processes in the Onyx River; Wright Valley, Antarctica: major ions, nutrients, trace metals. Geochim Cosmochim Acta 69(4):839–850

39. Hartmann J (2009) Bicarbonate fluxes and CO_2 consumption by chemical weathering on the Japanese Archipelago: application of multi-lithological model framework. Chem Geol 265:237–271

40. Hasnain SI, Thayyen RJ (1999) Controls on the major ion chemistry of Dokriani glacier meltwaters, Ganga basin, Garhwal Himalaya, India. J Glaciol 45:87–92

41. Henriksen A, Mill WA, Kot M, Rzychon D, Wathne BM (1997) Critical loads of acidity to surface waters. A case study from Polish Tatra Mountains. Acid rain research. NIVA Report 29, p 34

42. Hirschmiller J, Grujic D, Bookhagen B, Coutand I, Huyghe P, Mugnier J-L, Ojha T (2014) What controls the growth of the Himalayan foreland fold-and-thrust belt? Geology 42(3):247–250

43. Hodson A, Porter P, Lowe A, Mumford P (2002) Chemical denudation and silicate weathering in Himalayan glacier basins: Batura Glacier, Pakistan. J Hydrol 262:193–208

44. Huh Y, Panteleyev G, Babich D, Zaitsev A, Edmund JM (1998) The fluvial geochemistry of the rivers of eastern Siberia: II. Tributaries of the Lena, Omoloy, Yana, Indigirka, Kolyma, and Anadyr draining the collisional/accretionary zone of the Verkhoyansk and Cherskiy ranges. Geochim Cosmochim Acta 62(12):2053–2075

45. ICIMOD (2011) Glacial lakes and glacial lake outburst floods in Nepal. Nepal, International Centre for Integrated Mountain Development, Kathmandu, p 99

46. ICIMOD-UNEP (2002) Inventory of glaciers, glacial lakes and glacial lake outburst floods monitoring and early warning systems in the Hindu Kush Himalayan Region—Nepal. International Center for Integrated Mountain Development and United Nations Environmental Program, pp 197

47. Inger S, Harris BW (1992) Tectonothermal evolution of the High Himalayan crystalline sequence, Langtang Valley, Northern Nepal. J Metamorph Geol 10:439–452

48. Ittekkot V (1988) Global trends in the nature of organic matter in river suspensions. Nature 332:436–438

49. Jenkins A, Sloan WT, Cosby BJ (1995) Stream chemistry in the middle hills and mountains of the Himalayas, Nepal. J Hydrol 166(1–2):61–79

50. Kamiyama K (1984) Lakes and sediments around Yala Glacier. In: Higuchi K (ed) Glacier studies in Langtang Valley. Data Center of Glacier Research, Water Research Institute, Nagoya University, Nagoya, pp 85–89

51. Kattelmann R (2003) Glacial lake outburst floods in Nepal Himalaya: a manageable hazard? Nat Hazard 28(1):145–154

52. Keene WC, Pszenny AAP, Galloway JN, Hawley ME (1986) Sea-salt corrections and interpretation of constituent ratios in marine precipitation. J Geophys Res 91(D6):6647–6658

53. Kirkbride MP (1993) The temporal significance of transitions from melting of calving termini at glaciers in the central Southern Alps of New Zealand. Holocene 3:232–240

54. Krishnaswami S, Singh SK (2005) Chemical weathering in the river basins of the Himalaya, India. Curr Sci 89(5):841–849

55. Lliboutry L, Morales B, Pautre A, Schneider B (1977) Glaciological problems set by the control of dangerous lakes in Cordillera Blanca, Peru. I. Historical failure of moraine dams, their causes and prevention. J Glaciol 18:239–254

56. Ludwig W, Amiotte Suchet P, Munhoven G, Probst JL (1998) Atmospheric CO_2 consumption by continental erosion: present-day controls and implications for the last glacial maximum. Global Planet Change 16–17:107–120

57. Lupker M, France-Lanord C, Galy V, Lave J, Gaillardet J, Gajurel AP, Guilmette C, Rahman M, Singh SK, Sinha R (2012) Predominant floodplain over mountain weathering of Himalayan sediment (Ganga basin). Geochim Cosmochim Acta 84:410–432

58. McDowell WH, Asbury CE (1994) Export of Carbon, Nitrogen, and Major ions from three tropical montane watersheds. Limnol Oceanogr 39(1):111–125

59. McDowell WH, Sanchez CG, Asbury CE, Ramos Perez CR (1990) Influence of sea salt aerosols and long range transport on precipitation chemistry at El Verde, Puerto Rico. Atmos Environ 24(11):2813–2821

60. Meybeck M, Ragu A (1997) River Discharges to the Oceans: An assessment of suspended solids, major ions and nutrients. Nairobi, UNEP/WHO, Environment of Information and Assessment Division, UNEP, p 245

61. Millot R, Gaillardet J, Dupre B, Allegre CJ (2002) The global control of silicate weathering rates and the coupling with physical erosion: new insights from rivers of the Canadian Shield. Earth Planet Sci Lett 196:83–98

62. Millot R, Gaillardet J, Dupre B, Allegre CJ (2003) Northern latitude chemical weathering rates: clues from the Mackenzie River Basin, Canada. Geochim Cosmochim Acta 67(7):1305–1329

63. Ming J, Xiao C, Du Z, Yang X (2013) An overview of black carbon deposition in high Asia glaciers and its impact on radiation balance. Adv Water Resour 55:80–87

64. Musselman RC, Hudnell L, Williams MW, Summerfeld RA (1996) Water chemistry of rocky mountain front range aquatic ecosystems.

Research Paper RM-RP-325, USDA Forest Service, Fort Collins, CO, USA, p 13

65. Navarre-Sitchler A, Brantley SL (2007) Basalt weathering across scales. Earth Planet Sci Lett 261:321–334

66. Pandey SK, Singh AK, Hasnain SI (1999) Weathering and geochemical processes controlling solute acquisition in Ganga headwater-Bhagirathi river, Garhwal Himalaya, India. Aquat Geochem 5(4):357–379

67. Quade J, English N, DeCelles PG (2003) Silicate versus carbonate weathering in the Himalaya: a comparison of the Arun and Seti river watersheds. Chem Geol 202(3–4):275–296

68. Reynolds B, Chapman PJ, French MC, Jenkin A, Wheater HS (1995) Major, minor and trace element chemistry of surface waters in the Everest region of Nepal. In: Tonnessen KA, Williams MW, Tranter M (eds) Biogeochemistry of seasonally snow-covered catchments. IAHS Publ no 228, pp 405–412

69. Richardson SD, Reynolds JM (2000) An overview of glacial hazards in the Himalayas. Quat Int 65/66(1):31–47

70. Riebe CS, Kirchner JW, Finkel RC (2004) Sharp decrease in long-term chemical weathering rates along an altitudinal transects. Earth Planet Sci Lett 218(3–4):421–434

71. Sakai A, Takeuchi N, Fujita K, Nakawo M (2000) Role of supraglacial ponds in the ablation process of a debris-covered glacier in the Nepal Himalayas. In: Proceedings of a workshop on debris covered glaciers held at Seattle, Washington, USA. IAHS Publ no 265, pp. 119–130

72. Sakai A (2012) Glacier lakes in the Himalayas: a review on formation and expansion processes. Global Environ Res 16:23–30

73. Sakai A, Fujita K (2010) Formation condition of supraglacial lakes on debris-covered glaciers in Himalayas. J Glaciol 56:177–181

74. Sakai A, Nakawo M, Fujita K (1998) Melt rate of ice cliffs on the Lirung Glacier, Nepal Himalayas. Bull Glacier Res 16:57–66

75. Scherler D, Bookhagen B, Strecker MR (2011) Spatially variable response of Himalayan glaciers to climate change affected by debris cover. Nat Geosci.

76. Sharp M, Tranter M, Brown GH, Skidmore M (1995) Rates of chemical denudation and CO_2 drawdown in glacier-covered Alpine Catchments. Geology 76:195–206

77. Stallard RF (2000) Tectonic Processes and Erosion. Int Geophys 72:195–229

78. Stallard RF, Edmond JM (1981) Geochemistry of the Amazon 1. Precipitation chemistry and the marine contribution to the dissolved load at the time of peak discharge. J Geophys Res 86:9844–9858

79. Stallard RF, Edmond JM (1983) Geochemistry of the Amazon 2. The influence of geology and weathering environment on the dissolved load. J Geophys Res 88(C14):9671–9688

80. Subramanian V, Ittekkot V (1991) Carbon transport by the Himalayan rivers. In: Degens ET, Kempe S, Richey JE (eds) Biogeochemistry of Major World Rivers, SCOPE. Wiley, Chichester, pp 157–168

81. SD, Survey Department (1984) Land system map, central development region, Nepal. Topographical Survey Branch, Survey Department. Ministry of Land Reform, HMG, Kathmandu, Nepal

82. Takeuchi N, Kohshima S (2000) Effect of debris cover on species composition of living organism in supraglacial lakes on a Himalayan glacier. In: Proceedings of a workshop on debris covered glaciers held at Seattle, Washington, USA. IAHS Publ no 264, pp. 267–275

83. Takeuchi N, Sakai A, Kohshima S, Fujita K, Nakawo M (2012) Variation in suspended sediment concentration of supraglacial lakes on debris covered area of the Lirung glacier in the Nepal Himalayas. Aires Glob Environ Res 16:95–104

84. Tartari GA, Tartari G, Mosello R (1998) Water chemistry of high altitude lakes in the Khumbu and Imja Khola Valleys (Nepalese Himalayas). Mem Ist Ital Idrobiol 57:51–76

85. R Development Core Team (2015). R: a language and environment for statistical computing Retrieved 2015, from http://www.R-project.org

86. Tipper ET, Bickle MJ, Galy A, West AJ, Pomies C, Chapman HJ (2006) The short term climatic sensitivity of carbonate and silicate weathering fluxes: insight from seasonal variations in river chemistry. Geochim Cosmochim Acta 70(11):2737–2754

87. Tranter M, Raiswell R (1991) The composition of the englacial and subglacial component in bulk meltwaters draining the Gornergletscher, Switzerland. J Glaciol 37:59–66

88. Tranter M, Brown G, Raiswell R, Sharp M, Gurnell A (1993) A conceptual model of solute acquisition by Alpine glacier meltwaters. J Glaciol 39:573–581

89. Wadham JL, Hodson AJ, Tranter M, Dowdeswell JA (1998) The hydrochemistry of meltwaters during the ablation season at a high Arctic, polythermal-based glacier, South Svalbard: I. The ablation season. Hydrol Process 12:1825–1849

90. Watanabe O, Takenaka S, Iida H, Kamiyama K, Thapa KB, Mulmi DD (1984) First results from Himalayan Glacier Boring Project in 1981–1982: Part I Stratigraphic analyses of full depth cores from Yala glacier, Langtang Himal, Nepal. In: Higuchi K (ed) Glacial studies in Langtang Valley. Data Centre for Glacier Research, Water Resources Institute, Nagoya University, Nagoya, Japan, pp 7–23

91. West AJ, Bickle MJ, Collins R, Brasington J (2002) Small-catchments perspective on Himalayan weathering. Geology 30(4):355–358

92. West AJ, Galy A, Bickle M (2005) Tectonic and climate controls on silicate weathering. Earth Planet Sci Lett 235(1–2):211–228

93. White AF, Blum AE (1995) Effects of Climate on chemical weathering in watersheds. Geochim Cosmochim Acta 59(9):1729–1747

94. Wolff-Boenisch W, Gabet E, Burbank DW, Langner H, Putkonen J (2009) Spatial variations in chemical weathering and CO_2 consumption in Nepalese High Himalayan catchments during monsoon season. Geochim Cosmochim Acta 73:3148–3170

95. WWF-Nepal (2005) An overview of glaciers, glacier retreat, and subsequent impacts in Nepal, India, and China. Nepal, Kathmandu, p 70

96. Yamada T, Sharma CK (1993) Glacier lakes outburst floods in the Nepal Himalaya. IAHS 218:319–330

8

LATITUDINAL TRENDS AND DRIVERS IN THE CO_2–CARBONIC ACID SYSTEM OF PAPAHĀNAUMOKUĀKEA MARINE NATIONAL MONUMENT

Andrea K. Kealoha[1] • Samuel E. Kahng[1] • Fred T. Mackenzie[2] • Simone R. Alin[3] • Randall K. Kosaki[4] • Russell E. Brainard[5] • Christopher D. Winn[1]

1 Hawaii Pacific University, 1164 Bishop Street, Honolulu, HI 96813, USA

2 University of Hawaii at Manoa, 1000 Pope Road, Honolulu, HI 96822, USA

3 NOAA Pacific Marine Environmental Laboratory, 7600 Sand Point Way NE, Seattle, WA 98115, USA

4 NOAA Papaha¯naumokua¯kea Marine National Monument, 1845 Wasp Blvd., Honolulu, HI 96818, USA

5 NOAA Coral Reef Ecosystem Division, 1845 Wasp Blvd., Honolulu, HI 96818, USA

ABSTRACT

Emissions of anthropogenic carbon dioxide (CO_2) to the atmosphere and the consequent effects of climate change and ocean acidification on coral reef ecosystems have motivated significant interest in describing and understanding the CO_2–carbonic acid system of diverse coral reef environments. Although numerous studies have been successful in monitoring reef metabolism both in the field and in the laboratory, physical and biological forcings produce distinct conditions among environments. Due to the geographic isolation and associated difficulties with measuring marine carbon chemistry in waters of the Papahānaumokuākea Marine National Monument (PMNM), relatively few studies have described the CO_2–carbonic acid system and carbonate saturation state gradients of these waters. Yet, PMNM remains one of the largest conservation areas in the world with extensive and diverse coral reef ecosystems that could offer valuable insight into our current and future understanding about regional and global impacts of ocean acidification. In order to provide a broad overview of latitudinal trends and features of the marine carbon system in PMNM waters, continuous measurements for surface seawater fugacity of CO_2 (fCO_2) and pH were collected during August 2011 and July 2012 cruises of the NOAA Ship *Hi'ialakai*. These measurements indicate that pH andfCO_2 are three times more variable in nearshore monument waters relative to open ocean transect measurements. This variability can be observed up to 50 km away from the nearest reef and is likely the result of the direct and significant impact of coral reef metabolism on marine carbon chemistry around the islands and atolls. The relationship between total alkalinity and dissolved inorganic carbon is consistent with net calcification which creates an alkalinity sink throughout PMNM waters. Additionally, our measurements show clear latitudinal trends in fCO_2, pH, and aragonite saturation state that are influenced by environmental forcings, including temperature, wind speed, and residence time of the water. Collectively, our results suggest that coral reefs located at the northernmost atolls of PMNM may be the most susceptible to the adverse impacts of climate change and ocean acidification.

KEYWORDS

Carbon, Ocean acidification, Coral reef, Papahānaumokuākea Marine National Monument

1 INTRODUCTION

Papahānaumokuākea Marine National Monument (PMNM), encompassing the Northwestern Hawaiian Islands (NWHI), is home to a diverse array of biological, ecological, and cultural resources. The islands and coral reef atolls that make up this unique archipelago comprise one of the largest marine reserves in the world, encompassing nearly 140,000 square miles. In addition to serving as a biodiversity hotspot (DeMartini and Friedlander 2004; Kane et al. 2013) and important habitat to protected species, such as the critically endangered Hawaiian monk seal and threatened green sea turtle, the nearly uninhabited monument is in relatively pristine condition and exhibits a predator-dominated ecosystem structure observed in few places around the world (Friedlander and DeMartini 2002). Although conservation efforts in the monument are ongoing, the risks associated with climate change, including seawater acidification, sea level rise, and ocean warming, are considered threats to the health and abundance of resources in this important marine ecosystem (Selkoe et al. 2008).

Relatively few studies describing the trends and dynamics of the CO_2–carbonic acid system in marine waters of the NWHI have been published (Sabine and Mackenzie 1995; Thompson et al. 2014). Although Grigg (1982) quantified reef accretion rates as a function of latitude by measuring the linear growth rate of corals, there have been no recent publications describing the latitudinal gradients and features of the marine carbon system along the archipelago. The goal of this study is to provide a current baseline understanding of the CO_2–carbonic acid system of the NWHI, identify latitudinal gradients and features, and describe the processes and mechanisms that are responsible for these patterns. These measurements may provide some insight into the resilience and/or vulnerability of reef habitats in the monument waters undergoing the effects of climatic change, as well as assess the potential biological impacts in response to ocean acidification (OA).

2 CLIMATE CHANGE AND OCEAN ACIDIFICATION

Oceanic uptake of CO_2 decreases pH and shifts the distribution of dissolved inorganic carbon (DIC) species by increasing the partial pressure of carbon dioxide (pCO_2) and decreasing carbonate (CO32−)(CO32−) ion concentration and aragonite saturation state (Ω_{ar}), a process known as OA (Caldeira and Wickett 2003; Raven et al. 2005). Given the CO_2 emission scenarios described by the IPCC, surface ocean pH is expected to decline by 0.07–0.33 pH units by the end of the twenty-first century (Bopp et al. 2013). The maximum projected drop in pH is equivalent to an approximate 150 % increase in [H^+] and 50 % decrease in [CO32−][CO32−](Orr et al. 2005). A decrease in [CO32−][CO32−] may decrease the rate at which marine calcifying organisms form calcium carbonate shells and tests ($CaCO_3$) (e.g., Broecker 1971; Kleypas et al. 1999a, b).

At the present time, $CaCO_3$ saturation states are highest in shallow, warm tropical waters and lowest in cold, high-latitude regions and at depth in the sea, reflecting the influence of temperature and pressure on $CaCO_3$solubility (Morse and Mackenzie 1990; Feely et al. 2003; Doney et al. 2009). However, the Ω_{ar} of the global ocean is projected to decline by 2.8 by the year 2100, which reflects a 17–35 % decrease in calcification since preindustrial times. Additionally, the isolines of high Ω_{ar} will shift equatorward, narrowing the latitudinal boundaries for optimal reef growth (Kleypas et al. 1999a, b). Although some species of corals may be resilient to certain environmental conditions that inhibit calcification (Jokiel et al. 2008; Cohen and Holcomb2009), seasonally extreme temperature, and low-light conditions, combined with a decrease in Ω_{ar} may put calcifying organisms in the high-latitude regions as the first to experience the adverse impacts of OA (Grigg1982; Kleypas et al. 1999a, b; Andersson and Mackenzie 2004; Andersson et al. 2005; Orr et al. 2005).

The relative abundance of calcifying organisms such as bryozoans and crustose coralline algae (CCA) increases toward the northwest in the Hawaiian Archipelago (Chave 1954; Schlanger and Konishi 1975; Agegian and Mackenzie 1989). These organisms contain a higher percentage of Mg calcite, the most soluble phase of $CaCO_3$. CCA are an important component for the framework of coral reefs and facilitate the growth of larger structures by acting like cement that glues the reef together. The increased susceptibility of CCA–OA has been demonstrated in numerous studies (Anthony et al. 2008; Manzello et al. 2008; Fabricius et al.2011), and a loss of these sensitive reef-building species could result in a decline in the overall health of a coral reef.

3 SEASONAL TRENDS IN CO_2–CARBONIC ACID SYSTEM VARIABILITY IN THE NORTH PACIFIC SUBTROPICAL OCEAN

Oceanographic conditions and carbon chemistry trends in the North Pacific subtropical region are well characterized (Takahashi et al. 2009; Nakano and Watanabe 2005; http://www.socat.info/). Summer latitudinal (15°–35°N) gradients in sea surface temperature (SST) are relatively small (24–28 °C), whereas winter temperatures (15–25 °C) decrease rapidly with increasing latitude. Wind speed in the summer lacks a defined latitudinal gradient, while during the winter, wind speed increases in the north. (http://www.socat.info/). The average peak-to-peak ΔpCO_2 ($pCO_{2sw} - pCO_{2air}$) is approximately 40 μatm and reflects the influence of temperature on CO_2 solubility. The spatiotemporal distribution of pH below 40°N is small (<0.03) (Nakano and Watanabe 2005). These seasonal and geographic trends cause the northern ocean region to switch from a source of CO_2 in the summer to a CO_2 sink in the winter (Takahashi et al. 2009).

At the Hawai'i Ocean Time-Series (HOT) station A Long-term Oligotrophic Habitat Assessment (ALOHA), 100 km north of Oahu, pCO_2 in the surface seawater is monitored by time series observations. Since these waters are oligotrophic and average peak-to-peak SSTs are 23–27 °C, annual variations in surface pCO_2(350–390 μatm), pH (8.055–8.082), and total alkalinity (TA) (2305–2335 μeq kg^{-1}) are relatively small. Similarly, daily temperature only varies by about ±0.07 °C (Church et al. 2009), producing a diel pCO_2 signal of ±1.5 ppm (Karl 1999; Takahashi et al. 2002; Sarmiento and Gruber 2006; Dore et al. 2003; Sutton et al. 2010;http://hahana.soest.hawaii.edu/hot/hot-dogs/bseries.html).

Although time series data collected at station ALOHA reveal that open ocean waters surrounding the Hawaiian Islands currently act as a net sink for CO_2 (Dore 2009), monitoring studies have observed concurrent increases in atmospheric CO_2, surface seawater pCO_2, and DIC at station ALOHA (Winn et al.1998; Takahashi et al. 2006). The uptake of anthropogenic CO_2 is likely the cause for both long-term increases in DIC and decreases in $CaCO_3$ saturation state and pH (−0.02 units per decade) (Dore et al.2003). Over time, this shift in the distribution of DIC species will decrease the buffering capacity and the ability of the ocean to absorb anthropogenic CO_2 (Revelle and Suess 1957).

4 CORAL REEF METABOLISM IN COASTAL ENVIRONMENTS

Shallow-water environments play an important role in the global carbon cycle (Andersson and Mackenzie2004; Andersson et al. 2005) and display complex carbon system dynamics, particularly in tropical and subtropical coral reef ecosystems. Because of the great physical and biological heterogeneity of these environments (Gattuso et al. 1996; Lantz et al. 2013), spatial and temporal variations in the parameters of the CO_2–carbonic acid system are large and often difficult to assess accurately over time (Drupp et al. 2013). In addition to physical and environmental forcings, organic and inorganic metabolic processes significantly alter the chemical properties of the water. Photosynthesis and respiration and calcification and dissolution are opposing forces for changes in pCO_2 (Smith and Key 1975) and other components of the carbonate system (Morse and Mackenzie 1990; Gattuso et al. 1993, 1996; Yates and Halley 2006). Thus, determination of the physical and biological forcings driving changes is often difficult.

The effect of increased marine productivity near islands, or downstream from an island, is termed the 'island mass effect' (Doty and Oguri 1956). This effect refers to an upwelling-induced, biological enhancement around oceanic islands (Barton 2001). The topographic disturbance of an island assists in the formation of eddies downstream. These eddies can enhance the supply of nutrients into the euphotic zone and increase surface seawater productivity, stimulating higher biological carbon uptake and strengthening the flux of CO_2into the ocean (Caldeira et al. 2002; Bakker et al. 2007).

Lack of freshwater, nutrient and suspended solid inputs by rivers, and human discharge make the NWHI an ideal location to study the natural physical, biological, and chemical forcings that influence variability in the CO_2–carbonic acid system across the archipelago. Furthermore, studies describing this chemical system in the NWHI are relatively scarce, despite the critical need to characterize such environments for future time series observations tracking the long-term impacts of OA. In order to understand the processes that influence the marine carbon system and quantify the magnitude of variability in carbon parameters both spatially and temporally within coral reef ecosystems and adjacent waters of the NWHI, continuous high-resolution measurements of surface seawater fCO_2 and pH were made during oceanographic cruises in the summer seasons of 2011 and 2012 around the islands and atolls of the NWHI.

5 DESCRIPTION OF NORTHWESTERN HAWAIIAN ISLANDS AREA

PMNM (Fig. 1) is comprised of a number of islands and coral reef atolls in a relatively pristine environment undisturbed by local anthropogenic activity (Desch et al. 2009). The NWHI chain extends from the island of

Nihoa (23°N, 161°W) in the south to Kure atoll (28°N, 178°W) in the north. Due to the large expanse of the Monument, various natural factors play a role in influencing the structure of NWHI reef ecosystems. While SSTs at the southern end of the chain vary annually by 4 °C (~23–27 °C), SSTs at the northern end may vary annually by up to 10 °C (18–29 °C) (Friedlander et al. 2005; Desch et al. 2009). Surface currents in the NWHI are highly variable in both speed and direction (Firing et al. 2006). In the mean, surface currents at the southern end of the chain flow eastward, while the northern atolls experience predominantly southwestward flow. Winter storms in the North Pacific also affect the spatial distribution of benthic organisms, as some corals and/or algae are sensitive to high-wave-energy events generated by these storms (Grigg et al. 2008; Desch et al. 2009; Rooney et al. 2008). Additionally, internal tidal-driven waves have been observed along the Hawaiian Ridge and have the potential to cause turbulent mixing near island boundaries, bringing up waters that have different chemical characteristics than those of the overlying surface seawaters (Rudnick et al. 2003). Collectively, these oceanographic influences can affect the formation and distribution of coral reef communities in the NWHI (Polovina et al. 2001; Rudnick et al. 2003; Firing et al. 2006; Friedlander et al.2005; Grigg et al. 2008; Rooney et al. 2008; Desch et al. 2009).

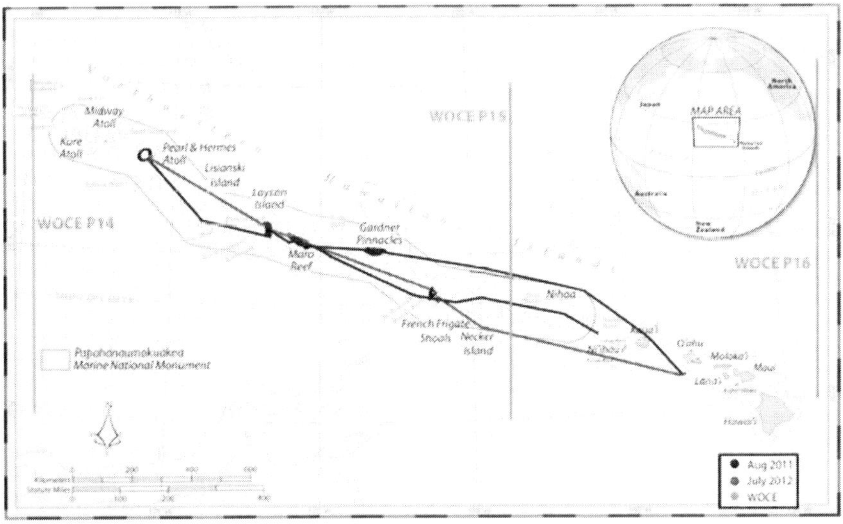

Figure 1. Map with boundaries of the PMNM (NOAA PMNM) and track of ship for August 2011 (*black*) and July 2012 (*red*) research cruises. Sections (20°–30°N) of WOCE legs P14, P15, and P16 are given in *white*

6 DATA COLLECTED AND QUALITY CONTROL

Continuous underway measurements of oceanic fCO_2, atmospheric fCO_2, pH, salinity, temperature, and oxygen were collected in surface seawaters of PMNM during August 2011 (7/24–8/21) and July 2012 (7/5–7/13) (see Fig. 1 for ship track). Meteorological data including wind speed and direction were recorded by the ship's instruments and are available on the National Oceanographic Data Center (NODC) website (http://www.nodc.noaa.gov/access/ accession numbers: 0075826, 0104266). Because the various instruments were sampled at different time intervals, the data were binned into the approximate 2-min sampling interval used by the underway system. These data include measurements in offshore waters throughout the monument, as well as shallow waters surrounding the islands and atolls.

Atmospheric and surface seawater fCO_2 were collected with a General Oceanics Model 8050 underwaypCO_2 system with a LI-COR, Inc., 7000 Infrared Gas (CO_2/H_2O) Analyzer (IRGA) located in the wet lab of the NOAA Ship *Hi'ialakai*. The instrument sampled air collected from the bridge and air equilibrated with seawater. Air samples were drawn in from both sides of the bridge where a sensor located 10 m above sea level analyzed wind speed and direction and used air from the upwind side least contaminated by the ship's stack gases. Seawater samples were collected approximately every 2 min from an intake port 5 m below the ship's waterline. These samples were collected with a ratio of 60 equilibrator samples to six atmospheric boundary layer samples (Pierrot et al. 2009). In order to calibrate the IRGA, the underway pCO_2 system measures four CO_2 standard gases every 2.5–3 h, and the differences between the certified values and the measured values are used to correct seawater and atmospheric CO_2 measurements. Offsets between certified and measured values for July 2012 indicate that on average the system was accurate to within ±2 μatm.

In order to calculate sea surface fCO_2 using mixing ratios determined by the IRGA, temperature and pressure inside the equilibrator, as well as SST, are required. Atmospheric pressure is also required to convert mixing ratios to partial pressures. Pressure inside the infrared analyzer is used to correct the signal for pressure effects. Temperature (within ±0.01 °C) and pressures (within ±0.2 mbars) were measured in order to achieve an accuracy of 0.2 μatm for atmospheric measurements and 2 μatm for surface seawater measurements (Dickson et al. 2007; Pierrot et al. 2009).

Measurements of SST were made using a digital oceanographic thermometer, Sea-Bird Electronic (SBE) 38, which is located at the intake port for the pCO_2 system. Sea surface salinity was measured and recorded by a SBE 21, and dissolved oxygen concentrations were determined using an Aanderra Optode 4330F, corrected for the influences of temperature and salinity. Both the salinity and oxygen sensors were located in the ship's wet lab.

pH was measured using a Honeywell Durafet (see Martz et al. 2010 for performance evaluation) pH electrode with a 2-min sampling interval. Duplicate water samples were collected daily for electrode calibration. Samples were collected in the ship's wet lab from the same inflow pipe used to measure $CO_2$and kept in 250-ml and 500-ml borosilicate glass bottles, which were allowed to overflow twice. Samples were then poisoned with 100–200 µl of saturated mercuric chloride ($HgCl_2$) solution to prevent further biological activity, and sealed with Apiezon grease. Rubber bands were affixed to ensure a tight seal. The samples were analyzed using spectrophotometric techniques (Clayton and Byrne 1993) and purified m-cresol purple dye (Liu et al. 2011) at 25 °C. A Lambda EZ210 spectrophotometer equipped with a temperature-controlled sample chamber was used for these analyses.

To test the accuracy of our laboratory measurements, certified reference materials (CRM) of a known TA, salinity, and DIC (Dickson et al. 2003) were used. Although CRMs are not available for pH, we calculated pH in CO_2 sys using the certified TA and DIC values (Pierrot et al. 2006). In order to limit CO_2 exchange, spectrophotometric pH was measured immediately after opening each bottle. Based on the offset between the calculated pH from the CRMs and our measured pH, the average difference was 0.006 pH unit with a standard deviation (SD) of 0.003 pH unit (Table S1). Thirty-three sets of duplicates were also analyzed for reproducibility and indicate the average difference between samples was 0.0002 pH unit (SD = 0.007) (Figure S1) (Dickson et al. 2007).

A total of 17 discrete samples were used for electrode calibration during July 2012. An average standard electrode potential (E_0) was calculated from spectrophotometric pH measurements on discrete samples corrected to in situ temperature. The average E_0 for July was −0.43485. This value, along with the voltage output from the electrode, was then applied in the calculation for pH over the entire transect.

7 RESULTS AND DISCUSSION

7.1 Coral Reef Metabolism and the Island Mass Effect

Net ecosystem calcification (NEC) over most coral reef environments is positive (i.e., NEC >0), while net organic production (NP) is near zero (Falter et al. 2011; Shamberger et al. 2011; Lantz et al. 2013). The resulting combination for net ecosystem production (NEP) causes coral reef environments to act as net sources of CO_2 to the atmosphere (Suzuki and Kawahata 2003). High-frequency carbon measurements collected during our study display more variability and consistently higher fCO_2 and lower pH values in surface seawaters than nearby transects in the open ocean, indicating the influence of coral reef metabolism and suggesting net calcification (i.e., net source of CO_2) over these reefs (Suzuki and Kawahata 1999; Borges et al. 2005; Fagan and Mackenzie 2007; Hofmann et al. 2011; Drupp et al. 2011, 2013). In order to describe quantitatively the magnitude of variability within our transects, we compared our underway $fCO_2$and pH to the underway pCO_2 and discrete pH data collected on the World Ocean Circulation Experiment (WOCE) transects located both west and east of the NWHI. The WOCE lines used for comparison here are P14 conducted in November–December 2007 (Fukasawa et al. 2008) and P16 conducted in February–March 2006 (Feely et al. 2008) (see Fig. 1).

7.1.1 August 2011 Transects

Although pH measurements were not made during the August 2011 cruise, these transects provided the greatest opportunity to examine latitudinal gradients in fCO_2 near the atolls. On the first transect traveling north, the surface seawaters of Nihoa, French Frigate Shoals (FFS), and Pearl and Hermes (P&H) were sampled. Traveling back south toward the main Hawaiian Islands (MHI), the Lisianski, Laysan, and Gardner Pinnacles surface seawaters was sampled. The fCO_2 data collected during the August 2011 transects clearly show an increase in the magnitude and variability of fCO_2 in shallow surface seawaters (denoted by the red circles) near the islands and atolls (Figs. 2, S6), while at further distances (denoted by the blue and green circles), lower values and less variability in fCO_2 is observed. The coefficient of variation (CV) was calculated for WOCE P16 (0.012), WOCE P14 (0.017), and NWHI August 2011 (0.034) and indicates two to three times higher variability in the NWHI measurements.

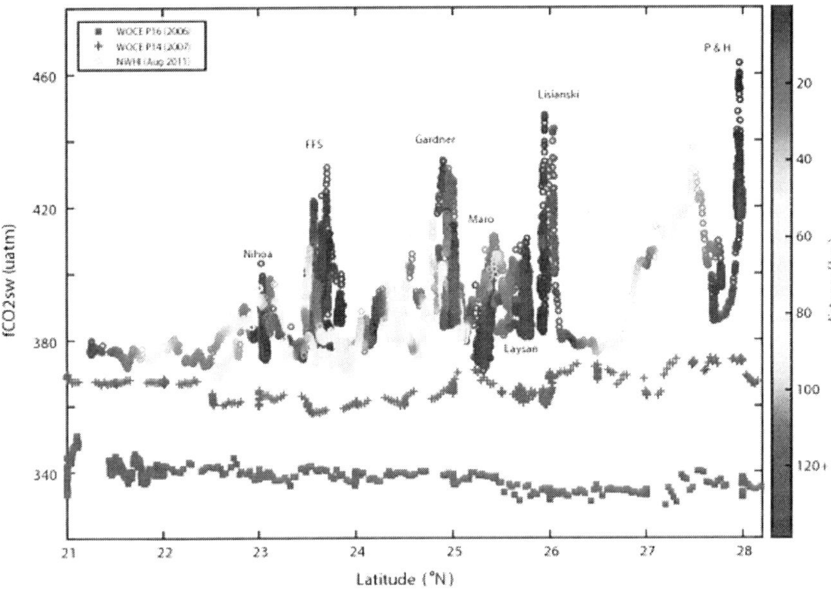

Figure 2. Underway surface seawater fCO_2 (at in situ *T*) during August 2011 (*circles*), WOCE P16 (*squares*), and WOCE P14 (*crosses*). The *colors* represent distance (km) from the nearest island or atoll

Whereas fCO_2 is relatively constant in surface seawaters of the WOCE transects, much more variability exists in our measurements within the monument waters, even while transiting between islands and at distances >50 km from the nearest reef (e.g., between 26°N and 28°N). Although this variability could be explained by turbulent mixing induced by eddies (Lumpkin 1998; Calil et al. 2008) or internal wave and tidal action along the island slope (Garrett and Munk 1979; Rudnick et al. 2003), surface seawaters would need to mix with waters at least 300 m below the surface (assuming a 50 % mixing ratio) to produce a 20–60 ppmfCO_2 increase in the open ocean near our transects. CTD profiles collected during our NWHI research cruises show that temperature and salinity at 300 m depth are generally lower by ~14 °C and 1.0 ppt, respectively, than in the surface seawater. These changes in salinity and temperature do not coincide with the fCO_2 increases observed on the NWHI transect. The increase in fCO_2 variability observed in our measurements throughout Monument waters therefore is probably primarily a result of photosynthesis/respiration and calcification/dissolution that occurs over and around the atolls within Papahānaumokuākea. The variability observed in our fCO_2 and pH measurements collected farther away from the islands is hypothesized to be a result of water parcels moving away from the reefs and not fully mixing with open ocean water.

There also appears to be an underlying trend in fCO_2 during the August 2011 cruises, with fCO_2 increasing with latitude near the atolls. Most coral reefs are balanced in terms of the uptake or contribution of DIC for photosynthesis and respiration (Falter et al. 2011; Shamberger et al. 2011; Lantz et al. 2013), respectively, and we can assume that organic metabolism in reefs of the NWHI are also balanced (Falter et al. 2011), since there is no land-derived inputs or evidence of upwelling in the northern region during the time of our study that would drive excess metabolism. If the latitudinal fCO_2 increase is not driven by net respiration, and net calcification produces CO_2, our 2011 data suggest that the northern reefs could be calcifying at a higher rate than reefs located at the southern end of the chain. Grigg (1982) found that net reef accretion of corals in the NWHI decreases with increasing latitude due to high-intensity storm events in the north, cooler water temperatures, less light availability, and subsidence of the Pacific plate. Therefore, other possible explanations for this trend were explored.

We have identified two main environmental factors that may contribute to the latitudinal increase in fCO_2during August 2011. First, there is a latitudinal increase in temperature (Figure S2). Normalizing fCO_2 with respect to temperature according to Takahashi et al. (1993, 2002) removes the effect of temperature on $fCO_2$for any given water sample (Fig. 3). Independent of temperature, fCO_2 at P&H is still ~20 μatm higher thanfCO_2 data measured in waters at lower latitudes indicating that temperature alone cannot account for all of the variability observed. Second, trade wind intensity and residence time of reef water also appear to play an important role in latitudinal carbon gradients.

The increase in fCO_2 of reef-associated surface seawaters over that of the open ocean during the cruise of August 2011 was coincident with a latitudinal decrease in trade wind intensity (Fig. 4) and a slight decrease in the flux of CO_2 to the atmosphere (Figure S5). Decreases in wind speed suppress air–sea exchange (Wanninkhof 1992; Fagan and Mackenzie 2007) and could decrease surface current velocities, which in turn may reduce the flushing of reef ecosystems and increase the residence time of waters within the lagoon. Additionally, islands located at the southern end of the chain (e.g., Nihoa and Gardner) and open atolls such as FFS and Maro Reef, are more exposed to flushing with the open ocean. The semi-enclosed northern atolls, including P&H and Midway, experience limited exchange between the lagoon and open ocean waters. Therefore, semi-enclosed bays and lagoons typically have longer residence times than fringing reef systems (Delesalle and Sournia 1992; Lowe et al. 2009), which contributes to greater magnitudes of change in carbon parameters (Shamberger et al. 2011) and increases fCO_2 through the processes of calcification and organic matter decomposition (Gattuso et al. 1996; Suzuki and Kawahata 1999; Kayanne et al. 2005).

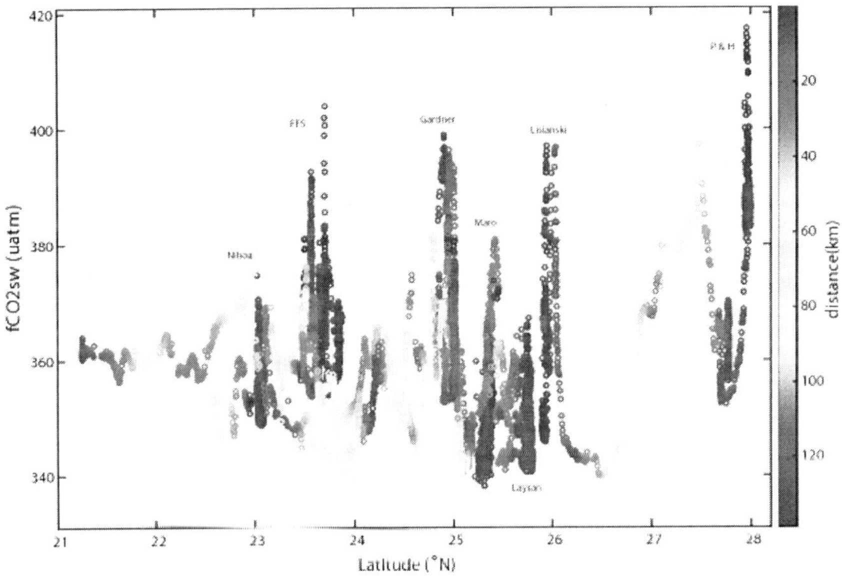

Figure 3. Underway surface seawater-normalized fCO_2 (at 25 °C) during August 2011. The *colors* represent distance (km) from the nearest island or atoll

7.1.2 July 2012 Transect

Both fCO_2 and pH data for surface seawaters were collected during the July 2012 cruise. July measurements indicated an increase in the variability (CV = 0.031) and magnitude of fCO_2 and pH in both nearshore and offshore waters of reef ecosystems (Fig. 5), relative to the WOCE transects. pH declined markedly (–0.06) in nearshore waters around FFS and also displayed a decreasing trend with increasing latitude. The latitudinal gradients in fCO_2, wind speed (Fig. 4) and CO_2 flux (Figure S5) are weak, in part, because only FFS and P&H were sampled. This increases our confidence that while biological processes are the dominant control on island-to-open ocean carbon gradients, latitudinal trends in wind speed and reef water residence time indirectly contribute to the build-up (or lack thereof) of CO_2.

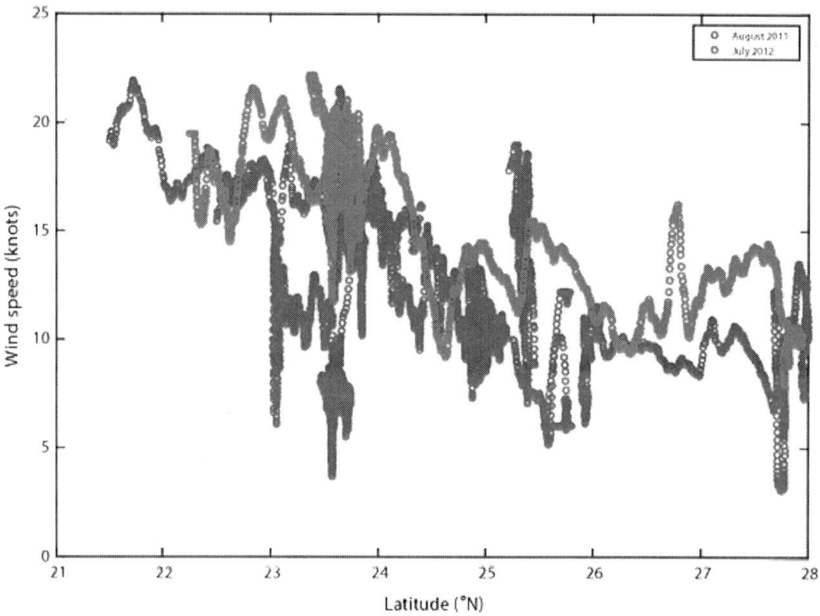

Figure 4. Wind speed (*knots*) as a function of latitude during August 2011 (*blue circles*) and July 2012 (*red circles*)

7.2 TA/DIC Relationship

Because both pH and pCO_2 were measured during the July cruise, we used these data to examine the latitudinal gradients in the TA/DIC ratio as well as in Ω_{ar}. Although TA and DIC can be computed from the fCO_2–pH pair, there are potentially significant errors associated with this calculation (Millero 2007). These errors are attributed to the strong covariance between pH and fCO_2 (Dickson and Riley 1978; Gray et al.2011). The uncertainty for our fCO_2 and our spectrophotometric pH calibration measurements was ±2 µatm and ±0.003 pH units, respectively. This combination of uncertainties can produce inaccuracies of 15–30 µmol/kg for DIC and alkalinity, respectively. However, since the calculation of both alkalinity and DIC is similarly affected by the covariance of pH and fCO_2, considerably greater confidence can be attributed to the TA/DIC ratio. In fact, doubling the uncertainties in both fCO_2 and pH described above yields a potential error of only 0.2 % in the ratio of these parameters.

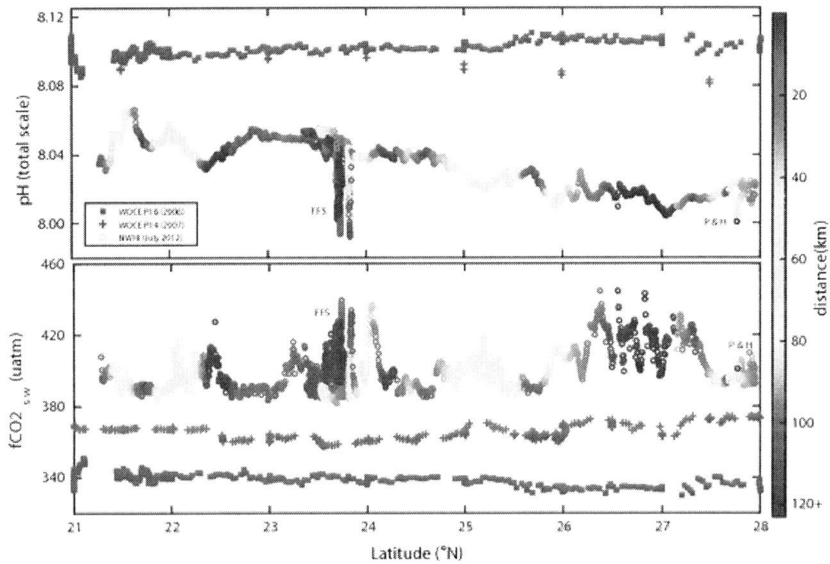

Figure 5. Underway surface seawater pH (*top*) and fCO_2 (*bottom*) during July 2012 (*circles*), WOCE P16 (*squares*) and WOCE P14 (*crosses*). Discrete pH samples were collected during the WOCE transects. The *colors* represent distance (km) from the nearest island or atoll. All values are measured at in situ temperature

The TA/DIC ratios computed from our underway data varied between 1.144 and 1.160 (Fig. 6), with the lowest values observed at FFS. The TA/DIC at station ALOHA during summer months, as well as the ratios on the available WOCE lines east and west of the island, are generally higher than the ratios we observed (Table 1). The relatively low TA/DIC ratios observed on the July transect are consistent with the higher

$fCO_2$and lower pH within the archipelago (Fig. 5), and are produced by TA decreasing at a higher rate than DIC. This pattern is consistent with net calcification that intensifies near the reefs and creates an alkalinity sink in the coastal waters surrounding the NWHI. A net increase in nDIC in the waters surrounding the NWHI is unlikely given that it would require net heterotrophy or net air–sea invasion of CO_2. Given the late summer timing of the NWHI surveys, these scenarios are implausible.

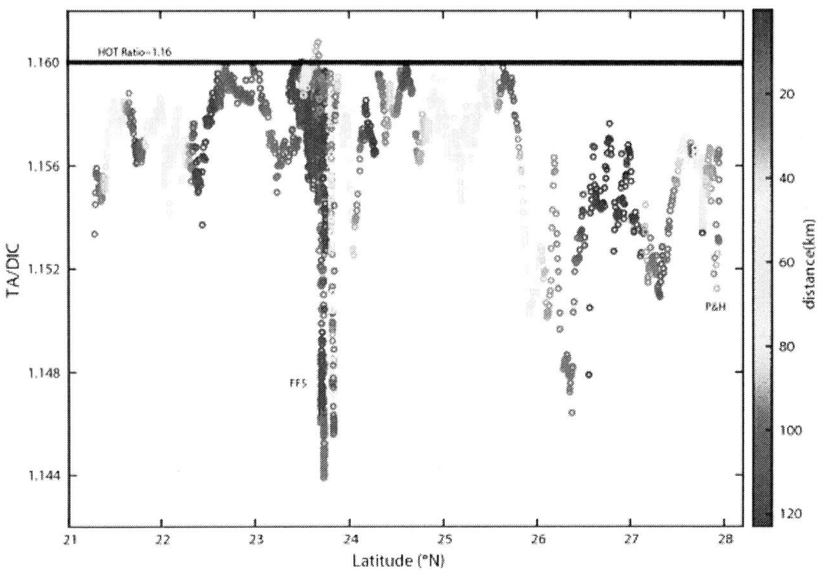

Figure6. Relationship between calculated TA and DIC with latitude during July 2012. The *black line* is the average TA/DIC over 1998–2012 at the Hawaii Ocean Time series station ALOHA. The *colors*represent distance (km) from the nearest island or atoll

Plotting changes in the TA relative to DIC can provide some insight into the relative impact of organic photosynthesis/respiration and $CaCO_3$ calcification/dissolution on the marine carbon system (Moulin et al.1985; Morse and Mackenzie 1990; Andersson and Gledhill 2013). Typical coral reef environments exhibit a slope of 0.95 when changes in TA are plotted versus changes in DIC (Andersson and Gledhill 2013; Andersson et al. 2014). However, studies have reported slopes ranging from 0.26 to 2.02 (Suzuki and Kawahata 2003; Lantz et al. 2013). A scatterplot of nTA versus nDIC for all of the data collected within the archipelago in July displays a remarkably consistent relationship with a least squares slope of 1.17 (Fig. 7). At the temperatures and salinities observed during the July transect, a slope of 1.22 is expected if the ocean remains in equilibrium with the atmosphere. The slightly lower slope observed through the NWHI suggests that the reaction pathway of the carbon system within the surface seawaters of the archipelago is predominantly governed by calcification and dissolution.

The net effect of the NWHI reef metabolism on the waters of the North Pacific Subtropical Gyre (NPSG) can be illustrated by comparing the average nTA and nDIC values for WOCE P-14 and P-16 (2300.9 and 1967.3, respectively) to the overall average for the NWHI (2227.1 and 1926.3, respectively) (Table 1). Assuming the WOCE data are representative of NPSG water unaffected by the Hawaiian Archipelago, the overall difference reveals a nTA/nDIC slope of approximately 1.80 which is close to the theoretical 2.00 value expected solely from net calcification without exchange with the overlying atmosphere.

7.3 Aragonite Saturation

As expected, Ω_{ar} showed the greatest range within the waters surrounding FFS, decreasing in shallow water near the reef due to calcification and increasing further away from the atoll as a result of dissolution and likely some mixing with open ocean water (Fig. 8). The decrease in pH with latitude also forces a latitudinal decrease in Ω_{ar}. Previous studies have reported average near-surface (~20 m depth) Ω_{ar} of 4.80 (Sabine1992; Sabine and Mackenzie 1995) and 3.61 (Thompson et al. 2014). Here, we report aragonite saturation levels ranging from 3.03 to 3.82 (mean = 3.44), with the lowest values measured in nearshore waters around FFS and in the northern end of the NWHI chain. Although our measurements provide greater coverage and data closer to shore, an overall decrease (1.36) in the mean value of Ω_{ar} over the past 20 years may be an indication of long-term change as a consequence of OA.

Table 1. Temperature, average normalized alkalinity, average normalized DIC, and average nTA/nDIC for WOCE P14 (1993, 2007), WOCE P15 (1994), WOCE P16 (1997, 2006), and this NWHI study (2012)

Program	Sample period	Temperature range	Average nTA	Average nDIC	Average nTA/nDIC
WOCE P14 1993	July	20.1–28.1	2298.0	1958	1.1736
WOCE P15 1994	October	21.3–27.8	2302.3	1962	1.1734
WOCE P16 1997	November	24.2–25.8	2301.5	1974.8	1.1654
WOCE P16 2006	March 2006	19.9–23.2	2302.2	1979.6	1.1630
WOCE P14 2007	October–November 2007	25.6–26.9	2300.6	1962	1.1726
Average			2300.9	1967.3	1.1696
This study	July 2012	25.1–27.3	2227.1	1926.3	1.1562
Difference			−73.8	−41.0	1.80

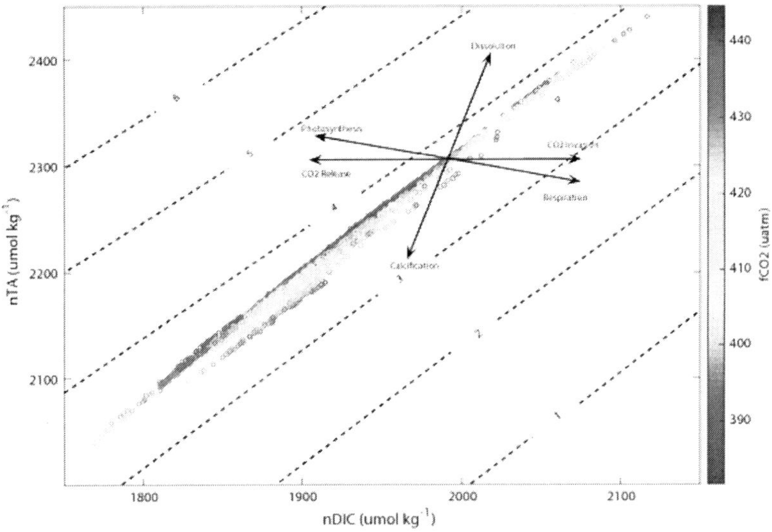

Figure7. TA plotted against DIC (slope = 1.17) for all data collected during July 2012. The *colors of the circles*represent fCO_2 concentration, and *dotted lines* represent Ω_{arag}

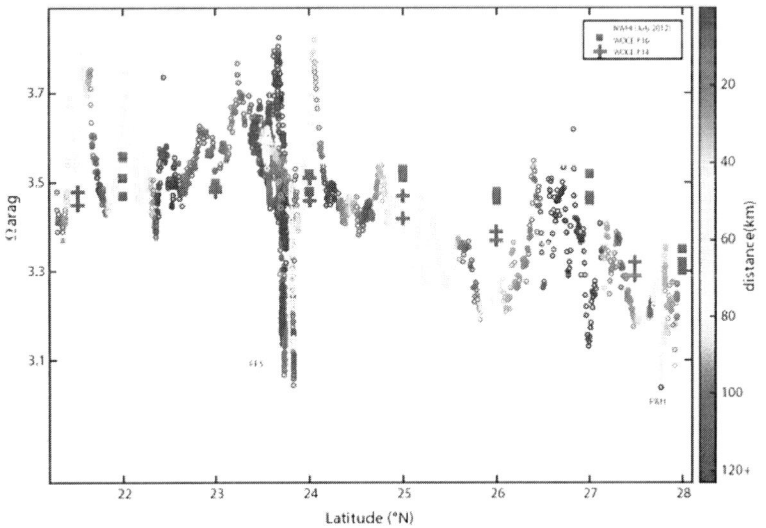

Figure 8. Aragonite saturation state during July 2012 (*blue circles*) calculated from underway fCO_2 and pH. WOCE P14 (*red triangles*) and WOCE P16 (*black boxes*) data are also shown. *Colors* represent distance (km) from the nearest island or atoll

Laboratory-derived studies have found that the calcification rate of corals decreases as carbonate saturation states decline (Gattuso et al. 1998), and previous research has shown that most coral reef communities exist in waters with $\Omega_{ar} > 3.3$ (Kleypas et al. 1999a, b). From a thermodynamic perspective, organisms expend more energy to calcify at lower saturation states, and at undersaturation, dissolution of $CaCO_3$ may exceed precipitation (Andersson et al. 2011; Mackenzie and Andersson 2013). A recent study by Shamberger et al. (2014) found that coral cover, diversity, and richness were maintained in waters with $\Omega_{ar} < 2.7$ at Rock Island, Palau. While this observation may have hopeful implications for some reefs in an OA world, northern reefs in the NWHI experience a combination of growth-inhibiting environmental factors (e.g., high temperature variability, strong storm events, less light availability) (Grigg 1982), in addition to low aragonite saturation states. Although the July 2012 transect only extended to P&H, a linear regression of the data yielded $-0.054\ \Omega_{ar}\ deg^{-1}$ °N, which would mean that reefs at Midway and Kure may experience average aragonite saturation states of 3.01 and 3.00, and chronically lower than those at P&H. The reef ecosystems at Midway and Kure may not have the capacity to maintain their growth rates under current or future OA conditions.

8 CONCLUSIONS

The observations presented in this study suggest that biological processes occurring within the coral reef ecosystems that comprise the NWHI significantly influence the CO_2–carbonic acid system in the surrounding coastal and open ocean environment. Our data demonstrate that considerably more variability in the seawater carbon system exists in and around the archipelago than in the surrounding open ocean. We interpret this spatial variability as being a consequence of the mixing of waters directly impacted by the coral reef ecosystem with open ocean waters being advected generally east to west across the NPSG. Our data also show that the impact of reef metabolism is not restricted to nearshore waters within the archipelago, but that the impact can be observed at distances up to 50 km from the nearest shallow water environments. In addition, our observations suggest that metabolic processes generate a significant decrease in the ratio of alkalinity to DIC relative to the open ocean waters both east and west of the archipelago. The change in this ratio is consistent with net calcification that decreases both alkalinity and DIC. These data taken together suggest that the reef ecosystems within the Pahapahānaumokuākea Marine National Monument have substantial impact on the inorganic carbon system in relatively vast regions (at least 10s of kilometers) around the archipelago.

Our data also show latitudinal variability within monument waters. There appeared to be gradients in $fCO_2$and pH that can be linked to latitudinal gradients in environmental factors, such as temperature, wind speed, and assumed atoll residence times that are produced, in part, by the configuration of the islands geomorphology. In addition, seawater aragonite saturation state values calculated from fCO_2 and pH during July 2012 reveal a decrease in nearshore waters as well as a latitudinal decrease in the surrounding open ocean. In addition to the lower Ω_{ar} values observed in this study, coral reefs of the northern atolls are subject to conditions less favorable for reef accretion due to higher wave energy during the winters, extreme seasonal variability in water temperatures, and lower solar insolation (Grigg 1982; Andersson and Mackenzie2004; Andersson et al. 2005; Orr et al. 2005; Rooney et al. 2008; Doney et al. 2009; Gove et al. 2013), relative to the southern reefs. Although we do not have winter carbon data in the NWHI, the environmental conditions of the North Pacific subtropical ocean would likely drive Ω_{ar} approximately one unit lower than the summer values observed in this study. Furthermore, the abundance of high-Mg-calcite-secreting organisms (i.e., coralline algae and bryozoa) increases in the northern Hawaiian waters (Chave 1954; Schlanger and Konishi 1975). These mineral compositions are more soluble than those of calcite and even aragonite. At the lower carbonate saturation states of northern waters, as anthropogenic CO_2 is added to surface ocean water, Mg calcite mineral phases are probably most at risk of dissolution, relative to organisms with primarily calcite and aragonite mineral compositions (Bischoff et al. 1987; Morse et al. 2006; Andersson et al. 2008; Manzello et al. 2008). Thus, the coral reef ecosystems of these northern atolls may be most susceptible to the adverse effects of climate change and OA. Additional surveying and monitoring of underway fCO_2 and pH in the Monument waters may help to quantify net calcification rates across the archipelago, measure the long-term effects of OA, and assist in the development of effective and informed management plans.

ACKNOWLEDGMENTS

The authors would like to thank Cathy Cosca and Geoff Lebon at NOAA Pacific Marine Environmental Laboratory. This work would not have been possible without their ongoing support through data management and mechanical expertise. We would also like to thank the NOAA *Hi'ialakai* ship crew including Michael Crumley and Verne Murakami for engineering and operational support, and Jackie Troller, Oscar Roig, and John Burns for their sampling support. Special thanks to Dr. Eric De Carlo, Coulson Lantz, and Bobby Thompson for the many fruitful discussions. This work was funded by NOAA's Office of National Marine Sanctuaries, through the Pahapahānaumokuākea Marine National Monument. This is PMEL Contribution Number 4253.

REFERENCES

1. Agegian CR, Mackenzie FT (1989) Calcareous organisms and sediment mineralogy on a mid-depth bank in the Hawaiian Archipelago. Pac Sci 43(1):56–66

2. Andersson AJ, Gledhill D (2013) Ocean acidification and coral reefs: effects on breakdown, dissolution, and net ecosystem calcification. Annu Rev Mar Sci 5:1.1–1.28

3. Andersson AJ, Mackenzie FT (2004) Shallow-water oceans: a source or sink of atmospheric CO_2? Front Ecol Environ 2(7):348–353

4. Andersson AJ, Mackenzie FT, Lerman A (2005) Coastal ocean and carbonate systems in the high CO_2 world of the Anthropocene. Am J Sci 305:875–918

5. Andersson AJ, Mackenzie FT, Bates NR (2008) Life on the margin: implications of ocean acidification on Mg-calcite, high latitude and cold-water marine calcifiers. Mar Ecol Prog Ser 373:265–273

6. Andersson AJ, Mackenzie FT, Gattuso J-P (2011) Effects of ocean acidification on benthic processes, organisms, and ecosystems. In: Gattuso J-P, Hansson L (eds) Ocean acidification. Oxford University Press, New York, pp 122–153

7. Andersson AJ, Yeakel KL, Bates NR, de Putron SJ (2014) Partial offsets in ocean acidification from changing coral reef biogeochemistry. Nat Clim Chang 4:56–61

8. Anthony KRN, Kline DI, Diaz-Pulido G, Dove S, Hoegh-Guldberg O (2008) Ocean acidification causes bleaching and productivity loss in coral reef builders. Proc Natl Acad Sci USA 105(45):17442–17446

9. Bakker CE, Nielsdottir MC, Morris PJ, Venables HJ, Watson AJ (2007) The island mass effect and biological carbon uptake for the subantarctic Crozet Archipelago. Deep Sea Res II 54:2174–2190

10. Barton ED (2001) Island wakes. In: Steele JH, Thorpe SA, Turekian KK (eds) Encyclopedia of ocean sciences, vol 5. Academic Press, Orlando, pp 1397–1402

11. Bischoff WD, Mackenzie FT, Bishop FC (1987) Stabilities of synthetic magnesian calcites in aqueous solution: comparison with biogenic material. Geochim Cosmochim Acta 51:581–589

12. Bopp L, Resplandy L, Orr JC, Doney SC, Dunne JP, Gehlen M, Halloran P, Heinze C, Ilyina T, Seferian R, Tjiputra J, Vichi M (2013) Multipe stressors of ocean ecosystems in the 21st century: projection with the CMIP5 models. Biogeosciences 10:6225–6245

13. Borges AV, Delille B, Frankignoulle M (2005) Budgeting sinks and sources of CO_2 in the coastal ocean: diversity of ecosystems counts. Geophys Res Let 32(14):L14601

14. Broecker WS (1971) Calcite accumulation rates and glacial–interglacial changes in ocean mixing. In: Turekian KK (ed) The late Cenozoic glacial ages. Yale University Press, New Haven, pp 239–265

15. Caldeira K, Wickett M (2003) Anthropogenic carbon and ocean pH. Nature 425(6956):365

16. Caldeira RMA, Groom S, Miller P, Pilgrim D, Nezlin NP (2002) Sea-surface signatures of the island mass effect phenomena around Madeira Island, Northeast Atlantic. Remote Sens Environ 80:336–360

17. Calil PH, Richards KJ, Jia Y, Bidigare RR (2008) Eddy activity in the lee of the Hawaiian Islands. Deep Sea Res Part II 55(10):1179–1194

18. Chave KE (1954) Aspects of the biogeochemistry of magnesium 1. Calcareous marine organisms. J Geol 62(3):266–283

19. Church MJ, Mahaffey C, Letelier RM, Lukas R, Zehr JP, Karl DM (2009) Physical forcing of nitrogen fixation and diazotroph community structure in the North Pacific subtropical gyre. Glob Biogeochem Cycles 23(2):GB2020

20. Clayton T, Byrne R (1993) Spectrophotometric seawater pH measurements: total hydrogen ion concentration scale calibration of m-cresol purple and at-sea results. Deep Sea Res Part I 40(10):2115–2129

21. Cohen AL, Holcomb M (2009) Why corals care about ocean acidification: uncovering the mechanism. Oceanography 22(4):118–127

22. Delesalle B, Sournia A (1992) Residence time of water and phytoplankton biomass in coral reef lagoons. Cont Shelf Res 12(7/8):939–949

23. DeMartini EE, Friedlander AM (2004) Spatial patterns of endemism in shallow-water reef fish populations of the Northwestern Hawaiian Islands. Mar Ecol Prog Ser 271:281–296

24. Desch A,Wynne T, Brainard R, Friedlander A, Christensen J (2009) Oceanographic and physical setting. A marine biogeographic assessment of the Northwestern Hawaiian Islands. NOAA technical memorandum NOS NCCOS 84. Office of National Marine Sanctuaries Papahānaumokuākea Marine National Monument, Silver Spring, p 363

25. Dickson AG, Riley JP (1978) The effect of analytical error on the evaluation of the components of the aquatic carbon-dioxide system. Mar Chem 6(1):77–85

26. Dickson AG, Afghan JD, Anderson GC (2003) Reference materials for oceanic CO_2 analysis: a method for the certification of total alkalinity. Mar Chem 80(2):185–197

27. Dickson A, Sabine C, Christian J (eds) (2007) Guide to best practices for ocean CO_2 measurements, vol 3. North Pacific Marine Science Organization, PICES Special Publication, Sidney, p 191

28. Doney S, Fabry V, Feely R, Kleypas J (2009) Ocean acidification: the other CO_2 problem. Annu Rev Mar Sci 1:169–192

29. Dore JE (2009) Physical and biogeochemical modulation of ocean acidification in the North Pacific. Proc Natl Acad Sci 106:12235–12240

30. Dore JE, Lukas R, Sadler DW, Karl DM (2003) Climate-driven changes to the atmospheric CO_2 sink in the subtropical North Pacific Ocean. Nature 124:754–757

31. Doty MS, Oguri M (1956) The island mass effect. J Cons Perm Int Explor 22:33–37

32. Drupp PS, De Carlo EH, Mackenzie FT, Bienfang P, Sabine CL (2011) Nutrient inputs, phytoplankton response, and CO_2 variations in a semi-enclosed subtropical embayment, Kaneohe Bay, Hawaii. Aquat Geochem 17:473–498

33. Drupp PS, De Carlo EH, Mackenzie FT, Sabine CL (2013) Comparison of CO_2 dynamics and air–sea exchange in different tropical reef environments. Aquat Geochem 19(5–6):371–397

34. Fabricius KE, Langdon C, Uthicke S, Humphrey C, Noonan S, De'ath G, Okazaki R, Muehllehner N, Glas MS, Lough JM (2011) Losers and winners in coral reefs acclimatized to elevated carbon dioxide concentrations. Nat Clim Chang 1(3):165–169

35. Fagan KE, Mackenzie FT (2007) Air–sea CO_2 exchange in a subtropical estuarine–coral reef system, Kaneohe Bay, Oahu, Hawai'i. Mar Chem 106:174–191

36. Falter JL, Atkinson MJ, Schar DW, Lowe RJ, Monismith SG (2011) Short-term coherency between gross primary production and community respiration in an algal-dominated reef flat. Coral Reefs 30(1):53–58

37. Feely RA, Sabine CL, Lee K, Berelson W, Kleypas J (2003) Impact of anthropogenic CO_2 on the $CaCO_3$ system in the oceans. Science 305:362–366

38. Feely RA, Sabine CL, Millero FJ, Langdon C, Dickson AG, Fine RA, Bullister JL, Hansell DA, Carlson CA, McNichos AP, Key RM, Byrne RH, Wanninkhof R (2008) Carbon dioxide, hydrographic, and chemical data obtained during the R/Vs Roger Revelle and Thomas

G. Thompson repeat hydrography cruises in the Pacific Ocean: CLIVAR CO_2 sections P16S_2005 (6 January–19 February 2005) and P16N_2006 (13 February–30 March, 2006). In: Kozyr A (ed) ORNL/CDIAC-155, NDP-090. Oak Ridge National Laboratory, Oak Ridge.

39. Firing J, Brainard RE, Firing E (2006) Ten years of shipboard ADCP measurements along the Northwestern Hawaiian Islands. Atoll Res Bull 543:347–364

40. Friedlander AM, DeMartini EE (2002) Contrasts in density, size, and biomass of reef fishes between the northwestern and the main Hawaiian Islands: the effects of fishing down apex predators. Mar Ecol Prog Ser 230:253–264

41. Friedlander A, Aeby G, Brainard R, Clark A, DeMartini E, Godwin S, Kenyon J, Kosaki R, Maragos J, Vroom, P (2005) The state of coral reef ecosystems of the Northwestern Hawaiian Islands. In: Waddell J (ed) The state of coral reef ecosystems of the United States and Pacific freely associated states: 2005. NOAA technical memorandum NOS NCCOS 11. NOAA/NCCOS Center for Coastal Monitoring and Assessment's Biogeography Team. Silver Spring, pp 270–311

42. Fukasawa M, Kawano, Murata, A, Uchida H, Doi T (2008) Carbon dioxide, hydrographic, and chemical data obtained during the R/V Mirai repeat hydrography cruise in the Pacific Ocean: CLIVAR CO_2 section P14_2007 (October 8–December 26, 2007. Oak Ridge National Laboratory, Oak Ridge.

43. Garrett C, Munk W (1979) Internal waves in the ocean. Annu Rev Fluid Mech 11(1):339–369

44. Gattuso JP, Pichon MM, Delesalle BB, Frankignoulle MM (1993) Community metabolism and air–sea CO_2fluxes in a coral reef ecosystem (Moorea, French Polynesia). Mar Ecol Prog Ser 96:259–267

45. Gattuso JP, Pichon M, Delesalle B, Canon C, Frankignoulle M (1996) Carbon fluxes in coral reefs. I. Lagrangian measurement of community metabolism and resulting air–sea CO_2 disequilibrium. Mar Ecol Prog Ser 145:109–121

46. Gattuso JP, Frankignoulle M, Wollast R (1998) Carbon and carbonate metabolism in coastal aquatic ecosystems. Annu Rev Ecol Syst 29(1):405–434

47. Gove JM, Williams GJ, McManus MA, Heron SF, Sandin SA, Vetter OJ, Foley DG (2013) Quantifying Climatological ranges and anomalies for Pacific coral reef ecosystems. PLoS One 8(4):e61974

48. Gray SEC, DeGrandpre MD, Moore TS, Martz TR, Friederich GE, Johnson KS (2011) Applications of in situ pH measurements for inorganic carbon calculations. Mar Chem 125(1):82–90

49. Grigg RW (1982) Darwin point: a threshold for atoll formation. Coral Reefs 1(1):29–34

50. Grigg RW, Polovina J, Friedlander AM, Rohmann SO (2008) Biology of coral reefs in the Northwestern Hawaiian Islands. In: Riegl BM, Dodge RE (eds) Coral reefs of the USA. Springer, Dordrecht, pp 573–594

51. Hofmann GE, Smith JE, Johnson KS, Send U, Levin LA, Micheli F, Paytan A, Price N, Peterson B, Takeshita Y, Matson PG, Crook E, Kroeker KJ, Gambi MC, Rivest EB, Frieder CA, Yu PC, Martz TR (2011) High-frequency dynamics of ocean pH: a multi-ecosystem comparison. PLoS One 6(12):e28983

52. Jokiel PL, Rodgers KS, Kuffner IB, Andersson AJ, Cox EF, Mackenzie FT (2008) Ocean acidification and calcifying reef organisms: a mesocosm investigation. Coral Reefs 27(3):473–483

53. Kane C, Kosaki R, Wagner D (2013) High levels of mesophotic reef fish endemism in the Northwestern Hawaiian Islands. Bull Mar Sci 90(2):693–703

54. Karl D (1999) A sea of change: biogeochemical variability in the North Pacific subtropical gyre. Ecosystems 2:181–214

55. Kayanne H, Hata H, Kudo S, Yamano H, Watanabe A, Ikeda Y, Nozaki K, Kato K, Negishi A, Saito H (2005) Seasonal and bleaching-induced changes in coral reef metabolism and CO_2 flux. Glob Biogeochem Cycles 19:GB3015

56. Kleypas JA, McManus JW, Meñez LAB (1999a) Environmental limits to coral reef development: where do we draw the line? Am Zool 39(1):146–159

57. Kleypas JA, Buddemeier RW, Archer D, Gattuso JP, Langdon C, Opdyke BN (1999b) Geochemical consequences of increased

atmospheric carbon dioxide on coral reefs. Science 284(5411):118–120

58. Lantz CA, Atkinson MJ, Winn CW, Kahng SE (2013) Dissolved inorganic carbon and total alkalinity of a Hawaiian fringing reef: chemical techniques for monitoring the effects of ocean acidification on coral reefs. Coral Reefs 33(1):105–115

59. Liu X, Patsavas MC, Byrne RH (2011) Purification and characterization of meta-cresol purple for spectrophotometric seawater pH measurements. Environ Sci Technol 45(11):4862–4868

60. Lowe RJ, Falter JL, Monismith SG, Atkinson MJ (2009) Wave-driven circulation of a coastal reef–lagoon system. J Phys Oceanogr 39(4):873–893

61. Lumpkin CF (1998) Eddies and currents of the Hawaiian Islands. Dissertation, UH Manoa

62. Mackenzie FT, Andersson AJ (2013) The marine carbon system and ocean acidification during phanerozoic time. Geochem Perspect 2(1):1–3

63. Manzello DP, Kleypas JA, Budd DA, Eakin CM, Glynn PW, Langdon C (2008) Poorly cemented coral reefs of the eastern tropical Pacific: possible insights into reef development in a high-CO_2 world. Proc Natl Acad Sci 105(30):10450–10455

64. Martz TR, Connery JG, Johnson KS (2010) Testing the honeywell durafet for seawater pH applications. Limnol Oceanogr Methods 8(5):172–184

65. Millero FJ (2007) The marine inorganic carbon cycle. Chem Rev 107(2):308–341

66. Morse JW, Mackenzie FT (1990) Geochemistry of sedimentary carbonates. Elsevier, Amsterdam

67. Morse JW, Andersson AJ, Mackenzie FT (2006) Initial responses of carbonate-rich shelf sediments to rising atmospheric pCO_2 and "ocean acidification": role of high Mg-calcites. Geochim Cosmochim Acta 70(23):5814–5830

68. Moulin E, Jordens A, Wollast R (1985) Influence of the aerobic bacterial respiration on the early dissolution of carbonates in coastal sediments. In: Greikan RV, Wollast R (eds) Progress in Belgian oceanographic research. University of Antwerp, Antwerp, pp 196–208

69. Nakano Y, Watanabe YW (2005) Reconstruction of pH in the surface seawater over the North Pacific basin for all seasons using temperature and chlorophyll-*a*. J Oceanogr 61:673–680

70. Orr J, Fabry V, Aumont O, Bopp L, Doney S, Feely R, Gnanadesikan A, Gruber N, Ishida A, Joos F, Key R, Lindsay K, Maier-Reimer E, Matear R, Monfray P, Mouchet A, Najjar R, Plattner G-K, Rodgers K, Sabine C, Sarmiento J, Schlitzer R, Slater R, Totterdell I, Weirig M-F, Yamanaka Y, Yool A (2005) Anthropogenic ocean acidification over the twenty-first century and its impact on calcifying organisms. Nature 437(7059):681–686

71. Pierrot D, Lewis E, Wallace DWR (2006) MS Excel program developed for CO_2 system calculations. Oak Ridge National Laboratory, Oak Ridge

72. Pierrot D, Neill C, Sullivan K, Castle R, Wanninkhof R, Luger H, Johannessen T, Olsen A, Feely R, Cosca C (2009) Recommendations for autonomous underway pCO_2 measuring system and data-reduction routines. Deep Sea Res II 56:512–522

73. Polovina JJ, Howell E, Kobayashi DR, Seki MP (2001) The transition zone chlorophyll front, a dynamic global feature defining migration and forage habitat for marine resources. Prog Oceanogr 49(1):469–483

74. Raven J, Caldeira K, Elderfield H, Hoegh-Guldberg O, Liss P, Riebesell U, Shepherd J, Turley C, Watson A (2005) Ocean acidification due to increasing atmospheric carbon dioxide. The Royal Society, London

75. Revelle R, Suess HE (1957) Carbon dioxide exchange between atmosphere and ocean and the question of an increase of atmospheric CO_2 during the past decades. Tellus 9:18–27

76. Rooney JJ, Wessel P, Hoeke R, Weiss J, Baker J, Parrish F, Fletcher C, Chojnacki J, Garcia M, Brainard R, Vroom P (2008) Geology and geomorphology of coral reefs in the Northwestern Hawaiian Islands. In: Riegl BM, Dodge RE (eds) Coral reefs of the USA. Coral reefs of the world, vol 1. Springer, Berlin, pp 515–567

77. Rudnick D, Boyd T, Brainard R, Carter G, Egbert G, Gregg M, Holloway P, Klymak J, Kunze E, Lee C (2003) From tides to mixing along the Hawaiian Ridge. Science 301(5631):355–357

78. Sabine C (1992) Geochemistry of particulate and dissolved inorganic carbon in the central North Pacific. Dissertation, UH Manoa

79. Sabine CL, Mackenzie FT (1995) Bank-derived carbonate sediment transport and dissolution in the Hawaiian Archipelago. Aquat Geochem 1(2):189–230

80. Sarmiento JL, Gruber N (2006) Ocean biogeochemical dynamics. Princeton University Press, Princeton

81. Schlanger SO, Konishi K (1975) The geographic boundary between the coral–algal and the bryozoan–algal limestone facies: a paleolatitude indicator. In: 9th international geological congress of sedimentology, nice, theme 1, sedimental indicators, pp 187–190

82. Selkoe KA, Halpern BS, Toonen R (2008) Evaluating anthropogenic threats to the Northwestern Hawaiian Islands. Aquat Conserv Mar Freshw Ecosyst 18(7):1149–1165

83. Shamberger KEF, Feely R, Sabine C, Atkinson M, De Carlo E, Mackenzie F (2011) Calcification and organic production on a Hawaiian coral reef. Mar Chem 127:64–75

84. Shamberger KEF, Cohen AL, Golbuu Y, McCorkle DC, Lentz SJ, Barkley HC (2014) Diverse coral communities in naturally acidified waters of a Western Pacific reef. Geophys Res Lett 41:499–504.

85. Smith SV, Key GS (1975) Carbon dioxide and metabolism in marine environments. Limnol Oceanogr 20(3):493–495

86. Sutton A, Sabine C, Maenner S, Musielewicz S, Bott R, Osborne J (2010) High-resolution ocean and atmosphere pCO_2 time-series measurements from mooring WHOTS_158W_23N. Oak Ridge National Laboratory, Oak Ridge.

87. Suzuki A, Kawahata H (1999) Partial pressure of carbon dioxide in coral reef lagoon waters: comparative study of atolls and barrier reefs in the Indo-Pacific oceans. J Oceanogr 55:731–745

88. Suzuki A, Kawahata H (2003) Carbon budget of coral reef systems: an overview of observations in fringing reefs, barrier reefs and atolls in the Indo-Pacific regions. Tellus 55B(2):428–444

89. Takahashi T, Olafsson J, Goddard JG, Chipman DW, Sutherland SC (1993) Seasonal-variation of CO_2 and nutrients in the high-latitude surface oceans—a comparative study. Glob Biogeochem Cycles 7(4):843–878

90. Takahashi T, Sutherland SC, Sweeney C, Poisson A, Metzl N, Tilbrook B, Bates N, Wanninkhof R, Feely RA, Sabine CL, Olafsson J, Nojiri Y (2002) Global sea–air CO_2 flux based on climatological surface ocean pCO_2, and seasonal biological and temperature effects. Deep Sea Res II 49:1601–1622

91. Takahashi T, Sutherland SC, Feely RA, Wanninkhof R (2006) Decadal change of surface water pCO_2 in the North Pacific: a synthesis of 35 years of observations. J Geophys Res 111.

92. Takahashi T, Sutherland SC, Wanninkhof R, Sweeney C, Feely RA, Chipman DW, Hales B, Friederich G, Chavez F, Sabine C, Watson A, Bakker DCE, Schuster U, Metzl N, Yoshikawa-Inoue H, Ishii M, Midorikawa T, Nojiri Y, Kortzinger A, Steinhoff T, Hoppema M, Olafsson J, Arnarson TS, Tilbrook B, Johannessen T, Olsen A, Bellerby R, Wong CS, Delille B, Bates NR, de Baar HJW (2009) Climatological mean and decadal change in surface ocean pCO_2, and net sea–air CO_2 flux over the global oceans. Deep Sea Res II 56:554–577

93. Thompson RW, Dickson AG, Kahng SE, Winn CD (2014) Nearshore carbonate dissolution in the Hawaiian Archipelago? Aquat Geochem.

94. Wanninkhof R (1992) Relationship between wind speed and gas exchange over the ocean. J Geophys Res 97(C5):7373–7382

95. Winn CD, Li Y-H, Mackenzie FT, Karl DM (1998) Rising surface ocean dissolved inorganic carbon at the Hawai'i Ocean time-series site. Mar Chem 60:33–47

96. Yates KK, Halley RB (2006) Diurnal variation in rates of calcification and carbonate sediment dissolution in Florida Bay. Estuaries Coasts 29(1):24–39

9

CLIMATE, NOT ATMOSPHERIC DEPOSITION, DRIVES THE BIOGEOCHEMICAL MASS-BALANCE OF A MOUNTAIN WATERSHED

Jared Heath 1• Jill S. Baron2

[1] Water Product Division, City of Fort Collins Utilities, Fort Collins, CO 80521, USA

[2] Fort Collins Science Center, United States Geological Survey, Fort Collins, CO 80526-8118, USA

ABSTRACT

Watershed mass-balance methods are valuable tools for demonstrating impacts to water quality from atmospheric deposition and chemical weathering. Owen Bricker, a pioneer of the mass-balance method, began applying mass-balance modeling to small watersheds in the late 1960s and dedicated his career to expanding the literature and knowledge of complex watershed processes. We evaluated long-term trends in surface-water chemistry in the Loch Vale watershed, a 660-ha. alpine/subalpine catchment located in Rocky Mountain National Park, CO, USA. Many changes in surface-water chemistry correlated with multiple drivers, including summer or monthly temperature, snow water equivalent, and the runoff-to-precipitation ratio. Atmospheric deposition was not a significant causal agent for surface-water chemistry trends. We observed statistically significant increases in both concentrations and fluxes of weathering products including cations, SiO_2, SO_4^{2-}, and ANC, and in inorganic N, with inorganic N being primarily of atmospheric origin. These changes are evident in the individual months June, July, and August, and also in the combined June, July, and August summer season. Increasingly warm summer temperatures are melting what was once permanent ice and this may release elements entrained in the ice, stimulate chemical weathering with enhanced moisture availability, and stimulate microbial nitrification. Weathering rates may also be enhanced by sustained water availability in high snowpack years. Rapid change in the flux of weathering products and inorganic N is the direct and indirect result of a changing climate from warming temperatures and thawing cryosphere.

KEYWORDS

Atmospheric deposition, Weathering, Climate change, Nitrogen, Snow water equivalent, Runoff-to-precipitation ratio

1 INTRODUCTION

Watershed mass-balance calculations have developed as an important technique for estimating processes resulting from interactions of water with biota, soils, and bedrock. Owen Bricker, a pioneer of the method, began applying mass-balance modeling to small watersheds in the late 1960s and dedicated his career to expanding the literature and knowledge of complex watershed processes (Bricker et al. 1968; Cleaves et al.1970; Katz et al. 1985; Baron and Bricker 1987; Simon et al. 2005). The mass-balance approach was developed to monitor the flux of water and associated dissolved and suspended chemicals from weathering and atmospheric deposition. Loch Vale Watershed (LVWS) in Rocky Mountain National Park, CO (Fig. 1), was instrumented in 1983 with Owen Bricker's help to address the response of an alpine and subalpine ecosystem to atmospheric deposition and climate (Baron 1992).

Previous LVWS mass-balance studies have provided insight into mineral weathering sources and rates, the fate of atmospherically deposited solutes, and internal biogeochemical cycling. Base cations, silica, and alkalinity are derived from weathering of biotite gneiss that is composed of the minerals apatite, quartz, plagioclase, microcline, biotite, microcrystalline calcite, and sillimanite within the watershed (Baron and Bricker 1987; Mast et al. 1990; Price et al. 2013). Price et al. (2013) found over their 24-year period of study that weathering of calcite accounts for 40–65 % of the contribution of Ca^{2+} to the watershed, while oligoclase and apatite account for 25 and 10–35 %, respectively. Wet deposition is a primary source for the relatively conservative species chloride (Cl^-), sulfate (SO_4^{2-}), and nitrate (NO_3^-) (Baron et al. 1995; Clow and Drever 1996; Baron and Campbell 1997; Campbell et al. 2000). Isotopic analysis of dissolved SO_4^{2-} in LVWS also identified an internal mineral source (Mast et al. 1990; Kester et al. 2003). Ammonium (NH_4^+) inputs from deposition exceed outputs and almost all NH_4^+ is rapidly nitrified or consumed (Campbell et al.1995, 2000; Williams et al. 1996).

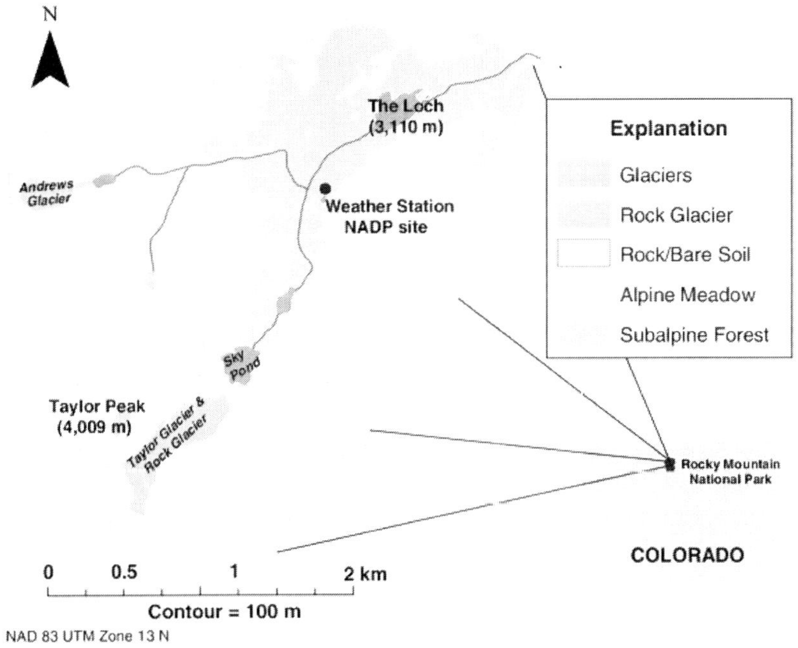

Figure 1. Map of 660-hectare Loch Vale Watershed located in Rocky Mountain National Park, CO

Recent studies in LVWS suggest that changes in climate and atmospheric deposition are altering the mass balance of major dissolved elements (Baron et al. 2009). Trends in wet deposition reported by the

National Atmospheric Deposition Program National Trends Network (NADP/NTN) show an increase in inorganic nitrogen (N) in precipitation throughout most of the United States since approximately 1980, while SO_4^{2-}has declined sharply (Lehmann et al. 2005). Increases in wet N deposition also occurred in the Rocky Mountains of the western United States, but there is no long-term (1984–2010) significant increase in wet N deposition in LVWS (Morris et al. 2012). The annual wet N deposition rate in LVWS exhibited high variability prior to 1995; from 1995 to 2010, wet N deposition was about 3.0 kg N ha^{-1} yr^{-1} (Morris et al. 2012). In spite of steady wet N deposition inputs, annual N export from LVWS has increased since 2000, suggesting either an internal source of N or that the inputs and outputs are not in annual balance (Baron et al. 2009). Wet SO_4^{2-} deposition has decreased over time, but increasing SO_4^{2-} concentrations in LVWS surface waters imply increased mineral weathering (Mast et al. 1990; Kester et al. 2003; Baron et al. 2009).

The Rocky Mountain climate has changed in recent years (Ray et al. 2008). Summer and autumn temperatures in the Front Range of Colorado showed statistically significant warming trends from 1991 to 2006 and the increase in summer temperatures has been associated with increasing July temperatures (Baron et al. 2009; McGuire et al. 2013). Mean and minimum July temperatures in LVWS increased by 0.22 and 0.18 °C yr^{-1} from 1991 to 2006, respectively, and this trend was statistically significant (Baron et al.2009). Warmer temperatures can increase weathering rates, influence the timing of snowmelt (Knowles et al.2006), and alter the proportion of precipitation that falls as snow or rain (Stewart et al. 2005).

Glacier and other ice feature ablation rates increase as climate warms. Alpine glaciers are retreating worldwide in response to warming (Fountain et al. 2012). Alpine catchments occupied by active glaciers may yield weathering rates greater than the global mean (Anderson et al. 1997), and changes to the cryosphere can influence the flux of water, weathering products, nutrients, and organic matter in mountainous environments (Baron et al. 2009; Bogdal et al. 2009; Saros et al. 2010; Singer et al. 2012). We evaluated long-term trends in surface-water chemistry in LVWS to try to apportion the cause of changes in solute flux over time. We compared the long-term surface-water chemical time series with the time series of atmospheric deposition, air temperature, precipitation, and snow water equivalent (SWE). We also addressed the possible indirect climate effects from warming-induced ice ablation.

2 SITE DESCRIPTION

Loch Vale watershed is a 660-hectare alpine/subalpine catchment in Rocky Mountain National Park, CO (Fig. 1), that was established to address the response of watershed-scale ecosystem processes to atmospheric deposition and climate change (Baron 1992; Baron et al. 2000; Campbell et al. 2000). It ranges in elevation from 4,009 m at the continental divide to 3,110 m at the outlet of The Loch, the lowest elevation subalpine lake. Fully 80 % of LVWS is exposed rock and talus, and 3 % of talus may cover rock glaciers (Baron 1992; Clow et al. 2003). Other land cover includes 1 % permanent snow and ice, 11 % alpine, 6 % old-growth conifer forest, and 1 % wetlands. The climate is high-mountain continental with 105 cm of mean annual (1984–2012) precipitation, about 20 cm of which falls during the summer months. Mean annual (1984–2012) temperature is 1.4 °C and summer temperatures average 11.6 °C (http://www.nrel.colostate.edu/projects/lvws/data.html). Long-term Natural Resource Conservation Service Snow Telemetry (SNOTEL) data at Bear Lake, the nearest snow measurement site, show that average peak SWE of 50 cm occurred on April 25 for the period 1979–2010 (http://www.wcc.nrcs.usda.gov/nwcc/site?sitenum=322&state=co). Discharge from 1984 to 2010 measured at The Loch outlet averaged 79 cm yr^{-1} (http://www.nrel.colostate.edu/projects/lvws/data.html).

3 METHODS

We compared trends in surface-water chemical concentrations and fluxes with trends in air temperature, precipitation amount, SWE, and wet deposition chemistry. Seasonal water sampling with weekly summer sample collection began in 1983. Weekly year-round sampling of surface waters at The Loch outlet began in 1991 and continues to present. Annual trends are presented from 1991 to 2010. We used the longer (1983–2010) record for June, July, and August and summer (combined June, July, and August). Since no trends were observed for other months or seasons, these results are not reported. All data were reported for water years (WY) October 1 through September 30.

Surface-water samples collected at The Loch outlet were analyzed following standard procedures (http://www.nrel.colostate.edu/projects/lvws/data.html). Data quality assurance procedures included field blanks, field duplicates, charge balance calculations, and comparisons between theoretical conductance and measured conductance (Allstott et al. 1999; Botte and Baron 2004; Richer and Baron 2011). Export was calculated by multiplying weekly concentrations of individual solutes by weekly discharge. Weekly export loads were summed for the year, summer, and June, July, and August to yield watershed outputs in kilograms per hectare per year (kg ha^{-1} yr^{-1}). Data reported here include the sum of base cations (Ca^{2+}, Mg^{2+}, Na^{+}, K^{+}), $SO_4{}^{2-}$, inorganic N, Cl^{-}, acid-neutralizing capacity (ANC), and dissolved silica (SiO_2).

Precipitation chemistry and amount have been measured at the National Atmospheric Deposition Program/National Trends Network (NADP/NTN) CO98 Loch Vale site (http://nadp.sws.uiuc.edu/sites/siteinfo.asp?id=CO98&net=NTN) located at an elevation of 3,159 m since 1983. Data reported in this study included precipitation amount, the sum of base cations, $SO_4{}^{2-}$, inorganic N, and Cl^{-}.

Daily temperature was measured at the Loch Vale Main Weather Station (http://co.water.usgs.gov/lochvale/data.html). Snow water equivalent was taken from the Bear Lake SNOTEL site (http://www.wcc.nrcs.usda.gov/nwcc/site?sitenum=322&state=co). Runoff-to-precipitation ratios (R:P) were calculated from measured discharge at The Loch outlet (http://www.nrel.colostate.edu/projects/lvws/data.html) and measured NADP precipitation at the CO98 site (http://nadp.sws.uiuc.edu/sites/siteinfo.asp?id=CO98&net=NTN). We assumed inter-annual comparability in catch efficiency. Precipitation measurements in snow-dominated mountain environments are always subject to uncertainty, but the analysis of overall error of NADP measurements from colocated samplers, including several mountainous sites, concluded the magnitude of error in sample volume is of minor environmental significance to data users (Wetherbee et al. 2005).

3.1 Trends in Long-Term Data

The Mann–Kendall test (MKT) was used to evaluate trends in wet deposition and surface-water chemical fluxes and concentrations. The MKT was performed using annual, summer, and June, July, and August data using a statistical package from Helsel et al. (2006). Statistical significance was determined to the 95 % confidence interval ($p \le 0.05$) and noted with an asterisk in tables and figures. In the "Results" and "Discussion" sections below, significance is used to denote statistical significance with 95 % confidence.

3.2 Kendall Rank Correlation Test

The Kendall rank correlation test was used to determine the statistical significance of correlations between climatic variables, including temperature, runoff-to-precipitation ratio (R:P), and peak SWE, with observed trends in water chemistry at The Loch outlet. Kendall's tau measures the monotonic strength of association between two independent nonparametric variables that exhibit skewness around a general trend (Helsel and Hirsch 2002). Correlations can either be described as positively correlated when Kendall's tau is positive, negatively correlated when Kendall's tau is negative, or no correlation when Kendall's tau is zero. Statistical significance was determined to the 95 % confidence interval ($p \leq 0.05$) and noted with an asterisk in tables and figures.

4 RESULTS

4.1 Wet Deposition Trends

Analysis of 1991–2010 annual wet deposition showed no significant temporal trends for cations or inorganic N (Table 1). There was, however, a significant decline of $SO_4{}^{2-}$, H^+, and Cl^- (Table 1). Annual $SO_4{}^{2-}$-deposition decreased by 0.12 kg ha^{-1} yr^{-1}, annual H^+ deposition decreased by 2.79×10^{-3} kg ha^{-1} yr^{-1}, and annual Cl^- deposition decreased by 0.02 kg ha^{-1} yr^{-1} (Table 1).

4.2 Surface-Water Chemistry Trends

Using annual resolution, there was no significant trend over time in the 1991–2010 fluxes of weathering materials or inorganic N, except for H^+, which decreased by 1.84×10^{-4} kg ha^{-1} yr^{-1} (Table 1). The fluxes of all solutes increased during the summer months from 1984 to 2010, however, some significantly (Fig. 2). The greatest flux increase was observed in June for the cations, $SO_4{}^{2-}$, and ANC.

Table 1. Annual wet deposition and water chemistry flux trends from Mann–Kendall trend test results for WY1991–2010 in LVWS, Rocky Mountain National Park, CO

Solute	Wet deposition			Loch outlet flux		
	Trend (kg ha^{-1} yr^{-1})	*P* value	Trend	Trend (kg ha^{-1} yr^{-1})	*P* value	Trend
Cations	0.08	0.098	Increasing	0.17	0.284	Increasing
Inorg. N	0.00	0.897	No trend	0.03	0.330	Increasing
SO_4^{2-}	−0.12	0.003	Decreasing*	0.16	0.098	Increasing
H^+	−2.79E−03	0.008	Decreasing*	−1.84E−04	0.002	Decreasing*
Cl^-	−0.02	0.007	Decreasing*	0.02	0.153	Increasing
SiO_2	n/a	n/a	n/a	0.15	0.284	Increasing
ANC	n/a	n/a	n/a	0.23	0.183	Increasing

Statistically significant trends ($p \leq 0.05$) are noted with an asterisk. Silica and ANC are not measured in wet deposition

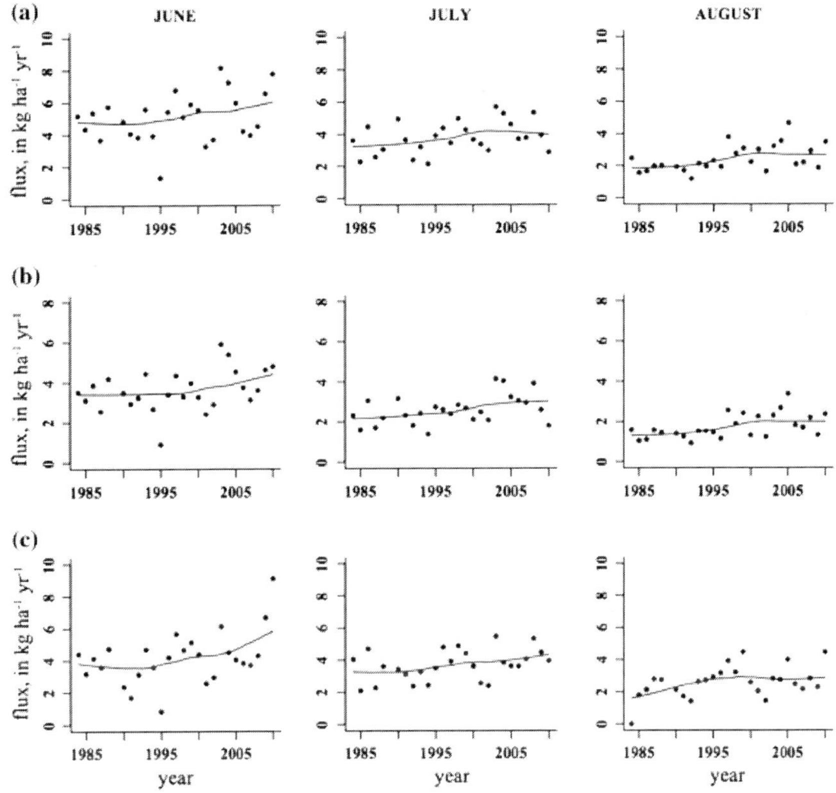

Figure 2. June, July, and August water chemistry flux in LVWS, Rocky Mountain National Park, CO, from 1984 to 2010 with smooth curve fitted by loess (locally weighted polynomial regression) for **a** cations, **b** $SO_4{}^{2-}$, and **c** ANC

Not all trends were significant. Cation flux trends were not significant for June and July, in spite of an increase in summer flux (Table 2). The summer total cation flux was driven primarily by an August increase of 0.05 kg ha^{-1} yr^{-1} (Fig. 2a; Table 2). The change in $SO_4{}^{2-}$ flux was insignificant in July, but strong and significant increases of 0.11 and 0.04 kg ha^{-1} yr^{-1} in June and August, respectively, provided a significant summer trend (Fig. 2b; Table 2). Acid-neutralizing capacity increased over time, with a summer increase of 0.13 kg ha^{-1} yr^{-1}, but the only significant flux trend occurred in July (Table 2). The summer increase in SiO_2flux of 0.20 kg ha^{-1} yr^{-1} was significant, but the trends were not significant for any individual month (Table 2). Finally, even though there was an increase in the amount of inorganic N and Cl^- flux over time, the only significant increase was for inorganic N in August (Table 2).

Table 2. Summer, June, July, and August water chemistry flux and concentration trends from Mann–Kendall test results for WY1984–2010 from LVWS, Rocky Mountain National Park, CO

Solute	Time period	Flux			Concentration		
		Trend (kg ha^{-1} yr^{-1})	*P* value	Trend	Trend (mg l^{-1} yr^{-1})	*P* value	Trend
Cations	Summer	0.13	0.028	Increasing*	0.02	0.005	Increasing*
	June	0.05	0.146	Increasing	0.02	0.004	Increasing*
	July	0.04	0.086	Increasing	0.02	0.004	Increasing*
	August	0.05	0.018	Increasing*	0.06	0.022	Increasing*
SO_4^{2-}	Summer	0.02	0.017	Increasing*	0.02	0.017	Increasing*
	June	0.11	0.012	Increasing*	0.02	0.028	Increasing*
	July	0.04	0.058	Increasing	0.02	0.007	Increasing*
	August	0.04	0.016	Increasing*	0.04	0.016	Increasing*
ANC	Summer	0.13	0.087	Increasing	0.02	0.004	Increasing*
	June	0.07	0.113	Increasing	0.02	0.001	Increasing*
	July	0.04	0.028	Increasing*	0.02	0.003	Increasing*
	August	0.03	0.103	Increasing	0.02	0.006	Increasing*
SiO_2	Summer	0.20	0.015	Increasing*	0.01	0.033	Increasing*
	June	0.09	0.086	Increasing	0.02	0.007	Increasing*
	July	0.02	0.402	Increasing	0.01	0.134	Increasing
	August	0.03	0.156	Increasing	0.03	0.156	Increasing
Inorg. N	Summer	0.02	0.058	Increasing	3.27E−03	0.086	Increasing
	June	4.53E−03	0.428	Increasing	2.58E−03	0.103	Increasing
	July	6.17E−03	0.158	Increasing	3.20E−03	0.04	Increasing*
	August	7.64E−03	0.009	Increasing*	0.01	0.009	Increasing*
Cl^-	Summer	0.01	0.052	Increasing	0.00	0.108	No trend
	June	4.68E−03	0.158	Increasing	1.50E−03	0.134	Increasing
	July	3.14E−03	0.086	Increasing	1.08E−03	0.139	Increasing
	August	2.56E−03	0.104	Increasing	2.56E−03	0.104	Increasing

Statistically significant trends ($p \leq 0.05$) are noted with an asterisk

The concentrations of nearly all solutes increased significantly in all summer months from 1984 to 2010 (Table 2). The exceptions were SiO_2, inorganic N, and Cl^-, whose trends were significant in some, but not all, months. A significant summer increase in SiO_2 concentration was driven by June values. Concentrations of inorganic N increased in all months, but the rate of increase, which was slight, was only significant in July and August. The increase in Cl^- concentrations over time was very small and insignificant (Table 2).

We compared the solute flux with environmental drivers to get an idea of causes for increasing flux over time from LVWS. All solute fluxes for all time periods were positively correlated with average maximum July temperature, R:P, and peak SWE at The Loch outlet (except for annual Cl^- flux, August $SO_4{}^{2-}$ flux, and all inorganic N flux with peak SWE), but not all correlations were significant (Table 3). The flux of different solutes displayed almost unique responses to different drivers. The annual cation flux correlated significantly with R:P, while June and July cation fluxes correlated significantly with peak SWE. August cation flux was significantly related to average maximum July temperatures. Annual $SO_4{}^{2-}$ flux correlated positively with average maximum July temperatures and R:P. Acid-neutralizing capacity and SiO_2 fluxes were significantly correlated with peak SWE annually and for June and July. July SiO_2 flux was also significantly correlated with R:P. Annual, June, and July inorganic N fluxes were significantly correlated with R:P, while the August inorganic N flux was most strongly correlated with maximum July temperatures. Chloride fluxes were not significantly related to any driver. None of the solutes showed a significant correlation with wet deposition (Table 3).

June flux values were not compared with July temperatures. Positive correlations are noted with a plus sign (+); negative correlations are noted with a minus sign (−); and no correlation is noted with a zero (0). Statistically significant trends ($p \leq 0.05$) are noted with an asterisk

4.3 Comparison of Flux in Extreme Years

We calculated the mean annual flux for all solutes for the warmest and coolest Julys, highest and lowest R:P years, and wettest (highest SWE) and driest (lowest SWE) years in order to gain insight into causal agents of change. Extremes were defined as the five lowest and five highest years for average maximum July temperatures, R:P, and peak SWE (Fig. 3; Table 4). The Mann–Whitney *U* test was used to determine whether the two extremes (lowest and highest) came from the same distribution.

There were greater annual fluxes of SiO_2, ANC, $SO_4{}^{2-}$, and cations in LVWS in years with the warmest July temperatures, but the increases were not statistically significant (Fig. 3a; Table 4). Years with the highest R:P produced more SiO_2, ANC, Cl^-, $SO_4{}^{2-}$, and cations compared to years with the lowest R:P, and the difference was significant for SiO_2 and cations (Fig. 3b). Years with the highest R:P had an average annual production of 18.67 (SD = 2.69) kg cations ha^{-1} yr^{-1} and 18.37 (SD = 0.97) kg SiO_2 ha^{-1} yr^{-1}. The years with the lowest R:P had an annual average production of 11.91 (SD = 1.38) kg cations ha^{-1} yr^{-1} and 12.79 (SD = 1.24) kg SiO_2 ha^{-1} yr^{-1} (Table 4b). In high R:P years, significantly less inorganic N was retained (0.20 (SD = 0.82) kg N ha^{-1} yr^{-1}) in the watershed than years when the R:P was low (1.61 (SD = 0.67) kg N ha^{-1} yr^{-1}). Years with the highest peak SWE resulted in greater net flux of cations, inorganic N, ANC, and SiO_2 in LVWS compared to years with the lowest peak SWE, but the differences were not statistically significant (Fig. 3c; Table 4).

Table 3. Correlations of annual, June, July, and August solute fluxes with average maximum July temperatures, runoff-to-precipitation ratios (R:P), peak snow water equivalent (SWE), and wet deposition using Kendall rank correlations for WY1991–2010

Solute flux	Season	Avg. Max July Temp.	R:P	Peak SWE	Wet deposition
Cations	Annual	+	+*	+	–
	June	N/A	+	+*	
	July	+	+	+*	
	August	+*	+	+	
SO_4^{2-}	Annual	+*	+*	+	–
	June	N/A	+	+	
	July	+	+	+	
	August	+	+	0	
ANC	Annual	+	+	+*	N/A
	June	N/A	+	+*	
	July	+	+	+*	
	August	+	+	+	
SiO_2	Annual	+	+	+*	N/A
	June	N/A	+	+*	
	July	+	+*	+*	
	August	+	+	+	
Inorg. N	Annual	+*	+*	+	+
	June	N/A	+*	+	
	July	+	+*	+	
	August	+*	+	–	
Cl^-	Annual	+	+	0	–
	June	N/A	+	+	
	July	+	+	+	
	August	+	+	+	

June flux values were not compared with July temperatures. Positive correlations are noted with a plus sign (+); negative correlations are noted with a minus sign (–); and no correlation is noted with a zero (0). Statistically significant trends ($p \leq 0.05$) are noted with an asterisk

Table 4. Comparison of the mean and standard deviation annual flux, in kg ha^{-1} yr^{-1}, for the five coolest versus warmest years, lowest runoff-to-precipitation ratio (R:P) versus highest R:P years, and driest (low peak SWE) versus wettest years (high peak SWE)

a)	Warmest maximum July temperatures						Coldest maximum July temperatures					
	Cations	SO_4^{2-}	Inorg.N	Cl^-	ANC	SiO_2	Cations	SO_4^{2-}	Inorg.N	Cl^-	ANC	SiO_2
Average	16.08	8.96	−0.42	0.71	16.45	15.56	12.16	5.47	−1.29	−0.03	14.57	11.94
SD	5.89	4.97	1.11	0.44	4.29	4.67	1.06	1.92	0.76	0.73	2.87	2.34
b)	**High R:P**						**Low R:P**					
	Cations	SO_4^{2-}	Inorg.N	Cl^-	ANC	SiO_2	Cations	SO_4^{2-}	Inorg.N	Cl^-	ANC	SiO_2
Average	18.67	10.00	−0.23	0.88	19.30	18.37	11.91	5.69	−1.61	0.28	16.81	12.79
SD	2.69	3.57	0.82	0.25	1.05	0.97	1.38	3.13	0.67	1.01	2.62	1.24
c)	**High peak SWE**						**Low peak SWE**					
	Cations	SO_4^{2-}	Inorg.N	Cl^-	ANC	SiO_2	Cations	SO_4^{2-}	Inorg.N	Cl^-	ANC	SiO_2
Average	15.05	−1.09	6.69	0.36	18.64	14.37	11.97	−1.10	6.41	0.57	13.49	11.03
SD	2.30	0.26	1.20	0.19	1.78	4.93	2.00	0.34	1.10	0.48	3.88	2.36

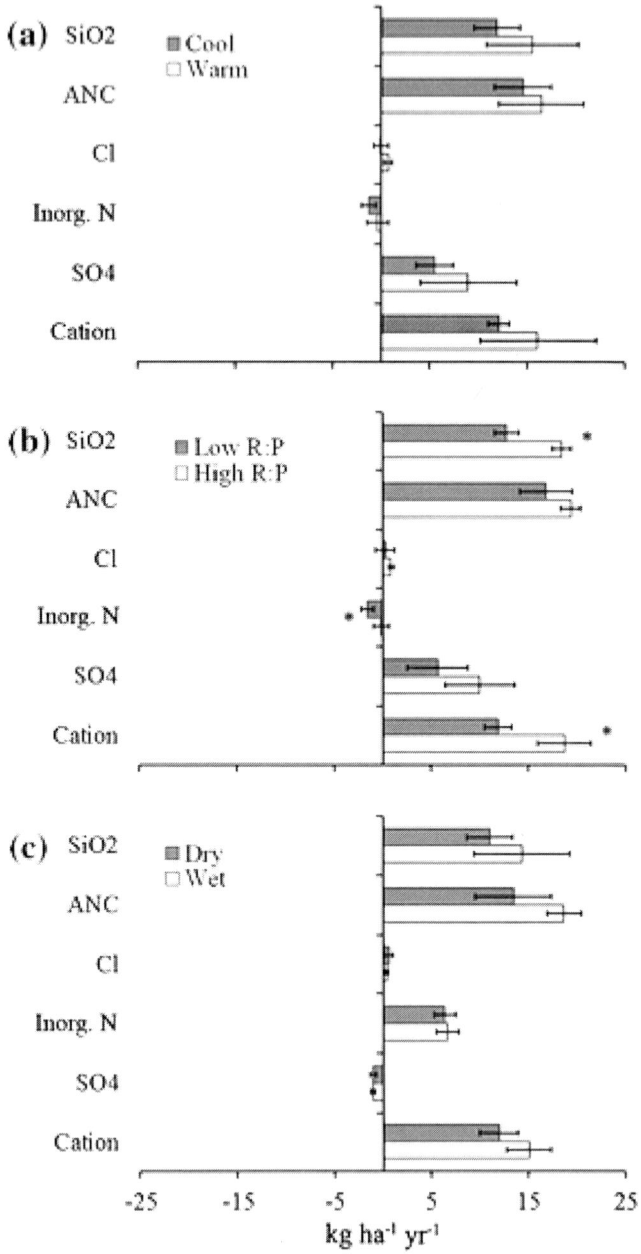

Figure 3. Comparison of means for the five most extreme years: **a** coolest versus warmest maximum July temperatures; **b** lowest R:P (runoff-to-precipitation ratio) versus highest R:P; and **c** driest (low peak SWE) versus wettest (high peak SWE). *Error bars* depict standard deviation

The R:P ratio is related to July maximum temperatures (Fig. 4). The average R:P was 0.77 from 1984 to 2010, suggesting slightly more than 23 % of precipitation was evaporated, transpired, or sublimated. Runoff approached or exceeded precipitation with R:P of 0.90, 0.89, 1.00, 1.16, 0.88, and 0.95, respectively, in water years 1998, 1999, 2000, 2003, 2005, and 2006 (Fig. 4a). Average maximum July temperatures were significantly and positively correlated with annual R:P (p = 0.034), suggesting that July temperatures influence the amount of discharge. In most years when the R:P was greater than average, the autumn (September, October, November) discharge was also greater than average (Fig. 4b).

5 DISCUSSION

Surface-water chemistry in LVWS is changing significantly during the summer months. This was not apparent using an annual time step, but is obvious in strong summer increases in weathering products and inorganic N. Annual fluxes in LVWS, as in other high-mountain watersheds with seasonal snowpacks, are dominated by snowmelt. Snowmelt drives the annual hydrograph and annual flux values, and while it may indirectly influence weathering rates, as discussed below, our observed changes in summer flux are too small to influence the annual totals.

There were significant increases in the summer fluxes and concentrations of cations, SiO_2, $SO_4{}^{2-}$, and ANC, and in inorganic N over time. Surprisingly, none of the trends were statistically correlated with wet atmospheric deposition. Instead, LVWS solute concentrations and flux were correlated with average maximum July temperature, R:P, and SWE. The greatest increases over time in concentrations and fluxes at The Loch outlet occurred in July and August. Hydrologically, this is commensurate with the receding hydrograph and return to base flow, as has been shown for other alpine catchments (e.g., Williams and Melack 1991). Late summer is also the time of maximum cryosphere thaw. We observed this with LVWS R:P ratios in warm years, and it matches theoretical constructs posed by Clow et al. (2003) and Williams et al. (2006), which suggest increasing contributions of glacier, rock glacier, and permafrost melt to mountain streams as the 0 °C isotherm increases in elevation.

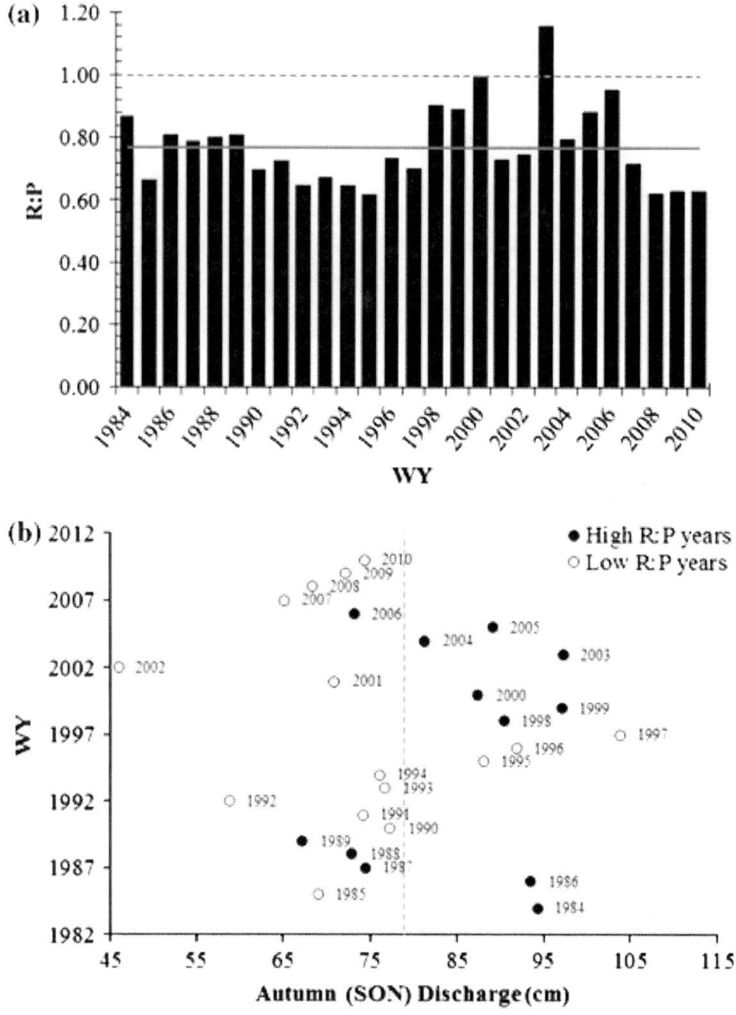

Figure 4. Time-series column plot of **a** the annual runoff-to-precipitation ratio (R:P) in LVWS for water years (WY) 1984 to 2010. The red line depicts the average annual R:P, while the *black dashed line* represents when runoff equals precipitation inputs; and **b** Autumn cumulative discharge for above average R:P years (*solid circles*) and below average R:P years (*hollow circles*). The *gray dashed line* represents average autumn discharge for LVWS

The increase in runoff in higher R:P years is much greater than can be accounted for entirely by glacial melt from small headwater glacial remnants. While some runoff may come from melting ice, high R:P may be more related to the snowmelt rate, with warm summer temperatures leading to more rapid melt and thus more runoff export instead of subsurface storage. The presence of liquid water in talus and block fields, or in newly thawed substrates, may enhance chemical weathering of calcite and apatite that are exposed by mechanical weathering (Mast et al. 1990, 1992; Price et al. 2013). Hall et al. (2002) suggested chemical weathering in cold environments is limited by moisture availability rather than temperature. Under this framework, it is possible that warm Julys increase the water available for mineral dissolution by thawing glacial ice or seasonal snow. In years of high SWE, seasonal snow persists longer into the summer, providing meltwater later in the summer season. Anderson et al. (1997) found chemical weathering in glacial environments to be relatively high, as have others in more recent studies (Williams et al. 2006; Saros et al.2010; Caine 2010). Cation, SiO_2, and ANC fluxes were positively correlated with peak SWE during June and July, the months that often correspond with spring snowmelt, suggesting a relationship between years with high snowfall and water chemical composition.

The source of some solutes coming from the cryosphere could include dry deposition in the form of dust or enhanced chemical weathering caused by prolonged moisture contact with rock as described above. Bulk snowpack chemistry data from the USGS Rocky Mountain Snowpack Chemistry Network (http://co.water.usgs.gov/projects/RM_snowpack/index.html) reveal cation concentrations in the LVWS snowpack have been increasing since 1993. A 2006 dust event covered much of the Rocky Mountains, including LVWS, resulting in a 10-fold increase in Ca^{2+} and 100-fold increase in ANC compared to early- and late-season snow (Rhoades et al. 2010). The concentration of cations in the LVWS snowpack in 2006 was 24.2 meq L^{-1}, while a similar dust event in 2008 resulted in a concentration of 26.6 meq L^{-1}. Neff et al. (2008) suggested that dust loads increased by 500 % following western settlement of the United States during the nineteenth century. In recent decades, the San Juan Mountains in southern Colorado experience up to seven large dust deposition events every year. Based on these studies, there is no doubt that episodic dust events add minerals to the snowpack and that the occurrence of these events is increasing. However, the importance of these events on LVWS water quality is unknown.

We previously hypothesized that increased late summer inorganic N flux could be the result of some combination of N released from melting ice and enhanced microbial nitrification rates (Baron et al. 2009). Recent research suggests both mechanisms are plausible. Saros et al.

(2010) found higher concentrations of nitrate (NO_3^-) in glacial meltwater compared to seasonal snow meltwater in the central and northern Rocky Mountains, and ice samples from two Front Range glaciers in 2012 contained considerable inorganic N (J.S. Baron, unpublished data). Even a small input of melting ice may affect water chemistry in warm years when rapid snowmelt leads to more runoff and export. Microbial nitrification also occurs in periglacial environments (Wynn et al. 2007; Saros et al. 2010; Osborne 2012). As with weathering rates, it appears moisture, not temperature, may govern microbial activity in cold environments. Additional evidence of microbial nitrification comes from comparison of isotopes in NO_3^- in snowpack, snowmelt, rainfall, and surface water. During snowmelt and throughout the summer, much of the inorganic N in stream water bears the signature of microbial nitrification, even at the highest elevations directly below ice (Campbell et al. 2000; E.K. Hall personal communication).

6 CONCLUSIONS

Watershed mass-balance methods have been valuable tools for demonstrating impacts to water quality from atmospheric deposition, leading to air quality policies that have greatly improved the ecological condition of US surface waters (Likens et al. 2001; Greaver et al. 2012). Mass-balance studies have also been fundamental to our understanding the role of mineral weathering on water quality (Bricker 1986; Price et al.2013). By exploring the change in mass balances over time in LVWS, we have discovered a change in summer water chemistry and export in response to warming summer temperatures and SWE. The connection between these drivers and the flux of weathering products and inorganic N is not necessarily direct. Warm summer temperatures and high SWE may stimulate mineral weathering with enhanced moisture availability, and warm summer temperatures may melt what was once permanent ice to release elements entrained therein. Moisture availability also appears to increase microbial nitrification, and all these processes likely contribute to summer season increases in LVWS chemical fluxes. There is an unknown but increasing influence of dust inputs. While long-term monitoring in instrumented watersheds has been criticized as unscientific, wasteful, and expensive, it is empirical studies such as these that increase our ability to understand and quantify cause and effect in times of rapid environmental change.

ACKNOWLEDGMENTS

Financial support was provided by the US Geological Survey and the National Park Service. Thanks to all who helped collect these data over many years, and to the National Park Service, especially Judy Visty, for long-term support. Alisa Mast, Stephanie Kampf, and Jason Price provided valuable comments that improved the paper. This paper is a product of the Western Mountain Initiative.

REFERENCES

1. Allstott EJ, Bashkin MJ, Baron JS (1999) Quality assurance report, Loch Vale Watershed Study, 1995–1998. USGS Open-File Report 99-111

2. Anderson SP, Drever JI, Humphrey NF (1997) Chemical weathering in glacial environments. Geology 25(5):399–402

3. Baron JS (ed) (1992) Biogeochemistry of a subalpine ecosystem: Loch Vale Watershed. Ecological Studies Series 90. Springer, New York

4. Baron J, Bricker OP (1987) Hydrologic and chemical flux in Loch Vale Watershed, Rocky Mountain National Park. In McKnight D, Averett RC (eds) Chemical quality of water and the hydrologic cycle. Lewis Publishers, Ann Arbor, MI, pp 141–156

5. Baron JS, Campbell DH (1997) Nitrogen fluxes in a high elevation Rocky Mountain basin. Hydrol Process 11(7):783–799

6. Baron JS, Allstott EJ, Newkirk BK (1995) Analysis of long term sulfate and nitrate budgets in a Rocky Mountain basin. In: Tonnesson KA, Williams MW, Tranter M (eds) Biogeochemistry of seasonally snow-covered catchments. IASH Publication No 228: International Association of Hydrologic Sciences Press, Institute of Hydrology, Wallingford, Oxfordshire, OX10 8BB, UK, pp 255–262

7. Baron JS, Rueth HM, Wolfe AP, Nydick KR, Allstott EJ, Minear JT, Moraska B (2000) Ecosystem responses to nitrogen deposition in the Colorado Front Range. Ecosystems 3(4):352–368

8. Baron JS, Schmidt TM, Hartman MD (2009) Climate-induced changes in high elevation stream nitrate dynamics. Glob Change Biol 15:1777–1789

9. Bogdal C, Schmid P, Zennegg M, Anselmetti FS, Scheringer M, Hungerbühler K (2009) Blast from the past: melting glaciers as a relevant source for persistent organic pollutants. Environ Sci Technol 43(21):8173–8177

10. Botte JA, Baron J (2004) Quality assurance report, Loch Vale Watershed, 1999–2002. USGS Open-File Report 2004–1306

11. Bricker OP (1986) Geochemical investigations of selected Eastern United States watersheds affected by acid deposition. J Geol Soc 143:621–626

12. Bricker OP, Godfrey AE, Cleaves ET (1968) Mineral-water interaction during the chemical weathering of silicates. In: Baker RA (ed) Advances in chemistry, vol 73. American Chemical Society, USA, pp 128–142

13. Caine N (2010) Recent hydrologic change in a Colorado alpine basin: an indicator of permafrost thaw? Ann Glaciol 51(56):130–134

14. Campbell DH, Baron JS, Tonnessen K, Brooks P, Schuster P (2000) Controls on nitrogen flux in alpine/subalpine watersheds. Water Resour Res 36(1):37–48

15. Campbell DH, Clow DW, Ingersoll GP, Mast MA, Spahr NE, Turk JT (1995) Processes controlling the chemistry of two snowmelt-dominated streams in the rocky mountains. Water Resour Res 31(11):2811–2821

16. Cleaves ET, Godfrey AE, Bricker OP (1970) Geochemical balance of a small watershed and its geomorphic implications. Geol Soc Am Bull 81(10):3015–3032

17. Clow DW, Drever JI (1996) Weathering rates as a function of flow through an alpine soil. Chem Geol 132:131–141

18. Clow DW, Schrott L, Webb R, Campbell DH, Torizzo A, Dornblaser M (2003) Ground water occurrence and contributions to streamflow in an alpine catchment. Colorado Front Range. Ground Water 41(7):937–950

19. Fountain AG, Campbell JL, Schuur EAG, Stammerjohn SE, Williams MW, Ducklow HM (2012) The disappearing cryosphere: impacts and ecosystem responses to rapid cryosphere loss. Bioscience 62(4):405–415

20. Greaver TL, Sullivan T, Herrick JD, Barber M, Baron J, Cosby B, Deerhake M, Dennis R, Dubois JJD, Goodale C, Herlihy A, Lawrence G, Liu L, Lynch J, Novak K (2012) Ecological effects from nitrogen and sulfur air pollution in the United States: what do we know? Front Ecol Environ 7:365–372

21. Hall K, Thorn CE, Matsuoka N, Prick A (2002) Weathering in cold regions: some thoughts and perspectives. Prog Phys Geogr 26(4):577–603

22. Helsel DR, Hirsch RM (2002) Statistical methods in water resources techniques of water resources investigations, Book 4, chapter A3. U.S. Geological Survey. http://pubs.usgs.gov/twri/twri4a3/

23. Helsel DR, Mueller DK, Slack JR (2006) Computer program for the Kendall family of trend tests. U.S. Geological Survey Scientific Investigations Report 2005–5275, 4 pp. http://pubs.usgs.gov/sir/2005/5275/pdf/sir2005-5275.pdf

24. Katz BG, Bricker OP, Kennedy MM (1985) Geochemical mass-balance relationships for selected ions in precipitation and stream water, Catoctin Mountains, MD. Am J Sci 285:931–962

25. Kester CL, Baron JS, Turk JT (2003) Isotopic study of sulfate sources and residence times in a subalpine watershed. Environ Geol 43(5):606–613

26. Knowles N, Dettinger MD, Cayan DR (2006) Trends in snowfall versus rainfall in the Western United States. J Clim 19:4545–4559

27. Lehmann CMB, Bowersox VC, Larson SM (2005) Spatial and temporal trends of precipitation chemistry in the United States. Environ Pollut 135(3):347–361

28. Likens GE, Butler TJ, Buso DC (2001) Long-and short-term changes in sulfate deposition: effects of the 1990 Clean Air Act Amendments. Biogeochemistry 52(1):1–11

29. Mast AM, Drever JI, Baron JS (1990) Chemical weathering in the Loch Vale watershed, Rocky Mountain National Park, Colorado. Water Resour Res 26(12):2971–2979

30. McGuire CR, Nufio CR, Bowers MD, Guralnick RP (2013) Elevation-dependent temperature trends in the Rocky Mountain Front Range: changes over a 56- and 20-year record. PlosOne 7(9):e44370.

31. Morris K, Mast A, Clow D, Wetherbee G, Baron J, Taipale C, Blett T, Gay D, Richer E (2012) 2010 monitoring and tracking wet nitrogen deposition at Rocky Mountain National Park: August 2012. Natural Resource Report NPS/NRSS/ARD/NRR—2012/562. National Park Service, Denver, CO

32. Neff JC, Ballantyne AP, Farmer GL, Mahowald NM, Conroy JL, Landry CC, Overpeck JT, Painter TH, Lawrence CR, Reynolds RL (2008) Increasing eolian dust deposition in the western United States linked to human activity. Nat Geosci 1:189–195

33. Osborne B (2012) The effects of temperature and moisture on alpine microbial processes across a gradient of soil development. Master's thesis, Colorado State University, Fort Collins

34. Price JR, Peresolak K, Brice RL, Tefend KS (2013) Temporal variability in the chemical weathering of Ca^{2+}-bearing phases in the Loch Vale Watershed, Colorado, USA: a mass-balance approach. Chem Geol 342:151–166

35. Ray AJ, Barsugli JJ, Averyt KB, Wolter K, Hoerling M (2008) Colorado climate change: a synthesis to support water resource management and adaptation, a report for the Colorado Water Conservation Board by the NOAA-CU Western Water Assessment. http://wwa.colorado.edu/publications/reports/WWA_ClimateChangeColoradoReport_2008.pdf

36. Rhoades C, Elder K, Greene E (2010) The influence of an extensive dust event on snow chemistry in the southern Rocky Mountains. Arct Antarct Alp Res 42(3):98–105

37. Richer EE, Baron JS (2011) Loch Vale Watershed long-term ecological research and monitoring program: quality assurance report, 2003–2009. USGS Open-File Report 2011–1137

38. Saros JE, Rose KC, Clow DW, Stephens VC, Nurse AB, Arnett HA, Stone JR, Williamson CE, Wolfe AP (2010) Melting high alpine

glaciers enrich lakes with reactive nitrogen. Environ Sci Technol 44(13):4891–4896

39. Simon NS, Bricker OP, Newell W, McCoy J, Morawe R (2005) The distribution of phosphorus in Popes Creek, VA, and in the Pocomoke River, MD: two watersheds with different land management practices in the Chesapeake Bay Basin. Water Air Soil Pollut 164:189–204

40. Singer GA, Fasching C, Wilhelm L, Niggemann J, Steier P, Dittmar T, Battin TJ (2012) Biogeochemically diverse organic matter in Alpine glaciers and its downstream fate. Nat Geosci 5(10):710–714

41. Stewart IT, Cayan DR, Dettinger MD (2005) Changes toward earlier stream flow timing across western North America. J Clim 18:1136–1155

42. Wetherbee GA, Latysh NE, Gordon JD (2005) Spatial and temporal variability of the overall error of National Atmospheric deposition program measurements determined by the USGS collocated-sampler program, water years 1989–2001. Environ Pollut 135(3):407–418

43. Williams MW, Melack JM (1991) Solute chemistry of snowmelt and runoff in an alpine basin. Sierra Nevada. Water Resour Res 27(7):1575–1588

44. Williams MW, Baron JS, Caine N, Sommerfeld R, Sanford R Jr (1996) Nitrogen saturation in the Rocky Mountains. Environ Sci Technol 30:640–646

45. Williams MW, Knauf N, Caine N, Lui F, Verplanck PL (2006) Geochemistry and source waters of rock glacier outflow, Colorado Front Range. Permafrost Periglac Process 17:13–33

46. Wynn PM, Hodson AJ, Heaton THE, Chenery SR (2007) Nitrate production beneath a high arctic glacier, Svalbard. Chem Geol 244(1–2):88–102

INDEX